1988

INTRODUCTION TO REAL ANALYSIS

INTRODUCTION TO REAL ANALYSIS

ROBERT G. BARTLE
DONALD R. SHERBERT

University of Illinois
Urbana-Champaign, Illinois

1807 175 YEARS OF PUBLISHING 1982

JOHN WILEY & SONS, INC.

New York • Chichester • Brisbane • Toronto • Singapore

Library of Congress Cataloging in Publication Data:

Bartle, Robert Gardner, 1927—
Introduction to real analysis

Includes index.
1. Mathematical analysis. 2. Functions of real variables. I. Sherbert, Donald R., 1935— . II. Title.

QA300.B294 515.8 81-15924
ISBN 0-471-05944-7 AACR2

Printed in the United States of America

10 9 8 7 6 5 4 3 2 1

To Carolyn and Janice

PREFACE

A careful study of real analysis is very important for any student who wants to use mathematics beyond the elementary manipulation of formulas to solve standard types of problems. In order to be able to modify techniques to different problems and to adapt concepts to new contexts, it is often necessary to develop a good understanding of the analysis of functions of a real variable.

As the use of mathematics in the areas of social science, life science, economics, and management science has grown in the last decade, it has become more important for students in these areas to study real analysis, along with their colleagues in physical sciences, engineering, computer science, and, of course, mathematics. Consequently, there is a need for an accessible, reasonably paced textbook that deals with the fundamental concepts and techniques of real analysis. In this book, which has been written to fill that need, we develop the material at a deliberate pace and are careful to provide examples to illustrate each idea. The pace is rather slow at first, but quickens somewhat in the later chapters, although in a manner that we have found to be consistent with the growing mathematical sophistication of the students.

To avoid undue complications in a first course, we restrict our attention here to functions of a single variable, leaving the study of functions of several variables to a later course. That distinguishes the present book from *The Elements of Real Analysis* (referred to in the book as *ERA*) by the first-named author, which deals with functions of several variables and which proceeds at a somewhat more rapid pace. However, the spirit and approach of the two books are similar.

In this book we initiate the study of limits by using sequences. This sequential approach has the distinct advantage of being very natural and accessible; it also relates well to numerical analysis and computer science. Of course, the topological concepts are not avoided, but they are kept somewhat in the background, at least initially.

In Chapter 1 we present a short summary of the set theoretical notions and notations that are used in the book. We strongly recommend that the existence of this chapter be noted, and that its content be briefly examined. However, any temptation to dwell on the formalism of set theory should be resisted, since such a study does not accurately preview the nature of real analysis. In particular, the uncountability of the set of real numbers should be noted without delving into the subtle complications found in the theory of cardinal numbers.

In Chapter 2 we present the properties of the real number system. The first few sections provide practice in elementary deductive thinking and the writing of proofs; they can be covered rather quickly, especially if the students have had some prior experience of this nature. The crucial completeness property of the real number system, introduced here through the Supremum Property, is enunciated in Section 2.4. This topic should be carefully discussed, for the use of suprema is essential in the following material. Fortunately, with increased familiarity of this use, students will achieve increased understanding.

In Chapter 3 we present a thorough treatment of real sequences and the associated limit concepts. This material is, of course, of the greatest importance. Fortunately, students find it rather natural. Chapter 4, by far the longest in the book, is also the heart of the book. Its initial sections dealing with the limits of functions rely heavily on the use of sequences. The presentation is almost completely parallel to the development of limits of sequences; this parallel approach reinforces the understanding of limits and permits the basic theorems on the limits of functions and on continuous functions to be covered rather quickly. (Some instructors may prefer to skip Section 4.3 in a first course.) The fundamental properties of continuous functions on intervals are developed in Section 4.6. The concept of compactness in terms of open covers is delayed to Section 4.9, since this notion is rather sophisticated for beginning students; some instructors may wish to delay this topic until later in the course.

The basic theory of the derivative is given in the first two sections of Chapter 5. Section 5.2 is essentially a sequence of consequences of the Mean Value Theorem. The fundamental theory of the Riemann integral is contained in the first three sections of Chapter 6. We introduce the integral by means of the upper and lower integrals, since this approach seems to be most readily understood. The later sections of Chapters 5 and 6 discuss various aspects of the derivative and the integral. They can be covered as time and interest dictate. We have included Chapter 7 on uniform convergence and Chapter 8 on infinite series because of their intrinsic importance and interest, and also to show how the material developed in the earlier chapters can be applied to numerical and analytical topics.

Throughout the book we have paid more attention to topics from numerical analysis and approximation theory than is usual. We have done so because of the increased importance of these topics to contemporary students and because a proper understanding of these topics is best found in the context of real analysis. At the same time, experience has shown that these topics enhance an understanding of "purely analytic" ideas as well.

Both of the authors have taught from preliminary versions of this book. It is our experience that the major topics can readily be covered in one semester provided that one does not get bogged down in the preliminary material. It is satisfying to see how the mathematical maturity of the students increases and how they learn to work comfortably with concepts that initially seem so mysterious to them. One should not feel, however, that one needs to cover every topic in one semester. There is ample material for the particular interests of the instructor and the students to be pursued.

We wish to thank all those who examined preliminary versions of the manuscript and who gave us their suggestions for improvement. We especially want to thank Professor Robert F. Geitz of Oberlin College for his very thorough reading of the manuscript. Finally we wish to express our appreciation to the staff of John Wiley and Sons for their competence and their patience.

Robert G. Bartle
Donald R. Sherbert

CONTENTS

INTRODUCTION TO REAL ANALYSIS

CHAPTER ONE

A GLIMPSE AT SET THEORY

The idea of a set is basic to all of mathematics, and all mathematical objects and constructions ultimately go back to set theory. In view of the fundamental importance of set theory, we shall present here a brief résumé of the set-theoretic notions that will be used frequently in this text. However, since the aim of this book is to present the *elements* (rather than the *foundations*) of real analysis, we adopt a rather pragmatic and naïve point of view. We shall be content with an informal discussion and shall regard the word "set" as understood and synonymous with the words "class", "collection", "aggregate", and "ensemble". No attempt will be made to define these terms or to present a list of axioms for set theory. A reader who is sophisticated enough to be troubled by our informal development should consult the references on set theory that are given at the end of this text. There he will learn how this material can be put on an axiomatic basis. He will find this axiomatization to be an interesting development in the foundations of mathematics. However, since we regard it to be outside the subject area of the present book, we shall not go through the details here.

The reader is strongly urged to read this introduction quickly to absorb the notations we shall employ. Unlike the later chapters, which must be *studied*, this introduction is to be considered background material. One should not spend much time on it.

SECTION 1.1 The Algebra of Sets

If A denotes a set and if x is an element, it is often convenient to write

$$x \in A$$

as an abbreviation for the statement that x is an **element** of A, or that x is a **member** of the set A, or that the set A **contains** the element x, or that x **is in** A. We shall not examine the nature of this property of being an element of a set any further. For most purposes it is possible to employ the naïve meaning of "membership", and an axiomatic characterization of this relation is not necessary.

If A is a set and x is an element which does *not* belong to A, we shall often write

$$x \notin A.$$

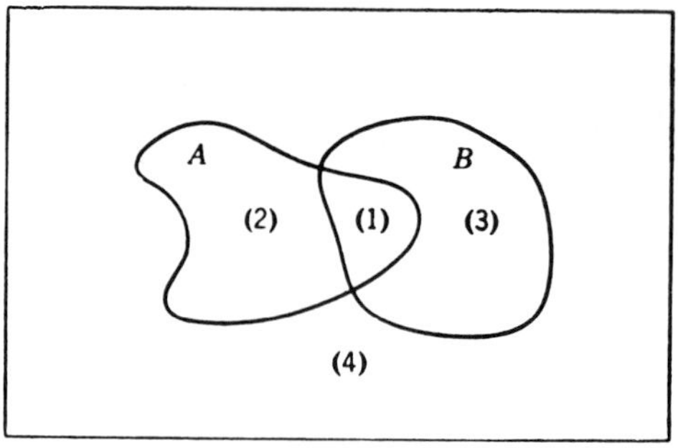

FIGURE 1.1.1 There are four possibilities for an element x.

In accordance with our naïve conception of a set, we shall require that exactly one of the two possibilities

$$x \in A, \qquad x \notin A,$$

holds for an element x and a set A.

If A and B are two sets and x is an element, then there are, in principle, four possibilities (see Figure 1.1.1):

(1) $x \in A$	and	$x \in B$;	(2) $x \in A$	and	$x \notin B$;
(3) $x \notin A$	and	$x \in B$;	(4) $x \notin A$	and	$x \notin B$.

If the second case cannot occur (that is, if every element of A is also an element of B), then we shall say that A is **contained** in B, or that B **contains** A, or that A is a **subset** of B, and we shall write

$$A \subseteq B \qquad \text{or} \qquad B \supseteq A.$$

If $A \subseteq B$ and there exists an element in B which is not in A, we say that A is a **proper subset** of B.

It should be noted that the statement that $A \subseteq B$ does not automatically preclude the possibility that A exhausts all of B. When this is true the sets A and B are "equal" in the sense we now define.

1.1.1 Definition. Two sets are **equal** if they contain the same elements. If the sets A and B are equal, we write $A = B$.

Thus in order to show that the sets A and B are equal we must show that the possibilities (2) and (3) mentioned above cannot occur. Equivalently, we must show that both $A \subseteq B$ and $B \subseteq A$.

The word "property" is not easy to define precisely. However, we shall not hesitate to use it in the usual (informal) fashion. If P denotes a property that is meaningful for a collection of elements, then we agree to write

$$\{x : P(x)\}$$

for the set of all elements x for which the property P holds. We usually read this as "the set of all x such that $P(x)$". It is often worthwhile to specify which elements we are testing for the property P. Hence we shall often write

$$\{x \in S : P(x)\}$$

for the subset of S for which the property P holds.

Examples. (a) If $\mathbf{N} := \{1, 2, 3, \ldots\}$ denotes† the set of natural numbers, then the set

$$\{x \in \mathbf{N} : x^2 - 3x + 2 = 0\}$$

consists of those natural numbers satisfying the stated equation. Now the only solutions of the quadratic equation $x^2 - 3x + 2 = 0$ are $x = 1$ and $x = 2$. Hence, instead of writing the above expression (since we have detailed information concerning all of the elements in the set under examination) we shall ordinarily denote this set by $\{1, 2\}$ thereby listing the elements of the set.

(b) Sometimes a formula can be used to abbreviate the description of a set. For example, the set of all even natural numbers could be denoted by $\{2x : x \in \mathbf{N}\}$, instead of the more cumbersome $\{y \in \mathbf{N} : y = 2x, x \in \mathbf{N}\}$.

(c) The set $\{x \in \mathbf{N} : 6 < x < 9\}$ can be written explicitly as $\{7, 8\}$, thereby exhibiting the elements of the set. Of course, there are many other possible descriptions of this set. For example:

$$\{x \in \mathbf{N} : 40 < x^2 < 80\},$$

$$\{x \in \mathbf{N} : x^2 - 15x + 56 = 0\},$$

$$\{7 + x : x = 0 \quad \text{or} \quad x = 1\}.$$

(d) In addition to the set of **natural numbers** (consisting of the elements denoted by 1, 2, 3, . . .) which we shall systematically denote by **N**, there are a few other sets for which we introduce a standard notation. The set of **integers** is

$$\mathbf{Z} := \{0, 1, -1, 2, -2, 3, -3, \ldots\}.$$

The set of **rational numbers** is

$$\mathbf{Q} := \{m/n : m, n \in \mathbf{Z} \quad \text{and} \quad n \neq 0\}.$$

†We shall often use the symbol := to mean that the symbol on the left is being *defined* by the expressions on the right.

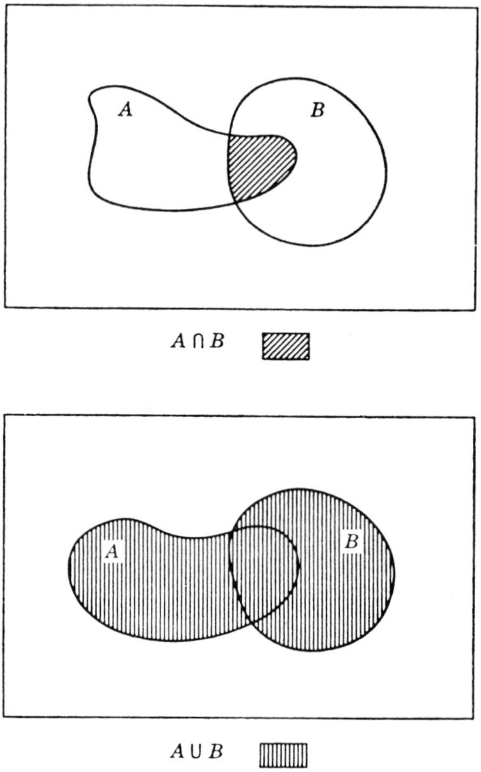

FIGURE 1.1.2 The intersection and union of two sets.

We shall treat the sets $\mathbf{N}$, $\mathbf{Z}$, and $\boldsymbol{Q}$ as if they are well understood and shall not reexamine their properties in much detail. Of basic importance for our later study is the set $\boldsymbol{R}$ of all real numbers which will be examined in Chapter 2.

Set Operations

We now introduce some methods of constructing new sets from given ones.

1.1.2 Definition. If A and B are sets, then their **intersection** is the set of all elements that belong to both A and B. We shall denote the intersection of the sets A, B by the symbol $A \cap B$, which is read "A intersect B". (See Figure 1.1.2.)

1.1.3 Definition. If A and B are sets, then their **union** is the set of all elements that belong either to A or to B or to both A and B. We shall denote the union of the sets A, B by the symbol $A \cup B$, which is read "A union B". (See Figure 1.1.2.)

We could also define $A \cap B$ and $A \cup B$ by

$$A \cap B := \{x : x \in A \quad \text{and} \quad x \in B\},$$

$$A \cup B := \{x : x \in A \quad \text{or} \quad x \in B\}.$$

In connection with the latter, it is important to realize that the word "or" is being used in the inclusive sense that is customary in mathematics and logic. In legal terminology this inclusive sense is sometimes indicated by "and/or".

We have tacitly assumed that the intersection and the union of two sets is again a set. Among other things this requires that there must exist a set that has no elements at all (for if A and B have no common elements, their intersection has no elements).

1.1.4 Definition. The set that has no elements is called the **empty** or the **void** set and will be denoted by the symbol $\emptyset$. If A and B are sets with no common elements (that is, if $A \cap B = \emptyset$), then we say that A and B are **disjoint** or that they are **non-intersecting**.

The next result gives some of the algebraic properties of the operations on sets that we have just defined. Since the proofs of these assertions are routine, we shall leave most of them to the reader as exercises.

1.1.5 Theorem. *Let A, B, C, be any sets, then*
(a) $A \cap A = A, \quad A \cup A = A;$
(b) $A \cap B = B \cap A, \quad A \cup B = B \cup A;$
(c) $(A \cap B) \cap C = A \cap (B \cap C), \quad (A \cup B) \cup C = A \cup (B \cup C);$
(d) $A \cap (B \cup C) = (A \cap B) \cup (A \cap C), \quad A \cup (B \cap C) = (A \cup B) \cap (A \cup C).$

These equalities are sometimes referred to as the *idempotent*, the *commutative*, the *associative*, and the *distributive properties*, respectively, of the operations of intersection and union of sets.

In order to give a sample proof, we shall prove the first equation in (d). Let x be an element of $A \cap (B \cup C)$, then $x \in A$ and $X \in B \cup C$. This means that $x \in A$, and either $x \in B$ or $x \in C$. Hence we either have (i) $x \in A$ and $x \in B$, or we have (ii) $x \in A$ and $x \in C$. Therefore, either $x \in A \cap B$ or $x \in A \cap C$, so $x \in (A \cap B) \cup (A \cap C)$. This shows that $A \cap (B \cup C)$ is a subset of $(A \cap B) \cup (A \cap C)$.

Conversely, let y be an element of $(A \cap B) \cup (A \cap C)$. Then, either (iii) $y \in A \cap B$, or (iv) $y \in A \cap C$. It follows that $y \in A$, and either $y \in B$ or $y \in C$. Therefore, $y \in A$ and $y \in B \cup C$ so that $y \in A \cap (B \cup C)$. Hence $(A \cap B) \cup (A \cap C)$ is a subset of $A \cap (B \cup C)$.

In view of Definition 1.1.1, we conclude that the sets $A \cap (B \cup C)$ and $(A \cap B) \cup (A \cap C)$ are equal.

In view of the relations in Theorem 1.1.5(c), we usually drop the parentheses and write merely

$$A \cap B \cap C, \qquad A \cup B \cup C.$$

It is possible to show that if $\{A_1, A_2, \ldots, A_n\}$ is a collection of sets, then there is a uniquely defined set A consisting of all elements which belong to *at least one* of the sets A_j, $j = 1, 2, \ldots, n$; and there exists a uniquely defined set B consisting of all elements which belong to *all* of the sets A_j, $j = 1, 2, \ldots, n$. Dropping the use of parentheses, we write

$$A = A_1 \cup A_2 \cup \cdots \cup A_n := \{x : x \in A_j \quad \text{for some } j\},$$

$$B = A_1 \cap A_2 \cap \cdots \cap A_n := \{x : x \in A_j \quad \text{for all } j\}.$$

Sometimes, in order to save space, we mimic the notation used for sums and employ a more condensed notation, such as

$$A = \bigcup_{j=1}^{n} A_j = \bigcup \{A_j : j = 1, 2, \ldots, n\},$$

$$B = \bigcap_{j=1}^{n} A_j = \bigcap \{A_j : j = 1, 2, \ldots, n\}.$$

Similarly, if for each j in a set J there is a set A_j, then $\bigcup\{A_j : j \in J\}$ denotes the set of all elements which belong to *at least one* of the sets A_j. In the same way, $\bigcap\{A_j : j \in J\}$ denotes the set of all elements which belong to *all* of the sets A_j for $j \in J$.

We now introduce another method of constructing a new set from two given ones.

1.1.6 Definition. If A and B are sets, then the **complement of B relative to A** is the set of all elements of A that do not belong to B. We shall denote this set by $A \setminus B$ (read "A minus B"), although the related notations $A - B$ and $A \sim B$ are sometimes used by other authors. (See Figure 1.1.3.)

In the notation introduced above, we have

$$A \setminus B := \{x \in A : x \notin B\}.$$

Sometimes the set A is understood and does not need to be mentioned explicitly. In this situation we refer simply to the *complement* of B and denote $A \setminus B$ by $\mathscr{C}(B)$.

Returning to Figure 1.1.1, we note that the elements x that satisfy (1) belong to $A \cap B$; those that satisfy (2) belong to $A \setminus B$; and those that satisfy (3) belong to $B \setminus A$.

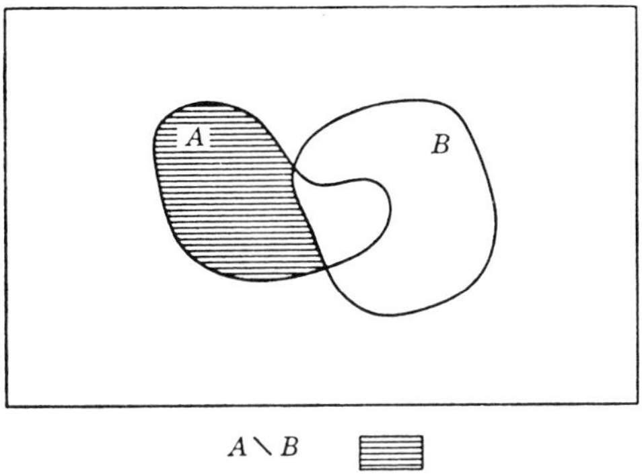

FIGURE 1.1.3 The relative complement.

We shall now state the *De Morgan laws* for three sets; a more general formulation will be given in the exercises.

1.1.7 Theorem. *If A, B, C, are any sets, then*

$$A \setminus (B \cup C) = (A \setminus B) \cap (A \setminus C),$$

$$A \setminus (B \cap C) = (A \setminus B) \cup (A \setminus C).$$

Proof. We shall carry out a demonstration of the first relation, leaving the second one to the reader. To establish the equality of the sets, we show that every element in $A \setminus (B \cup C)$ is contained in both $(A \setminus B)$ and $(A \setminus C)$ and conversely.

If x is in $A \setminus (B \cup C)$, then x is in A but x is not in $B \cup C$. Hence x is in A, but x is neither in B nor in C. (Why?) Therefore, x is in A but not B, and x is in A but not C. That is, $x \in A \setminus B$ and $x \in A \setminus C$, showing that $x \in (A \setminus B) \cap (A \setminus C)$.

Conversely, if $x \in (A \setminus B) \cap (A \setminus C)$, then $x \in (A \setminus B)$ and $x \in (A \setminus C)$. Thus $x \in A$ and both $x \notin B$ and $x \notin C$. It follows that $x \in A$ and $x \notin (B \cup C)$, so that $x \in A \setminus (B \cup C)$.

Since the sets $(A \setminus B) \cap (A \setminus C)$ and $A \setminus (B \cup C)$ contain the same elements, they are equal by Definition 1.1.1. Q.E.D.

Cartesian Product

We now define the Cartesian product of two sets.

1.1.8 Definition. If A and B are two non-void sets, then the **Cartesian product** $A \times B$ of A and B is the set of all ordered pairs (a, b) with $a \in A$ and $b \in B$. (See Figure 1.1.4 on the next page.)

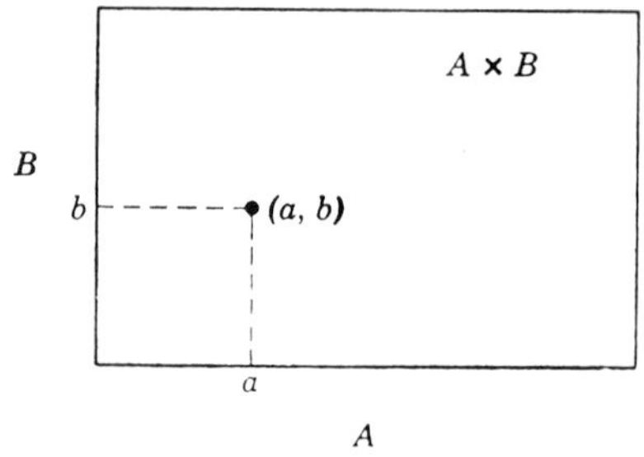

FIGURE 1.1.4 The Cartesian product.

Thus if $A = \{1, 2, 3\}$ and $B = \{4, 5\}$, then the set $A \times B$ is the set whose elements are the ordered pairs

$$(1, 4), (1, 5), (2, 4), (2, 5), (3, 4), (3, 5).$$

We may visualize the set $A \times B$ as the set of six points in the plane with the coordinates that we have just listed.

We often draw a diagram (such as Figure 1.1.4) to indicate the Cartesian product of two sets A, B. However, it should be realized that this diagram may be somewhat of a simplification. For example, if $A := \{x \in \boldsymbol{R} : 1 \leqslant x \leqslant 2\}$ and $B := \{x \in \boldsymbol{R} : 0 \leqslant x \leqslant 1 \text{ or } 2 \leqslant x \leqslant 3\}$, then instead of a rectangle, we should have a drawing like Figure 1.1.5.

Exercises for Section 1.1

1. Draw a diagram to represent each of the sets mentioned in Theorem 1.1.5.
2. Prove part (c) of Theorem 1.1.5.
3. Prove the second part of (d) of Theorem 1.1.5.
4. Prove that $A \subseteq B$ if and only if $A \cap B = A$.
5. Show that the set D of all elements that belong either to A or B but not to both is given by

$$D = (A \setminus B) \cup (B \setminus A).$$

   This set D is often called the **symmetric difference** of A and B. Represent it by a diagram.
6. Show that the symmetric difference D, defined in the preceding exercise, is also given by $D = (A \cup B) \setminus (A \cap B)$.
7. If $B \subseteq A$, show that $B = A \setminus (A \setminus B)$.
8. Given sets A and B, show that the sets $A \cap B$ and $A \setminus B$ are disjoint and that $A = (A \cap B) \cup (A \setminus B)$.
9. If A and B are any sets, show that $A \cap B = A \setminus (A \setminus B)$.
10. If $\{A_1, A_2, \ldots, A_n\}$ is a collection of sets, and if E is any set, show that

$$E \cap \bigcup_{j=1}^{n} A_j = \bigcup_{j=1}^{n} (E \cap A_j), \qquad E \cup \bigcup_{j=1}^{n} A_j = \bigcup_{j=1}^{n} (E \cup A_j).$$

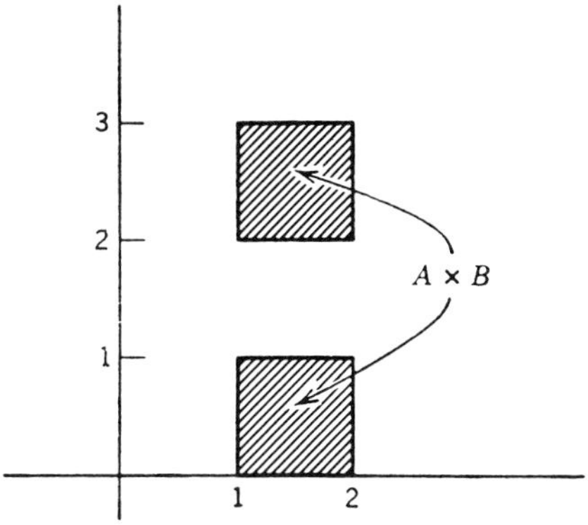

FIGURE 1.1.5 The Cartesian product of *A* and *B*.

11. If $\{A_1, A_2, \ldots, A_n\}$ is a collection of sets, and if E is any set, show that

$$E \cap \bigcap_{j=1}^{n} A_j = \bigcap_{j=1}^{n} (E \cap A_j), \qquad E \cup \bigcap_{j=1}^{n} A_j = \bigcap_{j=1}^{n} (E \cup A_j).$$

12. Let E be a set and $\{A_1, A_2, \ldots, A_n\}$ be a collection of sets. Establish the De Morgan laws:

$$E \setminus \bigcap_{j=1}^{n} A_j = \bigcup_{j=1}^{n} (E \setminus A_j), \qquad E \setminus \bigcup_{j=1}^{n} A_j = \bigcap_{j=1}^{n} (E \setminus A_j).$$

Note that if $E \setminus A_j$ is denoted by $\mathscr{C}(A_j)$, these relations take the form

$$\mathscr{C}\left(\bigcap_{j=1}^{n} A_j\right) = \bigcup_{j=1}^{n} \mathscr{C}(A_j), \qquad \mathscr{C}\left(\bigcup_{j=1}^{n} A_j\right) = \bigcap_{j=1}^{n} \mathscr{C}(A_j).$$

13. Let J be any set and, for each $j \in J$, let A_j be contained in X. Show that

$$\mathscr{C}\left(\bigcap\{A_j : j \in J\}\right) = \bigcup\{\mathscr{C}(A_j) : j \in J\},$$

$$\mathscr{C}\left(\bigcup\{A_j : j \in J\}\right) = \bigcap\{\mathscr{C}(A_j) : j \in J\}.$$

14. If B_1 and B_2 are subsets of B and if $B = B_1 \cup B_2$, then

$$A \times B = (A \times B_1) \cup (A \times B_2).$$

SECTION 1.2 Functions

We now turn to a discussion of the fundamental notion of a *function* or *mapping*. It will be seen that a function is a special kind of a set, although there are other visualizations that are often suggestive. All of the later sections will be concerned with various types of functions, but they will usually be of less abstract nature than considered in the present introductory section.

To the mathematician of a century ago the word "function" ordinarily meant a definite formula, such as

$$f(x) := x^2 + 3x - 5,$$

which associates to each real number x another real number $f(x)$. The fact that certain formulas, such as

$$g(x) := \sqrt{x - 5},$$

do not give rise to real numbers for all real values of x was, of course, well-known but was not regarded as sufficient grounds to require an extension of the notion of function. Probably one could arouse controversy among those mathematicians as to whether the absolute value

$$h(x) := |x|$$

of a real number is an "honest function" or not. For, after all, the definition of $|x|$ is given "in pieces" by

$$|x| := \begin{cases} x, & \text{if} \quad x \geqslant 0, \\ -x, & \text{if} \quad x < 0. \end{cases}$$

As mathematics developed, it became increasingly clear that the requirement that a function be a formula was unduly restrictive and that a more general definition would be useful. It also became evident that it is important to make clear distinction between the function itself and the values of the function. The reader probably finds himself in the position of the mathematician of a century ago in these two respects due to no fault of his own. We propose to bring him up to date with the current usage, but we shall do so in two steps. Our first revised definition of a function would be:

A function f from a set A to a set B is a rule of correspondence that assigns to each x in a certain subset D of A, a uniquely determined element $f(x)$ of B.

Certainly, the explicit formulas of the type mentioned above are included in this tentative definition. The proposed definition allows the possibility that the function might not be defined for certain elements of A and also allows the consideration of functions for which the sets A and B are not necessarily real numbers (but might even be desks and chairs—or cats and dogs).

However suggestive the proposed definition may be, it has a significant defect: it is not clear. There remains the difficulty of interpreting the phrase "rule of correspondence". Doubtless the reader can think of phrases that will satisfy him better than the above one, but it is not likely that he can dispel the fog entirely. The most satisfactory solution seems to be to define "function" entirely in terms of sets and the notions introduced in the preceding section. This has the disadvantage of being more artificial and losing some of the intuitive content of the earlier description, but the gain in clarity outweighs these disadvantages.

The key idea is to think of the graph of the function: that is, a collection of ordered pairs. We notice that an arbitrary collection of ordered pairs cannot be the graph of a function, for once the first member of the ordered pair is named, the second is uniquely determined.

1.2.1 Definition. Let A and B be sets (which are not necessarily distinct). A **function from** A **to** B is a set f of ordered pairs in $A \times B$ with the property that if (a, b) and (a, b') are elements of f, then $b = b'$. The set of all elements of A that can occur as first members of elements in f is called the **domain** of f and will be denoted $D(f)$. The set of all elements of B that can occur as second members of elements f is called the **range** of f (or the **set of values** of f) and will be denoted by $R(f)$. In case $D(f) = A$, we often say that f **maps** A **into** B (or is a **mapping** of A into B) and write $f : A \to B$. (See Figure 1.2.1.)

If (a, b) is an element of a function f, then it is customary to write

$$b = f(a) \qquad \text{or} \qquad f : a \mapsto b$$

instead of $(a, b) \in f$. We often refer to the element b as the **value** of f at the point a, or the **image under** f of the point a.

Transformations and Machines

Aside from graphs, we can visualize a function as a *transformation* of part of the set A into part of B. In this phraseology, when $(a, b) \in f$, we think of f as taking the element a from the subset $D(f)$ of A and "transforming" or "mapping"

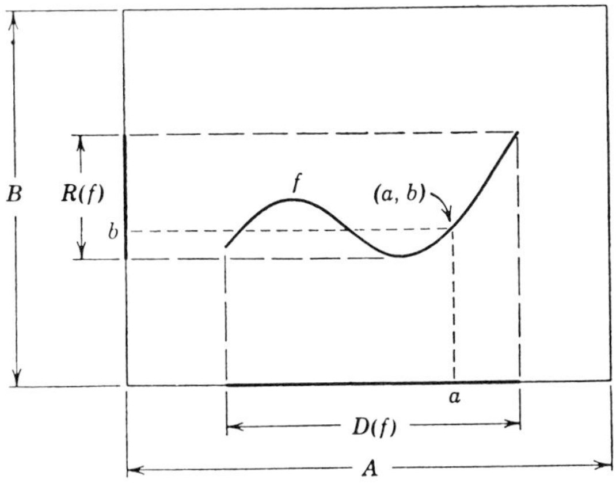

FIGURE 1.2.1 A function as a graph.

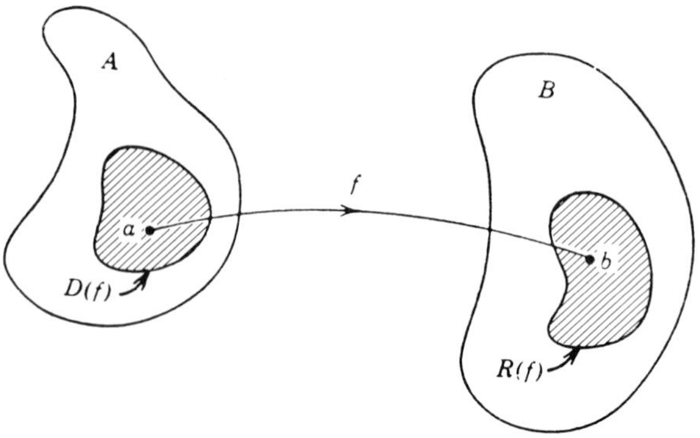

FIGURE 1.2.2 A function as a transformation.

it into an element $b := f(a)$ in the subset $R(f)$ of B. We often draw a diagram such as Figure 1.2.2. We frequently use this geometrical representation of a function even when the sets A and B are not subsets of the plane.

There is another way of visualizing a function: namely, as a *machine* which will accept elements of $D(f)$ as inputs and yield corresponding elements of $R(f)$ as outputs. If we take an element x from $D(f)$ and put it into f, then out comes the corresponding value $f(x)$. If we put a different element y of $D(f)$ into f, we get $f(y)$ [which may or may not differ from $f(x)$]. If we try to insert something which does not belong to $D(f)$ into f, we find that it is not accepted, for f can operate only on elements belonging to $D(f)$. (See Figure 1.2.3.)

This last visualization makes clear the distinction between f and $f(x)$: the first is the machine, the second is the output of the machine when we put x into it. Certainly it is useful to distinguish between a machine and its outputs. Only a fool would confuse a meat grinder with ground meat; however, enough people have confused functions with their values that it is worthwhile to make a modest effort to distinguish between them notationally.

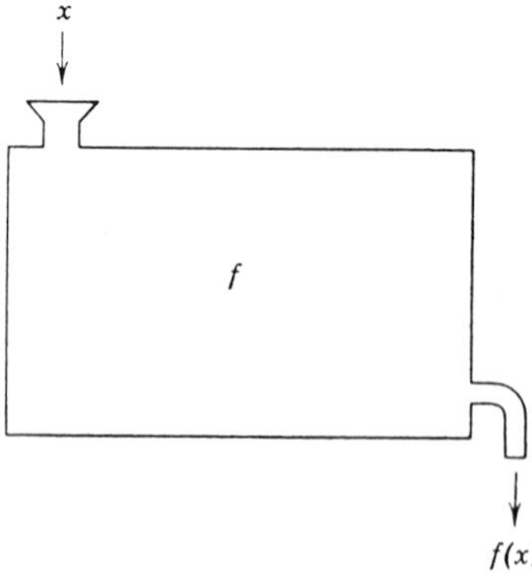

FIGURE 1.2.3 A function as a machine.

Restrictions and Extensions of Functions

If f is a function with domain $D(f)$ and D_1 is a subset of $D(f)$, it is often useful to define a new function f_1 with domain D_1 by $f_1(x) := f(x)$ for all $x \in D_1$. This function f_1 is called the *restriction of f to the set D_1*. In terms of Definition 1.2.1, we have

$$f_1 := \{(a, b) \in f : a \in D_1\}.$$

Sometimes we write $f_1 = f \mid D_1$ to denote the restriction of the function f to the set D_1.

A similar construction (that appears less artificial) is the notion of an "extension". If g is a function with domain $D(g)$ and $D_2 \supseteq D(g)$, then any function g_2 with domain D_2 such that $g_2(x) = g(x)$ for all $x \in D(g)$ is called an *extension of g to the set D_2*.

Composition of Functions

We now want to "compose" two functions by first applying f to each x in $D(f)$ and then applying g to $f(x)$ whenever possible [that is, when $f(x)$ belongs to $D(g)$]. In doing so, some care needs to be exercised concerning the domain of the resulting function. For example, if f is defined on $\boldsymbol{R}$ by $f(x) := x^3$ and if g is defined for $x \geqslant 0$ by $g(x) := \sqrt{x}$, then the composition $g \circ f$ can be defined only for $x \geqslant 0$, and for these real numbers it is to have the value $\sqrt{x^3}$.

1.2.2 Definition. Let f be a function with domain $D(f)$ in A and range $R(f)$ in B and let g be a function with domain $D(g)$ in B and range $R(g)$ in C. (See Figure 1.2.4.) The **composition** $g \circ f$ (note the order!) is the function from A to C given by

$$g \circ f := \{(a, c) \in A \times C: \text{there exists an element } b \in B \text{ such that } (a, b) \in f \text{ and } (b, c) \in g\}.$$

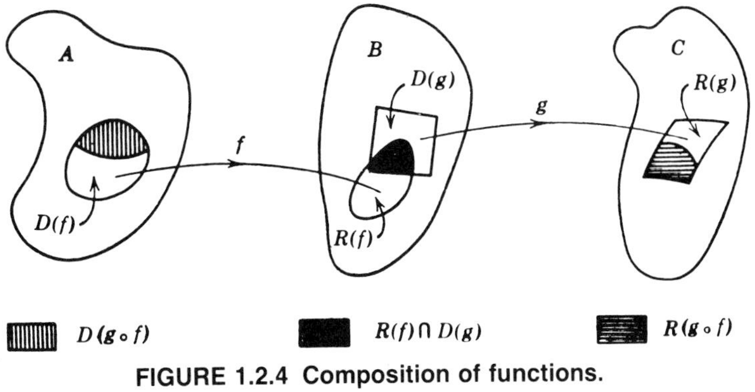

FIGURE 1.2.4 Composition of functions.

The reader should verify that the composition of two functions, as defined in Definition 1.2.2, is in fact a function.

We note that if f and g are functions and if $x \in D(f)$, then in order that g be defined at the element $f(x)$, it must be the case that $f(x)$ is an element of $D(g)$. Thus we see that the domain of the composition $g \circ f$ is the set

$$D(g \circ f) := \{x \in D(f) : f(x) \in D(g)\}.$$

For an element x in $D(g \circ f)$, the value of $g \circ f$ at x is given by $g \circ f(x) := g(f(x))$. It follows that the range of $g \circ f$ is the set

$$R(g \circ f) := \{g(f(x)) : x \in D(g \circ f)\}.$$

1.2.3 Examples. (a) Let f, g be functions whose values at the real number x are the real numbers given by

$$f(x) := 2x, \qquad g(x) := 3x^2 - 1.$$

Since $D(g)$ is the set $\boldsymbol{R}$ of all real numbers and $R(f) \subseteq D(g)$, the domain $D(g \circ f)$ is also $\boldsymbol{R}$ and $g \circ f(x) = 3(2x)^2 - 1 = 12x^2 - 1$. On the other hand, $D(f \circ g) = \boldsymbol{R}$, but $f \circ g(x) = 2(3x^2 - 1) = 6x^2 - 2$.

(b) If h is the function with $D(h) := \{x \in \boldsymbol{R} : x \geqslant 1\}$ defined by

$$h(x) := \sqrt{x - 1},$$

and if f is as in part (a), then $D(h \circ f) = \{x \in \boldsymbol{R} : 2x \geqslant 1\} = \{x \in \boldsymbol{R} : x \geqslant \frac{1}{2}\}$ and $h \circ f(x) = \sqrt{2x - 1}$. Also $D(f \circ h) = \{x \in \boldsymbol{R} : x \geqslant 1\}$ and $f \circ h(x) = 2\sqrt{x - 1}$. If g is the function in part (a), then $D(h \circ g) = \{x \in \boldsymbol{R} : 3x^2 - 1 \geqslant 1\} = x \in \boldsymbol{R} : x \leqslant -\sqrt{\frac{2}{3}}$ or $x \geqslant \sqrt{\frac{2}{3}}\}$ and $h \circ g(x) = \sqrt{3x^2 - 2}$. Also $D(g \circ h) = \{x \in \boldsymbol{R} : x \geqslant 1\}$ and $g \circ h(x) = 3x - 4$. (Note that the formula expressing $g \circ h$ has meaning for values of x other than those in the domain of $g \circ h$.)

(c) Let F, G be the functions with domains $D(F) := \{x \in \boldsymbol{R} : x \geqslant 0\}$, and $D(G) := \boldsymbol{R}$, such that the values of F and G at a point x in their domains are

$$F(x) := \sqrt{x}, \qquad G(x) := -x^2 - 1.$$

Then $D(G \circ F) = \{x \in \boldsymbol{R} : x \geqslant 0\}$ and $G \circ F(x) = -x - 1$, whereas $D(F \circ G) = \{x \in D(G) : G(x) \in D(f)\}$. This last set is void as $G(x) < 0$ for all $x \in D(G)$. Hence the function $F \circ G$ is not defined at any point, so $F \circ G$ is the "void function".

Injective and Inverse Functions

We now give a way of constructing a new function from a given one in case the original function does not take on the same value twice.

1.2.4 Definition. Let f be a function with domain $D(f)$ in A and range $R(f)$ in B. We say that f is **injective** or **one-one** if, whenever (a, b) and (a', b) are elements of f, then $a = a'$. If f is injective we may say that f is an **injection**.

In other words, f is injective if and only if the two relations $f(a) = b$ and $f(a') = b$ imply that $a = a'$. Alternatively, f is injective if and only if when a, a' are in $D(f)$ and $a \neq a'$, then $f(a) \neq f(a')$.

We claim if f is injective from A to B, then the set of ordered pairs in $B \times A$ obtained by interchanging the first and second members of ordered pairs in f yields a function g that is also injective.

We omit the proof of this assertion, leaving it as an exercise; it is a good test for the reader. The connections between f and g are:

$$D(g) = R(f), \qquad R(g) = D(f),$$

$$(a, b) \in f \quad \text{if and only if} \quad (b, a) \in g.$$

This last statement can be written in the more usual form:

$$b = f(a) \quad \text{if and only if} \quad a = g(b).$$

1.2.5 Definition. Let f be an injection with domain $D(f)$ in A and range $R(f)$ in B. If $g := \{(b, a) \in B \times A : (a, b) \in f\}$, then g is an injection with domain $D(g) = R(f)$ in B and with range $R(g) = D(f)$ in A. The function g is called the function **inverse** to f and is denoted by f^{-1}.

The inverse function can be interpreted from the mapping point of view. (See Figure 1.2.5.) If f is injective, it maps distinct elements of $D(f)$ into distinct elements of $R(f)$. Thus, each element b of $R(f)$ is the image under f of a unique element a in $D(f)$. The inverse function f^{-1} maps the element b into this unique element a.

1.2.6 Examples. (a) Let F be the function with domain $D(F) := \mathbf{R}$, the set of all real numbers, and range in $\mathbf{R}$ such that the value of F at the real number x is $F(x) := x^2$. (In other words, F is the function $\{(x, x^2) : x \in \mathbf{R}\}$.) It is readily

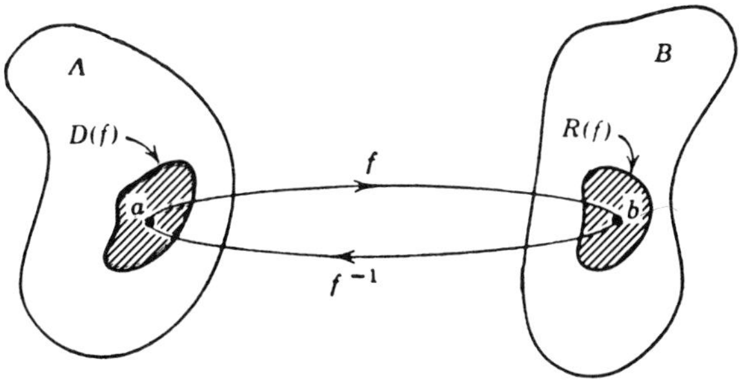

FIGURE 1.2.5 The inverse function.

seen that F is not one-one; in fact, the ordered pairs $(2, 4)$, $(-2, 4)$ both belong to F. Since F is not one-one, it does not have an inverse.

(b) Let f be the function with domain $D(f) := \{x \in \mathbf{R} : x \geq 0\}$ and $R(f) \subseteq \mathbf{R}$ whose value at x in $D(f)$ is $f(x) := x^2$. Note that f is the restriction to $D(f)$ of the function F in part (a). In terms of ordered pairs, $f := \{(x, x^2) : x \in \mathbf{R}, x \geq 0\}$. Unlike the function F in part (a), f is injective, for if $x^2 = y^2$ with x, y in $D(f)$, then $x = y$. (Why?) Therefore, f has an inverse function g with $D(g) = R(f) = \{x \in \mathbf{R} : x \geq 0\}$ and $R(g) = D(f) = \{x \in \mathbf{R} : x \geq 0\}$. Furthermore, $y = x^2 = f(x)$ if and only if $x = g(y)$. This inverse function g is ordinarily called the **positive square root function** and is denoted by

$$g(y) := \sqrt{y}, \qquad y \in \mathbf{R}, \qquad y \geq 0.$$

(c) If f_1 is the function $\{(x, x^2) : x \in \mathbf{R}, x \leq 0\}$, then as in (b), f_1 is one-one and has domain $D(f_1) := \{x \in \mathbf{R} : x \leq 0\}$ and range $R(f_1) := \{x \in \mathbf{R} : x \geq 0\}$. Note that f_1 is the restriction to $D(f_1)$ of the function F of part (a). The function g_1 inverse to f is called the **negative square root function** and is denoted by

$$g_1(y) := -\sqrt{y}, \qquad y \in \mathbf{R}, \qquad y \geq 0,$$

so that $g_1(y) \leq 0$.

(d) The sine function F introduced in trigonometry with $D(F) := \mathbf{R}$ and $R(F) := \{y \in \mathbf{R} : -1 \leq y \leq +1\}$ is well-known not to be injective (for example, $\sin 0 = \sin 2\pi = 0$). However, if we let f be its restriction to the set $D(f) := \{x \in \mathbf{R} : -\pi/2 \leq x \leq +\pi/2\}$, then f is injective. It therefore has an inverse function g with $D(g) = R(f)$ and $R(g) = D(f)$. Also, $y = \sin x$ with $x \in D(f)$ if and only if $x = g(y)$. The function g is called the (principal branch of) the **inverse sine function** and is often denoted by

$$g(y) := \text{Arc} \sin y \qquad \text{or} \qquad g(y) = \text{Sin}^{-1} y.$$

Surjective and Bijective Functions

1.2.7 Definition. Let f be a function with domain $D(f) \subseteq A$ and range $R(f) \subseteq B$. We say that f is **surjective**, or that f **maps onto** B, in case the range $R(f) = B$. If f is surjective, we may say that f is a **surjection.**

In defining a function it is important to specify the domain of the function and the set in which the values are taken. Once this has been done it is possible to inquire whether or not the function is surjective. For example, the function $f(x) := x^2$ is a surjection of $\mathbf{R}$ onto $\{x \in \mathbf{R} : x \geq 0\}$ but not of $\mathbf{R}$ onto $\mathbf{R}$.

1.2.8 Definition. A function f with domain $D(f) \subseteq A$ and range $R(f) \subseteq B$ is said to be **bijective** if (i) it is injective (that is, it is one-one), and (ii) it is surjective (that is, it maps $D(f)$ onto B). If f is bijective, we may say that f is a **bijection**.

Direct and Inverse Images

Let $f : A \to B$ be a function with domain A and range in B. We do *not* assume that f is injective.

1.2.9 Definition. If E is a subset of A, then the **direct image** of E under f is the subset $f(E)$ of B given by

$$f(E) := \{f(x) : x \in E\}.$$

If H is a subset of B, then the **inverse image** of H under f is the subset $f^{-1}(H)$ of A given by

$$f^{-1}(H) := \{x : f(x) \in H\}.$$

We emphasize that f need not be injective so that the inverse function f^{-1} need not exist. However, in the case that f^{-1} does exist, then the set $f^{-1}(H)$ can be viewed as the direct image of H under f^{-1}, as well as the inverse image of H under f. In either case, the set denoted by $f^{-1}(H)$ is unambiguously determined.

1.2.10 Examples. (a) If $f : \boldsymbol{R} \to \boldsymbol{R}$ is defined by $f(x) := 3x + 4$, then the direct image of the set $E := \{x : -1 \leqslant x \leqslant 2\}$ is the set $f(E) = \{y : 1 \leqslant y \leqslant 10\}$. The set $H := \{y : -2 \leqslant y \leqslant 7\}$ has the inverse image $f^{-1}(H) = \{x : -2 \leqslant x \leqslant 1\}$ as the reader should verify. Since f is injective and has an inverse function given by $f^{-1}(y) = (y - 4)/3$, the set $f^{-1}(H)$ can also be thought of as the direct image of H under f^{-1}.

(b) Let $h : \boldsymbol{R} \to \boldsymbol{R}$ be defined by $h(x) := x^2 + 1$, $x \in \boldsymbol{R}$, and let $E := \{x : 0 \leqslant x \leqslant 2\}$. Then the direct image of E under h is the set $h(E) = \{y : 1 \leqslant y \leqslant 5\}$. The inverse image of $h(E)$ is the set $h^{-1}(h(E)) = \{x : -2 \leqslant x \leqslant 2\}$ which is *not* the same as E. Note that h is not injective; for example, we have $h^{-1}(\{5\}) = \{2, -2\}$, that is, $h(2) = h(-2) = 5$.

Sequences

Functions that have the set $\boldsymbol{N}$ of natural numbers as domain play a special role in analysis. For such functions we have special terminology and notation, which we now introduce.

1.2.11 Definition. A **sequence** in a set S is a function whose domain is the set $\boldsymbol{N}$ of natural numbers and whose range is contained in the set S.

For a sequence $\varphi : \boldsymbol{N} \to S$, the value of φ at n in $\boldsymbol{N}$ is often denoted by φ_n instead of $\varphi(n)$. The sequence itself is often given by the notation $(\varphi_n : n \in \boldsymbol{N})$ or more simply by (φ_n). For example, the sequence in $\boldsymbol{R}$ designated by $(\sqrt{n} : n \in \boldsymbol{N})$ is the same as the function $\varphi : \boldsymbol{N} \to \boldsymbol{R}$ defined by $\varphi(n) := \sqrt{n}$.

It is important to distinguish between a sequence $(\varphi_n : n \in \mathbf{N})$ and its range $\{\varphi_n : n : n \in \mathbf{N}\}$. The terms of a sequence should be viewed as having an order induced by the order of the natural numbers, whereas the range of a sequence is simply a set. For example, the terms of the sequence $((-1)^n : n \in \mathbf{N})$ alternate between -1 and 1, but the range of this sequence is the set $\{1, -1\}$ consisting of two elements.

Sequences will be studied at considerable length in Chapter 3.

Exercises for Section 1.2

1. Let $A := B := \mathbf{R}$ and consider the subset $C := \{(x, y) : x^2 + y^2 = 1\}$ of $A \times B$. Is this set a function with domain in $\mathbf{R}$ and range in $\mathbf{R}$?
2. Give an example of two functions f, g on $\mathbf{R}$ to $\mathbf{R}$ such that $f \neq g$, but such that $f \circ g = g \circ f$.
3. Prove that if f is an injection from A to B, then $f^{-1} := \{(b, a) : (a, b) \in f\}$ is a function. Then prove it is an injection.
4. Suppose f is an injection. Show that $f^{-1} \circ f(x) = x$ for all x in $D(f)$ and $f \circ f^{-1}(y) = y$ for all y in $R(f)$.
5. Let f and g be functions and suppose that $g \circ f(x) = x$ for all x in $D(f)$. Show that f is an injection and that $R(f) \subseteq D(g)$ and $R(g) \supseteq D(f)$.
6. Let f, g be functions such that

$$g \circ f(x) = x \qquad \text{for all } x \text{ in } D(f),$$

$$f \circ g(y) = y \qquad \text{for all } y \text{ in } D(g).$$

   Prove that $g = f^{-1}$.
7. Let f be the function on $\mathbf{R}$ to $\mathbf{R}$ given by $f(x) := x^2$, and let $E := \{x \in \mathbf{R} : -1 \leqslant x \leqslant 0\}$ and $F := \{x \in \mathbf{R} : 0 \leqslant x \leqslant 1\}$. Show that $E \cap F = \{0\}$ and $f(E \cap F) = \{0\}$ while $f(E) = f(F) = \{y \in \mathbf{R} : 0 \leqslant y \leqslant 1\}$. Hence $f(E \cap F)$ is a proper subset of $f(E) \cap f(F)$. Now delete 0 from E and F.
8. If f, E, F are as in Exercise 7, find the sets $E \setminus F$ and $f(E) \setminus f(F)$ and show that it is *not* true that $f(E \setminus F) \subseteq f(E) \setminus f(F)$.
9. Show that if $f : A \to B$ is an injection of A into B and if $H \subseteq B$, then the inverse image of H under f coincides with the direct image of H under the inverse function f^{-1}.
10. If f and g are as in Definition 1.2.2, show that $D(g \circ f) = f^{-1}(D(g))$.
11. Show that if $f : A \to B$ and E, F are subsets of A, then $f(E \cup F) = f(E) \cup f(F)$ and $f(E \cap F) \subseteq f(E) \cap f(F)$.
12. Show that if $f : A \to B$ and G, H are subsets of B, then $f^{-1}(G \cup H) = f^{-1}(G) \cup f^{-1}(H)$ and $f^{-1}(G \cap H) = f^{-1}(G) \cap f^{-1}(H)$.
13. Show that if $f : A \to B$ is injective and H is a subset of B, then $f(f^{-1}(H)) = H$. Also if E is a subset of A, then $f^{-1}(f(E)) = E$. Give examples to show that these equalities need not hold in general.
14. Show that if $f : A \to B$ and $g : B \to C$ and H is a subset of C, then we have $(g \circ f)^{-1}(H) = f^{-1}(g^{-1}(H))$.

SECTION 1.3 Mathematical Induction

Mathematical induction is an important method of proof that will be used frequently throughout this book. It is a method used to establish the validity of statements that are given in terms of the natural numbers. Though its utility is restricted to this rather special context, mathematical induction is an indispensable tool in all branches of mathematics. Since many induction proofs follow the same formal lines of argument, we shall often state only that a result follows from mathematical induction and leave it to the reader to supply the necessary details. In this section we shall state the Principle of Mathematical Induction and give several examples to illustrate how inductive proofs proceed.

We shall assume familiarity with the set of natural numbers

$$\mathbf{N} := \{1, 2, 3, \ldots\},$$

with the usual arithmetic operations of addition and multiplication and with the meaning of one natural number being less than another. We shall also assume the following fundamental property of $\mathbf{N}$.

1.3.1 Well-Ordering Property of *N*. *Every non-empty subset of* $\mathbf{N}$ *has a least element.*

A more detailed statement of this property is as follows: if S is a subset of $\mathbf{N}$ and if $S \neq \emptyset$, then there exists an element $m \in S$ such that $m \leqslant k$ for all $k \in S$.

On the basis of the Well-Ordering Property, we shall derive a version of the Principle of Mathematical Induction expressed in terms of subsets of $\mathbf{N}$. The property described in this version is sometimes referred to as the "hereditary" property of $\mathbf{N}$.

1.3.2 Principle of Mathematical Induction. *Let* S *be a subset of* $\mathbf{N}$ *that possesses the properties:*

(1) $1 \in S$;

(2) *if* $k \in S$, *then* $k + 1 \in S$.

Then $S = \mathbf{N}$.

Proof. Suppose to the contrary that $S \neq \mathbf{N}$. Then the set $\mathbf{N} \setminus S$ is not empty, and therefore by the Well-Ordering Property it contains a least element. Let m denote the least element of $\mathbf{N} \setminus S$. Because $1 \in S$ by hypothesis (1), we know that $m \neq 1$. Therefore $m > 1$ so that $m - 1$ is also a natural number. Since $m - 1 < m$ and since m is the least element of $\mathbf{N}$ such that $m \notin S$, it must be the case that $m - 1$ is in S.

We now apply the hypothesis (2) to the element $k = m - 1$ in S, and infer that $k + 1 = (m - 1) + 1 = m$ is in S. This conclusion contradicts the statement that m is not in S. Since m was obtained by assuming that $\mathbf{N} \setminus S$ was not empty, we are forced to the conclusion that $\mathbf{N} \setminus S$ is empty. Therefore we have shown that $S = \mathbf{N}$. Q.E.D.

The Principle of Mathematical Induction is often set forth in the framework of properties or statements about natural numbers. If $P(n)$ denotes a meaningful statement about $n \in \mathbf{N}$, then $P(n)$ may be true for some values of n and false for others. For example, if $P(n)$ is the statement: "$n^2 = n$", then $P(1)$ is true while $P(n)$ is false for all $n \neq 1$, $n \in \mathbf{N}$. In this context, the Principle of Mathematical Induction can be formulated as follows.

For each $n \in \mathbf{N}$, let $P(n)$ be a statement about n. Suppose that:

(1′) $P(1)$ is true;
(2′) if $P(k)$ is true, then $P(k + 1)$ is true.

Then $P(n)$ is true for all $n \in \mathbf{N}$.

The connection with the preceding version of mathematical induction, given in 1.3.2, is made by letting $S := \{n \in \mathbf{N}: P(n) \text{ is true}\}$. Then the conditions (1) and (2) of 1.3.2 correspond exactly to the conditions (1′) and (2′), respectively. The conclusion that $S = \mathbf{N}$ in 1.3.2 corresponds to the conclusion that $P(n)$ is true for all $n \in \mathbf{N}$.

The following examples illustrate how the Principle of Mathematical Induction is employed as a method of proving assertions about the natural numbers.

1.3.3 Examples. (a) For each $n \in \mathbf{N}$, the sum of the first n natural numbers is given by

$$1 + 2 + \cdots + n = \tfrac{1}{2}n(n + 1).$$

To prove this formula, we let S be the set of all $n \in \mathbf{N}$ for which the formula is true. We must verify that conditions (1) and (2) of 1.3.2 are satisfied. If $n = 1$, then we have $1 = \frac{1}{2}\cdot 1 \cdot (1 + 1)$ so that $1 \in S$. Thus condition (1) is satisfied. Next we *assume* that $k \in S$ and wish to infer from this assumption that $k + 1 \in S$. If $k \in S$, then we have

$$1 + 2 + \cdots + k = \tfrac{1}{2}k(k + 1).$$

If we add $k + 1$ to both sides of the assumed equality, we obtain

$$\begin{aligned} 1 + 2 + \cdots + k + (k + 1) &= \tfrac{1}{2}k(k + 1) + (k + 1) \\ &= \tfrac{1}{2}(k + 1)(k + 2). \end{aligned}$$

Since this is the stated formula for $n = k + 1$, we conclude that $k + 1 \in S$. Hence condition (2) of 1.3.2 is satisfied. Therefore by the Principle of Mathematical Induction, we conclude that $S = \mathbf{N}$ and the formula is valid for all $n \in \mathbf{N}$.

(b) For each $n \in \mathbf{N}$, the sum of the squares of the first n natural numbers is given by the formula

$$1^2 + 2^2 + \cdots + n^2 = \tfrac{1}{6}n(n+1)(2n+1).$$

To prove the validity of the formula, we first note that it is true for the case $n = 1$ since $1^2 = \frac{1}{6}\cdot 1\cdot 2\cdot 3$. If we assume it is true for k, then adding $(k+1)^2$ to both sides of the assumed equality gives

$$1^2 + 2^2 + \cdots + k^2 + (k+1)^2 = \tfrac{1}{6}k(k+1)(2k+1) + (k+1)^2$$

$$= \tfrac{1}{6}(k+1)(2k^2 + k + 6k + 6) = \tfrac{1}{6}(k+1)(k+2)(2k+3).$$

The validity of the formula for all $n \in \mathbf{N}$ follows by mathematical induction.

(c) Consider the statement "$n + 5 = n$" for $n \in \mathbf{N}$. If we let S be the set of all natural numbers for which the statement is true, then the assumption that $k \in S$ implies that $k + 1 \in S$. Therefore condition (2) of 1.3.2 is satisfied. However, since the statement is false for $n = 1$, condition (1) is not satisfied. Thus, we may not use mathematical induction to conclude that $n + 5 = n$ for all $n \in \mathbf{N}$.

(d) The inequality $2^n \leqslant (n+1)!$ can be established by mathematical induction as follows. We first observe that it is true for $n = 1$. Then if we assume that $2^k \leqslant (k+1)!$, it follows that

$$2^{k+1} = 2\cdot 2^k \leqslant 2(k+1)! \leqslant (k+2)(k+1)! = (k+2)!$$

Thus, if the inequality holds for k, it holds for $k + 1$. Therefore, by mathematical induction, the inequality is true for all $n \in \mathbf{N}$.

(e) If $r \in \boldsymbol{R}$, $r \neq 1$, and $n \in \mathbf{N}$, then

$$1 + r + r^2 + \cdots + r^n = \frac{1 - r^{n+1}}{1 - r}.$$

This is the formula for the sum of the terms in a geometric progression. It can be established on the basis of mathematical induction as follows. First, if $n = 1$, we have $1 + r = (1-r^2)/(1-r)$, so that the formula is valid in this case. If we assume the truth of the statement for $n = k$ and add the term r^{k+1} to both sides, then we get

$$1 + r + \cdots + r^k + r^{k+1} = \frac{1 - r^{k+1}}{1 - r} + r^{k+1} = \frac{1 - r^{k+2}}{1 - r},$$

which is the formula for the case $n = k+1$. It follows from the Principle of Mathematical Induction that the formula is true for all $n \in \mathbf{N}$.

Actually this result can be proved without employing mathematical induction. If we let $s_n := 1 + r + \cdots + r^n$, then $rs_n = r + r^2 + \cdots + r^{n+1}$ so that

$$(1 - r)s_n = s_n - rs_n = 1 - r^{n+1}.$$

If we solve for s_n, we obtain the stated formula.

(f) Careless use of the Principle of Mathematical Induction can lead to obviously absurd conclusions. The reader is invited to find the error in the "proof" of the following "theorem".

If n is any natural number, and if the maximum of two natural numbers p and q is n, then $p = q$. (Consequently, if p and q are any two natural numbers, then ($p = q$.) "*Proof*" Let S be the set of natural numbers for which the assertion is true. Then $1 \in S$ because if p and q are in $\mathbf{N}$ and if their maximum is 1, then $p = q = 1$. If we assume that $k \in S$ and that the maximum of the natural numbers p and q is $k + 1$, then the maximum of $p - 1$ and $q - 1$ is k. Therefore $p - 1 = q - 1$ since $k \in S$, and hence we conclude that $p = q$. Thus, $k + 1 \in S$ and we conclude that the assertion is true for all $n \in \mathbf{N}$.

(g) There are statements that are true for many natural numbers, but that are not true for *all* of them. For example the formula $p(n) = n^2 - n + 41$ gives a prime number for $n = 1, 2, \ldots, 40$. However, $p(41)$ is evidently not a prime number.

There is another version of the Principle of Mathematical Induction that is sometimes quite useful. It is often referred to as the "Principle of Strong Induction" even though it is in fact equivalent to the previous version. We shall leave it to the reader to establish the equivalence of these two versions.

1.3.4 Principle of Strong Induction. *Let S be a subset of $\mathbf{N}$ such that $1 \in S$, and if $\{1, 2, \ldots, k\} \subseteq S$ then $k + 1 \in S$. Then $S = \mathbf{N}$.*

Exercises for Section 1.3

1. Prove that $1^3 + 2^3 + \cdots + n^3 = [\frac{1}{2}n(n + 1)]^2$ for all $n \in \mathbf{N}$.
2. Prove that $n < 2^n$ for all $n \in \mathbf{N}$.
3. Prove that the sum of the cubes of any three consecutive natural numbers n, $n + 1$, $n + 2$ is divisible by 9.
4. Conjecture a formula for the sum of the first n odd natural numbers $1 + 3 + \cdots + (2n - 1)$, and check your conjecture by using mathematical induction.
5. Prove the following variation of 1.3.2. Let S be a non-empty subset of $\mathbf{N}$ such that for some $n_0 \in \mathbf{N}$ it is true that (a) $n_0 \in S$; and (b) if $k \geq n_0$ and if $k \in S$, then $k + 1 \in S$. Then S contains the set $\{n \in \mathbf{N}: n \geq n_0\}$.
6. Prove that $2^n < n!$ for all $n \geq 4$, $n \in \mathbf{N}$. (See Exercise 5.)
7. For which natural numbers is it true that $n^2 < 2^n$? Prove your assertion. (See Exercise 5.)
8. Let S be a subset of $\mathbf{N}$ such that (a) 2^k is in S for all $k \in \mathbf{N}$, and (b) if $k \in S$, and $k \geq 2$, then $k - 1 \in S$. Prove that $S = \mathbf{N}$.
9. Let the sequence (x_n) be defined as follows: $x_1 := 1$, $x_2 := 2$, and $x_{n+2} := \frac{1}{2}(x_{n+1} + x_n)$ for $n \in \mathbf{N}$. Use the Principle of Strong Induction 1.3.4 to show that $1 \leq x_n \leq 2$ for all $n \in \mathbf{N}$.

SECTION 1.4 Infinite Sets

The purpose of this section is very restricted; it is to introduce the terms "finite", "infinite", and "countable". We shall use these notions to contrast the set of rational numbers and the set of irrational numbers by showing that the former set is countably infinite but the latter set is not. The general theories of infinite sets, cardinal numbers, and ordinal numbers are extensive and fascinating in their own right, but we shall not pursue these studies. It turns out that very little exposure to these topics is essential for the material in this text.

1.4.1 Definition A set B is **finite** if it is empty or if there is a bijection with domain B and range in an initial segment $\{1, 2, \ldots, n\}$ of $\mathbf{N}$. If there is no such function, the set is **infinite**. If there is a bijection of B onto $\mathbf{N}$, then the set B is **denumerable** (or **enumerable**). If a set is either finite or denumerable, it is said to be **countable**.

When there is an injective (or one-one) function with domain B and range C, we sometimes say that B can be put into *one-one correspondence* with C. By using this terminology, we rephrase Definition 1.4.1 and say that a set B is finite if it is empty or can be put into one-one correspondence with a subset of an initial segment of $\mathbf{N}$. We say that B is denumerable if it can be put into one-one correspondence with all of $\mathbf{N}$.

It will be noted that, by definition, a set B is either finite or infinite. However, it may be that, owing to the description of the set, it may not be a trivial matter to decide whether the given set B is finite or infinite.

The subsets of $\mathbf{N}$ denoted by $\{1, 3, 5\}$, $\{2, 4, 5, 8, 10\}$, $\{2, 3, \ldots, 100\}$, are finite since, although they are not initial segments of $\mathbf{N}$, they are contained in initial segments of $\mathbf{N}$ and hence can be put into one-one correspondence with subsets of initial segments of $\mathbf{N}$. The set E of even natural numbers

$$E := \{2, 4, 6, 8, \ldots\}$$

and the set O of odd natural numbers

$$O := \{1, 3, 5, 7, \ldots\}$$

are not initial segments of $\mathbf{N}$. However, since they can be put into one-one correspondence with all of $\mathbf{N}$ (how?), they are both denumerable.

Even though the set $\mathbf{Z}$ of all integers

$$\mathbf{Z} := \{\ldots, -2, -1, 0, 1, 2, \ldots\},$$

contains the set $\mathbf{N}$, it may be seen that $\mathbf{Z}$ is a denumerable set. (How?)

We now state some theorems without proof. At first reading it is probably best to accept them without further examination; on a later reading, however, the reader will do well to attempt to provide proofs for these statements. In doing so, he will find the inductive property of the set $\boldsymbol{N}$ of natural numbers to be useful.

1.4.2 Theorem. *A set B is countable if and only if there is an injection with domain B and range in $\boldsymbol{N}$.*

1.4.3 Theorem. *Any subset of a finite set is finite. Any subset of a countable set is countable.*

1.4.4 Theorem. *The union of a finite collection of finite sets is a finite set. The union of a countable collection of countable sets is a countable set.*

It is a consequence of the second part of Theorem 1.4.4 that the set $\boldsymbol{Q}$ of all rational numbers forms a countable set. (We recall that a rational number is a fraction m/n, where m and n are integers and $n \neq 0$.) To see that $\boldsymbol{Q}$ is a countable set we form the sets

$$
\begin{aligned}
A_0 &:= \{0\}, \\
A_1 &:= \left\{\frac{1}{1}, -\frac{1}{1}, \frac{2}{1}, -\frac{2}{1}, \frac{3}{1}, -\frac{3}{1}, \ldots\right\}, \\
A_2 &:= \left\{\frac{1}{2}, -\frac{1}{2}, \frac{2}{2}, -\frac{2}{3}, \frac{3}{2}, -\frac{3}{2}, \ldots\right\}, \\
&\ldots\ldots\ldots\ldots\ldots\ldots\ldots\ldots \\
A_n &:= \left\{\frac{1}{n}, -\frac{1}{n}, \frac{2}{n}, -\frac{2}{n}, \frac{3}{n}, -\frac{3}{n}, \ldots\right\}, \\
&\ldots\ldots\ldots\ldots\ldots\ldots\ldots\ldots
\end{aligned}
$$

Note that each of the sets A_n is countable and that their union is all of $\boldsymbol{Q}$. Hence Theorem 1.4.4 asserts that $\boldsymbol{Q}$ is countable. In fact, we can enumerate $\boldsymbol{Q}$ by the "diagonal procedure":

$$
0, \frac{1}{1}, -\frac{1}{1}, \frac{1}{2}, \frac{2}{1}, -\frac{1}{2}, \frac{1}{3}, \ldots
$$

By using this type of argument, the reader should be able to construct a proof of Theorem 1.4.4.

The Uncountability of *R* and *I*

Despite the fact that the set of rational numbers is countable, the entire set $\boldsymbol{R}$ of real numbers is not countable. In fact, the set $\boldsymbol{I}$ of real numbers x satisfying

$0 \leqslant x \leqslant 1$ is not countable. To demonstrate this, we shall use the elegant "diagonal" argument of G. Cantor. We assume it is known that every real number x with $0 \leqslant x \leqslant 1$ has a decimal representation in the form $x = 0.a_1a_2a_3 \cdots$, where each a_k denotes one of the digits 0, 1, 2, 3, 4, 5, 6, 7, 8, 9. It is to be realized that certain real numbers have two representations in this form (for example, the rational number $\frac{1}{10}$ has the two representations

$$0.1000 \cdots \qquad \text{and} \qquad 0.0999 \cdots).$$

We could decide in favor of one of these two representations, but it is not necessary to do so. Since there are infinitely many rational numbers in the interval $0 \leqslant x \leqslant 1$ (why?), the set $\boldsymbol{I}$ cannot be finite. We shall now show that it is not denumerable. Suppose that there is an enumeration $x_1, x_2, x_3 \ldots$ of all real numbers satisfying $0 \leqslant x \leqslant 1$ given by

$$x_1 = 0.a_1a_2a_3 \cdots$$

$$x_2 = 0.b_1b_2b_3 \cdots$$

$$x_3 = 0.c_1c_2c_3 \cdots$$

$$\cdots\cdots\cdots\cdots$$

Now let $y_1 = 2$ if $a_1 \geqslant 5$ and let $y_1 = 7$ if $a_1 \leqslant 4$; let $y_2 = 2$ if $b_2 \geqslant 5$ and let $y_2 = 7$ if $b_2 \leqslant 4$; and so on. Consider the number y with decimal representation

$$y = 0.y_1y_2y_3 \cdots$$

which clearly satisfies $0 \leqslant y \leqslant 1$. The number y is not one of the numbers with two decimal representations, since $y_n \neq 0, 9$. At the same time $y \neq x_n$ for any n (since the nth digits in the decimal representations for y and x_n are different). Therefore, any denumerable collection of real numbers in this interval will omit at least one real number belonging to this interval. Therefore, this interval is not a countable set.

The fact that the set $\boldsymbol{R}$ of real numbers is uncountable can be combined with the fact that the set $\boldsymbol{Q}$ of rational numbers is countable to conclude that the set $\boldsymbol{R} \setminus \boldsymbol{Q}$ is uncountable. Indeed, since the union of two countable sets is a countable set by Theorem 1.4.4, the assumed countability of $\boldsymbol{R} \setminus \boldsymbol{Q}$ would imply that $\boldsymbol{R}$, the union of $\boldsymbol{Q}$ and $\boldsymbol{R} \setminus \boldsymbol{Q}$, would be a countable set. From this contradiction, we infer that the set of irrational numbers $\boldsymbol{R} \setminus \boldsymbol{Q}$ is an uncountable set.

Suppose that a set A is infinite; we shall *suppose* that there is a one-one correspondence with a subset of A and all of $\boldsymbol{N}$. In other words, we *assume that every infinite set contains a denumerable subset.* This assertion is a weak form of the so-called Axiom of Choice, which is one of the usual axioms of set theory. After the reader has digested the contents of this book, he may turn to an axiomatic treatment of the foundations that we have been discussing in a somewhat informal fashion. However, for the moment he would do well to take the above statement as a temporary axiom. It can be replaced later by a more far-reaching axiom of set theory.

Exercises for Section 1.4

1. Exhibit a one-one correspondence between the set E of even natural numbers and the set $\mathbf{N}$.
2. Exhibit a one-one correspondence between the set O of odd natural numbers and the set $\mathbf{N}$.
3. Exhibit a one-one correspondence between $\mathbf{N}$ and a proper subset of $\mathbf{N}$.
4. If A is contained in some initial segment $\{1, 2, \ldots, n\}$ of $\mathbf{N}$, use the well-ordering property of $\mathbf{N}$ to define a bijection of A onto some initial segment of $\mathbf{N}$.
5. Give an example of a countable collection of finite sets whose union is not finite.
6. Use the fact that every infinite set has a denumerable subset to show that every infinite set can be put into one-one correspondence with a proper subset of itself.
7. Show that if the set A can be put into one-one correspondence with a set B, then B can be put into one-one correspondence with A.

CHAPTER TWO

THE REAL NUMBERS

In this chapter we shall discuss the properties of the real number system $\boldsymbol{R}$. Although it would be possible to present a formal construction of this system on the basis of a more primitive set (such as the set $\boldsymbol{N}$ of natural numbers or the set $\boldsymbol{Q}$ of rational numbers), we shall not do so. Instead, we shall exhibit a list of fundamental properties that are associated with the real numbers and show how other properties can be deduced from the ones assumed.

The real number system can be described as a "complete ordered field". However, for the sake of clarity, we prefer not to state all the properties of the real number system at once. Instead, we first introduce, in Section 2.1, the "algebraic" properties (often called the "field" properties) that are based on the two operations of addition and multiplication. We next introduce, in Section 2.2, the "order" properties and some of their consequences along with several inequalities that illustrate the use of these properties. The notion of absolute value, which is based on the order properties, is briefly discussed in Section 2.3. In Section 2.4, we make the final step by adding the "completeness" property to the algebraic and order properties of $\boldsymbol{R}$.

This procedure may appear somewhat piecemeal, but there are a number of properties involved and it is well to take a few at a time. Furthermore, the proofs required in the preliminary stages are different in nature from some of the later proofs. Finally, since there are several other ways of discussing the notion of "completeness", we wish to have this property isolated from the other assumptions.

Part of the purpose of Sections 2.1, 2.2, and 2.3 is to provide examples of proofs of elementary theorems that are derived from explicitly stated assumptions. Students who have not had much exposure to formal proofs can gain experience before proceeding to more complicated arguments. However, students who are familiar with the axiomatic method and the technique of proofs can go to Section 2.4 after a cursory look at the earlier sections.

In Section 2.5 we establish the Nested Interval Theorem and the important Bolzano-Weierstrass Theorem. There is also a brief discussion of binary and decimal representations of real numbers based on nested intervals. We conclude the chapter with a brief introduction to open sets and closed sets in $\boldsymbol{R}$, given in Section 2.6.

SECTION 2.1 The Algebraic Properties of $\boldsymbol{R}$

In this section we shall discuss the "algebraic structure" of the real number system. This is done by first giving a list of basic properties of addition and multiplication. This list embodies all the essential algebraic properties of $\boldsymbol{R}$ in the sense that all other such properties can be deduced as theorems. In the terminology of abstract algebra, the system of real numbers is a "field" with respect to addition and multiplication. The properties listed in 2.1.1 are known as the "field axioms".

By a **binary operation** on a set F we mean a function B with domain $F \times F$ and range in F. Thus, a binary operation associates with each ordered pair (a,b) of elements of the set F a unique element $B(a,b)$ in F. However, instead of using the notation $B(a,b)$, we use the conventional notations of $a + b$ and $a \cdot b$ (or merely ab) when discussing the properties of addition and multiplication. Examples of other binary operations can be found in the exercises.

2.1.1 Algebraic Properties of $\boldsymbol{R}$ On the set $\boldsymbol{R}$ of real numbers there are two binary operations, denoted by $+$ and $\cdot$ and called **addition** and **multiplication**, respectively. These operations satisfy the following properties:

(A1) $a + b = b + a$ for all a, b in $\boldsymbol{R}$ (*commutative property of addition*);
(A2) $(a + b) + c = a + (b + c)$ for all a, b, c in $\boldsymbol{R}$ (*associative property of addition*);
(A3) there exists an element 0 in $\boldsymbol{R}$ such that $0 + a = a$ and $a + 0 = a$ for all a in $\boldsymbol{R}$ (*existence of a zero element*);
(A4) for each a in $\boldsymbol{R}$ there exists an element $-a$ in $\boldsymbol{R}$ such that $a + (-a) = 0$ and $(-a) + a = 0$ (*existence of negative elements*);
(M1) $a \cdot b = b \cdot a$ for all a, b in $\boldsymbol{R}$ (*commutative property of multiplication*);
(M2) $(a \cdot b) \cdot c = a \cdot (b \cdot c)$ for all a, b, c in $\boldsymbol{R}$ (*associative property of multiplication*);
(M3) there exists an element 1 in $\boldsymbol{R}$ *distinct from* 0 such that $1 \cdot a = a$ and $a \cdot 1 = a$ for all a in $\boldsymbol{R}$ (*existence of a unit element*);
(M4) for each $a \neq 0$ in $\boldsymbol{R}$ there exists an element $1/a$ in R such that $a \cdot (1/a) = 1$ and $(1/a) \cdot a = 1$ (*existence of reciprocals*);
(D) $a \cdot (b + c) = (a \cdot b) + (a \cdot c)$ and $(b + c) \cdot a = (b \cdot a) + (c \cdot a)$ for all a, b, c in $\boldsymbol{R}$ (*distributive property of multiplication over addition*).

These properties should be familiar to the reader. The list is not "minimal" since, for convenience, we have included a few redundancies. For example, the second assertions in (A3) and (A4) follow from the first assertions by using (A1).

To illustrate the basic nature of the listed properties, we now deduce several elementary (but important) consequences. We begin by proving that the elements 0 and 1, whose existence were asserted in (A3) and (M3), are in fact unique.

2.1.2 Theorem. (a) *If z and a are elements of* **R** *such that $z + a = a$, then $z = 0$.*

(b) *If u and $b \neq 0$ are elements of* **R** *such that $u \cdot b = b$, then $u = 1$.*

Proof. (a) The hypothesis is that $z + a = a$. We add the element $-a$, whose existence is given in (A4), to both sides. Then using (A4), (A2), (A4), and (A3) in succession, we obtain

$$0 = a + (-a) = (z + a) + (-a)$$
$$= z + \big(a + (-a)\big) = z + 0 = z.$$

The proof of part (b) is left as an exercise. Note that the hypothesis $b \neq 0$ is crucial. Q.E.D.

We next show that, for a given element a in **R**, the element $-a$ and the element $1/a$ (when $a \neq 0$) are uniquely determined.

2.1.3 Theorem. (a) *If a and b are elements of* **R** *such that $a + b = 0$, then $b = -a$.*

(b) *If $a \neq 0$ and b are elements of* **R** *such that $a \cdot b = 1$, then $b = 1/a$.*

Proof. (a) If $a + b = 0$, then we add $-a$ to both sides to get

$$(-a) + (a + b) = (-a) + 0.$$

We now use (A2) on the left side and (A3) on the right side to obtain $\big((-a) + a\big) + b = -a$. Now apply (A4) and (A3) on the left side, to get $b = -a$.

The proof of part (b) is left as an exercise. Note that the hypothesis $a \neq 0$ is used. Q.E.D.

If we view the preceding properties in terms of solving equations, we note that (A4) and (M4) enable us to solve the equations $a + x = 0$ and $a \cdot x = 1$ (when $a \neq 0$) for x, and Theorem 2.1.3 implies that the solutions are unique. The next theorem shows that the right-hand sides of these equations can be any element of **R**.

2.1.4 Theorem. *Let a, b be arbitrary elements of* **R**. *Then:*

(a) *the equation $a + x = b$ has the unique solution $x = (-a) + b$;*

(b) *if $a \neq 0$, the equation $a \cdot x = b$ has the unique solution $x = (1/a) \cdot b$.*

Proof. Since $a + \big((-a) + b\big) = \big(a + (-a)\big) + b = 0 + b = b$ by properties (A2), (A4), (A3), we see that $x = (-a) + b$ is a solution of the equation $a + x = b$. To establish that it is the only solution, supppose that x_1 is any solution of this equation; then

$$a + x_1 = b.$$

If we add $-a$ to both sides, we get

$$(-a) + (a + x_1) = (-a) + b.$$

Use (A3), (A4), and (A2), to get

$$\begin{aligned} x_1 = 0 + x_1 &= (-a + a) + x_1 \\ &= (-a) + (a + x_1) = (-a) + b. \end{aligned}$$

Hence,

$$x_1 = (-a) + b.$$

The proof of part (b) is left as an exercise. Q.E.D.

In the three theorems established so far, we have considered the properties of addition and multiplication separately. To examine the interplay between the two operations, we must employ the distributive property (D). This is illustrated in the next theorem.

2.1.5 Theorem. *If a is any element of $\mathbf{R}$, then:*
(a) $a \cdot 0 = 0$;
(b) $(-1) \cdot a = -a$;
(c) $-(-a) = a$;
(d) $(-1) \cdot (-1) = 1$.

Proof. (a) From (M3) we know that $a \cdot 1 = a$. Then (D) and (A3) give

$$\begin{aligned} a + a \cdot 0 &= a \cdot 1 + a \cdot 0 \\ &= a \cdot (1 + 0) = a \cdot 1 = a. \end{aligned}$$

Since 0 is unique, by Theorem 2.1.2(a), we conclude that $a \cdot 0 = 0$.

(b) We use (D), in conjunction with (M3), (A4) and part (a), to obtain

$$\begin{aligned} a + (-1) \cdot a &= 1 \cdot a + (-1) \cdot a \\ &= \big(1 + (-1)\big) \cdot a = 0 \cdot a = 0. \end{aligned}$$

Thus, from Theorem 2.1.3(a), we conclude that $(-1) \cdot a = -a$.

(c) By (A4) we have $(-a) + a = 0$. According to the uniqueness assertion of Theorem 2.1.3(a), it follows that $a = -(-a)$.

(d) In part (b), substitute $a = -1$. Then

$$(-1) \cdot (-1) = -(-1).$$

Hence, the assertion follows from part (c) with $a = 1$. Q.E.D.

We conclude our formal deductions from the field properties with the following important results.

2.1.6 Theorem. *Let a, b, c, be elements of* $\boldsymbol{R}$.
(a) *If $a \neq 0$, then $1/a \neq 0$ and $1/(1/a) = a$.*
(b) *If $a \cdot b = a \cdot c$ and $a \neq 0$, then $b = c$.*
(c) *If $a \cdot b = 0$, then either $a = 0$ or $b = 0$.*

Proof. (a) We are given $a \neq 0$, so that $1/a$ exists. If $1/a = 0$, then $1 = a \cdot (1/a) = a \cdot 0 = 0$, contrary to (M3). Thus $1/a \neq 0$ and since $(1/a) \cdot a = 1$, the uniqueness asserted in Theorem 2.1.3(b) implies $1/(1/a) = a$.

(b) If we multiply both sides of the equation $a \cdot b = a \cdot c$ by $1/a$ and apply the associative property (M2), we get

$$((1/a) \cdot a) \cdot b = ((1/a) \cdot a) \cdot c.$$

Thus, $1 \cdot b = 1 \cdot c$ which is the same as $b = c$.

(c) It suffices to assume $a \neq 0$ and deduce that $b = 0$. (Why?) Since $a \cdot b = 0 = a \cdot 0$, we apply part (b) to the equation $a \cdot b = a \cdot 0$ to conclude that $b = 0$ if $a \neq 0$. Q.E.D.

The above theorems represent a small, but important, sample of the algebraic properties of the real numbers. Many additional consequences of the field properties of $\boldsymbol{R}$ can be deduced, and some additional ones are given in the exercises.

The operation of **subtraction** is defined by $a - b := a + (-b)$ for a, b in $\boldsymbol{R}$. Similarly, **division** is defined for a, b in $\boldsymbol{R}$, $b \neq 0$, by $a/b := a \cdot (1/b)$. In the following, we shall use this customary notation for subtraction and division and freely use the familiar properties of these operations without further elaboration. Similarly, from now on we shall generally drop the use of the dot to denote multiplication and write ab for $a \cdot b$. As usual, we shall write a^2 for aa, a^3 for $(a^2)a$; more generally, for $n \in \boldsymbol{N}$, we define $a^{n+1} := (a^n) \cdot a$. We agree to adopt the convention that $a^0 = 1$ and $a^1 = a$ for any $a \in \boldsymbol{R}$. Also, if $a \neq 0$, the notation a^{-1} will be used for $1/a$ and if $n \in \boldsymbol{N}$ we write a^{-n} for $(1/a)^n$ when it is convenient to do so.

Rational Numbers

Elements of $\boldsymbol{R}$ that can be written in the form b/a where $a, b \in \boldsymbol{Z}$ and $a \neq 0$ are called **rational numbers.** The set of all rational numbers in $\boldsymbol{R}$ will be denoted by the standard notation $\boldsymbol{Q}$. The sum and product of two rational numbers is again a rational number. (Prove this.) Moreover, the field properties listed at the beginning of this section can be shown to hold for $\boldsymbol{Q}$.

The fact that there are elements of $\boldsymbol{R}$ that are not in $\boldsymbol{Q}$ is not immediately apparent. In the sixth century B.C. the ancient Greek Pythagoreans discovered that the diagonal of a square with unit sides could not be expressed as a ratio of integers. In view of the Pythagorean Theorem for right triangles, this implies that the square of no rational number can equal 2. This discovery had considerable impact on the development of Greek mathematics. One consequence is that elements of $\boldsymbol{R}$ that are not in $\boldsymbol{Q}$ became known as **irrational numbers.** Though this terminology is unfortunate, it is quite standard and we shall adopt it.

We close this section with a proof of the fact that there does not exist a rational number whose square is 2.

2.1.7 Theorem. *There does not exist a rational number r such that $r^2 = 2$.*

Proof. Suppose, on the contrary, that p and q are integers such that $(p/q)^2 = 2$. It may be assumed that p and q have no common integer factors other than 1. (Why?) Since $p^2 = 2q^2$, we see that p^2 is even. This implies that p is even (for if $p = 2k + 1$ is odd, then $p^2 = 4k^2 + 4k + 1 = 2(2k^2 + 2k) + 1$ is also odd). Therefore, q must be an odd integer. However, $p = 2m$ for some integer m because p is even, and hence $4m^2 = 2q^2$ so that $2m^2 = q^2$. As above, it follows that q is an even integer, and we have arrived at a contradiction to the fact that no integer is both even and odd. Q.E.D.

Exercises for Section 2.1

1. Prove part (b) of Theorem 2.1.2.
2. Prove part (b) of Theorem 2.1.3.
3. Solve the following equations, justifying each step by referring to an appropriate property or theorem.
   (a) $2x + 5 = 8$.
   (b) $x^2 = 2x$.
   (c) $(x - 1)(x + 2) = 0$.
4. Prove that if $a, b, \in \boldsymbol{R}$, then
   (a) $-(a + b) = (-a) + (-b)$.
   (b) $(-a)\cdot(-b) = a\cdot b$.
   (c) $1/(-a) = -(1/a)$ if $a \neq 0$.
5. If $a \in \boldsymbol{R}$ satisfies $a\cdot a = a$, prove that either $a = 0$ or $a = 1$.
6. If $a \neq 0$ and $b \neq 0$, show that $1/(ab) = (1/a)\cdot(1/b)$.
7. Use the argument in the proof of Theorem 2.1.7 to show that there does not exist a rational number s such that $s^2 = 6$.
8. Modify the argument in the proof of Theorem 2.1.7 to show that there does not exist a rational number t such that $t^2 = 3$.
9. Show that if $\xi \in \boldsymbol{R}$ is irrational and $r \neq 0$ is rational, then $r + \xi$ and $r\xi$ are irrational.
10. If x and y are rational numbers, show that $x + y$ and xy are rational.

11. If x and y are irrational numbers, show that it does *not* follow that $x + y$ and xy are irrational.
12. Let B be a binary operation on $\boldsymbol{R}$. We say that B:
(i) is **commutative** if $B(a,b) = B(b,a)$ for all a, b in $\boldsymbol{R}$;
(ii) is **associative** if $B(a, B(b,c)) = B(B(a,b), c)$ for all a, b, c in $\boldsymbol{R}$;
(iii) has an **identity** if there exists an element $e \in \boldsymbol{R}$ such that $B(a,e) = a = B(e,a)$ for all $a \in \boldsymbol{R}$.
Determine which of these properties hold for the following binary operations, defined for all a, $b \in \boldsymbol{R}$, by:
(a) $B_1(a,b) := \frac{1}{2}(a + b)$; (b) $B_2(a,b) := \frac{1}{2}(ab)$;
(c) $B_3(a,b) := a - b$; (d) $B_4(a,b) := 1 + ab$.
13. A binary operation B on $\boldsymbol{R}$ is said to be **distributive over addition** if it satisfies $B(a,b + c) = B(a,b) + B(a,c)$ for all a, b, c in $\boldsymbol{R}$. Which (if any) of the binary operations given in Exercise 12 are distributive over addition?
14. Use induction to show that if $a \in \boldsymbol{R}$ and m, $n \in \boldsymbol{N}$, then $a^{m+n} = a^m a^n$ and $(a^m)^n = a^{mn}$.

SECTION 2.2 The Order Properties of R

In this section we shall introduce the "order properties" of $\boldsymbol{R}$. This entails the notions of positivity and inequalities. As with the algebraic structure of the system of real numbers, we proceed by isolating several basic order properties from which the other order properties follow. The simplest way to do this is to make use of the notion of "positivity".

2.2.1 The Order Properties of R There is a non-empty subset P of $\boldsymbol{R}$, called the set of **strictly positive real numbers**, that satisfies the following properties:
(i) If a,b belong to P, then $a + b$ belongs to P.
(ii) If a,b belong to P, then ab belongs to P.
(iii) If a belongs to $\boldsymbol{R}$, then exactly one of the following holds:

$$a \in P, \quad a = 0, \quad -a \in P.$$

The first two properties insure the compatibility of order with the operations of addition and multiplication, respectively. Condition (iii) is called the **Trichotomy Property** since it divides $\boldsymbol{R}$ into three distinct types of elements. It states that the set $\{-a : a \in P\}$ of **strictly negative** real numbers has no elements in common with P, and, moreoever, the set $\boldsymbol{R}$ is the union of three disjoint sets.

2.2.2 Definition. If a is in P, we say that a is a **strictly positive** real number and write $a > 0$. If a is either in P or is 0, we say that a is a **positive** real number and write $a \geqslant 0$. If $-a$ is in P, we say that a is a **strictly negative** real number and write $a < 0$. If $-a$ is either in P or is 0, we say that a is a **negative** real number and write $a \leqslant 0$.

Remark. It should be noted that, according to the terminology just introduced, the number 0 is *both* positive and negative; it is the only number with this dual status. This terminology is not completely standard, but it will prove to be a convenience. Some authors reserve the term "positive" for the elements of the set P and use the term "non-negative" for the elements of $P \cup \{0\}$.

We can now introduce the notion of inequality between elements of $\boldsymbol{R}$ in terms of the set P of positive elements.

2.2.3 Definition. Let a,b be elements of $\boldsymbol{R}$.
(i) If $a - b \in P$, then we write $a > b$ or $b < a$.
(ii) If $a - b \in P \cup \{0\}$, then we write $a \geqslant b$ or $b \leqslant a$.

For notational convenience, we shall write

$$a < b < c$$

to mean that both

$$a < b \qquad \text{and} \qquad b < c$$

are satisfied. Similarly, if $a \leqslant b$ and $b \leqslant c$, we shall write $a \leqslant b \leqslant c$. Also, if $a \leqslant b$ and $b < d$, we shall write $a \leqslant b < d$.

Properties of the Order

We shall now establish some of the basic properties of the order relation on $\boldsymbol{R}$. These are the familiar "rules of inequalities" that the reader has used in earlier mathematics courses. They will be used frequently in later sections.

2.2.4 Theorem. *Let a, b, c be elements of* $\boldsymbol{R}$.
(a) *If $a > b$ and $b > c$, then $a > c$.*
(b) *Exactly one of the following holds:* $a > b$, $a = b$, $a < b$.
(c) *If $a \geqslant b$ and $b \geqslant a$, then $a = b$.*

Proof. (a) If $a - b \in P$ and $b - c \in P$, then 2.2.1(i) implies that $(a - b) + (b - c) = a - c$ belongs to P. Hence, $a > c$.

(b) By the Trichotomy Property 2.2.1(iii), exactly one of the following possibilities occurs: $a - b \in P$, $a - b = 0$, $-(a - b) = b - a \in P$.

(c) If $a \neq b$, then $a - b \neq 0$, so from part (b) we have either $a - b \in P$ or $b - a \in P$; that is, either $a > b$ or $b > a$. In either case, one of the hypotheses is contradicted. Therefore we must have $a = b$. Q.E.D.

It is natural to expect that the natural numbers are strictly positive. We show next how this positivity is derived from the basic properties given in 2.2.1.

The key observation is that the square of any non-zero real number is strictly positive.

2.2.5 Theorem. (a) *If* $a \in \mathbf{R}$ *and* $a \neq 0$, *then* $a^2 > 0$.
(b) $1 > 0$.
(c) *If* $n \in \mathbf{N}$, *then* $n > 0$.

Proof. (a) By the Trichotomy Property if $a \neq 0$, then either $a \in P$ or $-a \in P$. If $a \in P$, then by 2.2.1(ii), we have $a^2 = a \cdot a \in P$. Similarly, if $-a \in P$, then by 2.2.1(ii), we have $(-a)(-a) \in P$. From 2.1.5(b) and 2.1.5(d) it follows that

$$(-a)(-a) = ((-1)a)((-1)a) = (-1)(-1) \cdot a^2 = a^2,$$

whence it follows that $a^2 \in P$. We conclude that if $a \neq 0$, then $a^2 > 0$.

(b) Since $1 = (1)^2$, part (a) implies that $1 > 0$.

(c) Use mathematical induction. The validity of the assertion with $n = 1$ is just part (b). If we suppose the assertion is true for the natural number k, then $k \in P$. Since $1 \in P$, we then have $k + 1 \in P$ by 2.2.1(i). Hence, the assertion is true for all natural numbers. Q.E.D.

The following properties relate the order in $\mathbf{R}$ to addition and multiplication. They provide some of the tools with which we work in dealing with inequalities.

2.2.6 Theorem. *Let* a, b, c, d *be elements of* $\mathbf{R}$.
(a) *If* $a > b$, *then* $a + c > b + c$.
(b) *If* $a > b$ *and* $c > d$, *then* $a + c > b + d$.
(c) *If* $a > b$ *and* $c > 0$, *then* $ca > cb$.
If $a > b$ *and* $c < 0$, *then* $ca < cb$.
(d) *If* $a > 0$, *then* $1/a > 0$.
If $a < 0$, *then* $1/a < 0$.

Proof. (a) The conclusion follows from the fact that $(a + c) - (b + c) = a - b > 0$.

(b) If $a - b \in P$ and $c - d \in P$, then $(a + c) - (b + d) = (a - b) + (c - d)$ also belongs to P by 2.2.1(i). Thus, $a + c > b + d$.

(c) If $a - b \in P$ and $c \in P$, then $ca - cb = c(a-b) \in P$ by 2.2.1(ii), and thus $ca > cb$ when $c > 0$.

On the other hand, if $c < 0$, then $-c \in P$ so that $cb - ca = (-c)(a - b) \in P$. Hence, $cb > ca$ when $c < 0$.

(d) If $a > 0$, then $a \neq 0$ (by the Trichotomy Property), so that $1/a \neq 0$ by 2.1.6(a). If $1/a < 0$, then part (c) with $c = 1/a$ implies that $1 = a(1/a) < 0$, contradicting 2.2.5(b). Therefore, we have $1/a > 0$ since the other two possibilities have been excluded.

Similarly, if $a < 0$, then the possibility $1/a > 0$ again leads to the contradiction $1 = a(1/a) < 0$. Q.E.D.

Combining 2.2.5(c) and 2.2.6(d), we see that the reciprocal $1/n$ of any natural number n is strictly positive. Consequently, rational numbers that have the form $m/n = m(1/n)$, where m, n are natural numbers, are strictly positive.

2.2.7 Theorem. *If a and b are in $\mathbf{R}$ and if $a > b$, then $a > \frac{1}{2}(a + b) > b$.*

Proof. Since $a > b$, it follows from 2.2.6(a) that $2a = a + a > a + b$ and also that $a + b > b + b = 2b$. Therefore we have

$$2a > a + b > 2b.$$

By 2.2.5(c), we have $2 > 0$; therefore, by 2.2.6(d), it follows that $\frac{1}{2} > 0$. We infer from 2.2.6(c) that

$$a = \tfrac{1}{2}(2a) > \tfrac{1}{2}(a + b) > \tfrac{1}{2}(2b) = b.$$

Hence $a > \frac{1}{2}(a + b) > b$. Q.E.D.

It is worth noting that the elementary order properties that have been established so far are enough to show that no smallest strictly positive real number can exist.

2.2.8 Corollary. *If $a \in \mathbf{R}$ and $a > 0$, then $a > \frac{1}{2}a > 0$.*

Proof. Take $b = 0$ in 2.2.7. Q.E.D.

2.2.9 Theorem. *If $a \in \mathbf{R}$ is such that $0 \leq a < \varepsilon$ for every strictly positive ε in $\mathbf{R}$, then $a = 0$.*

Proof. Suppose to the contrary that $a > 0$. Then it follows from Corollary 2.2.8 that $a > \frac{1}{2}a > 0$. If we take $\varepsilon_0 := \frac{1}{2}a$, then we obtain $a > \varepsilon_0 > 0$ so that it is false that $a < \varepsilon$ for every $\varepsilon > 0$. Since the supposition that $a > 0$ leads to a contradiction, we conclude that $a = 0$. Q.E.D.

The product of two strictly positive real numbers is again strictly positive. However, the strict positivity of a product of two real numbers does not imply that the separate factors are strictly positive. The correct conclusion is that the factors must have the same sign (both strictly positive or both strictly negative), as we show next.

2.2.10 Theorem. *If $ab > 0$, then either* (i) *$a > 0$ and $b > 0$, or* (ii) *$a < 0$ and $b < 0$.*

Proof. First we note that $ab > 0$ implies that $a \neq 0$ and $b \neq 0$ (since if either a or b is 0 then their product would be 0). From the Trichotomy Property, either $a > 0$ or $a < 0$. If $a > 0$, then $1/a > 0$ by 2.2.6(d) and therefore

$$b = 1 \cdot b = ((1/a)a)b = (1/a)(ab) > 0.$$

Similarly, if $a < 0$, then $1/a < 0$ so that $b = (1/a)(ab) < 0$. Q.E.D.

2.2.11 Corollary. *If $ab < 0$, then either* (i) $a < 0$ *and* $b > 0$, *or* (ii) $a > 0$ *and* $b < 0$.

Inequalities

We now show how the order properties established in this section can be used to "solve" certain inequalities. The reader should justify each of the steps.

2.2.12 Examples. (a) Determine the set A of all real numbers x such that $2x + 3 \leq 6$.

†We note that $x \in A \Leftrightarrow 2x + 3 \leq 6 \Leftrightarrow 2x \leq 3 \Leftrightarrow x \leq \frac{3}{2}$. Therefore, we have $A = \{x \in \boldsymbol{R} : x \leq \frac{3}{2}\}$.

(b) Determine the set $B := \{x \in \boldsymbol{R} : x^2 + x > 2\}$.

We rewrite the inequality so that Theorem 2.2.10 can be applied. Note that $x \in B \Leftrightarrow x^2 + x - 2 > 0 \Leftrightarrow (x - 1)(x + 2) > 0$. Therefore, we either have (i) $x - 1 > 0$ and $x + 2 > 0$, or we have (ii) $x - 1 < 0$ and $x + 2 < 0$. In case (i) we must have both $x > 1$ and $x > -2$, which is satisfied if and only if $x > 1$. In case (ii) we must have both $x < 1$ and $x < -2$, which is satisfied if and only if $x < -2$. We conclude that $B = \{x \in \boldsymbol{R} : x > 1\} \cup \{x \in \boldsymbol{R} : x < -2\}$.

(c) Determine the set $C := \{x \in \boldsymbol{R} : (2x + 1)/(x + 2) < 1\}$. We note that $x \in C \Leftrightarrow (2x + 1)/(x + 2) - 1 < 0 \Leftrightarrow (x - 1)/(x + 2) < 0$. Therefore we have either (i) $x - 1 < 0$ and $x + 2 > 0$, or (ii) $x - 1 > 0$ and $x + 2 < 0$ (why?). In case (i) we must have both $x < 1$ and $x > -2$, which is satisfied if and only if $-2 < x < 1$. In case (ii), we must have both $x > 1$ and $x < -2$, which is impossible. We conclude that $C = \{x \in \boldsymbol{R} : -2 < x < 1\}$.

The following examples illustrate the use of the order properties of $\boldsymbol{R}$ in establishing inequalities. The reader should verify the steps in the arguments by identifying the properties that are employed. These results are of interest in their own right and will frequently be used in later work. It should be noted that the existence of square roots of strictly positive real numbers has not been formally established as yet; however, we assume their existence here for the purpose of discussing the examples. (The existence of square roots will be discussed in Section 2.4.)

† The symbol $\Leftrightarrow$ should be read "if and only if".

2.2.13 Examples. (a) If $0 < a < b$, then $a^2 < b^2$ and $\sqrt{a} < \sqrt{b}$.

The first assertion is proved by noting that the strict positivity of $b - a$ and $b + a$ implies that $b^2 - a^2 = (b - a)(b + a) > 0$. Thus, $b^2 > a^2$.

The second assertion follows from the identity $(\sqrt{b} - \sqrt{a})(\sqrt{b} + \sqrt{a}) = b - a$ because the right-hand side $b - a$ is strictly positive and the factor $\sqrt{b} + \sqrt{a}$ in the left-hand side is strictly positive. This forces the other factor $\sqrt{b} - \sqrt{a}$ to be strictly positive. Hence $\sqrt{b} > \sqrt{a}$.

(b) If a and b are strictly positive real numbers, then their **arithmetic mean** is $\frac{1}{2}(a + b)$ and their **geometric mean** is $\sqrt{ab}$. The **Arithmetic-Geometric Mean Inequality** for a, b is:

$$\sqrt{ab} \leqslant \tfrac{1}{2}(a + b) \tag{1}$$

with equality occurring if and only if $a = b$.

To prove this, note that if $a > 0$, $b > 0$, and $a \neq b$, then it follows from part (a) that $\sqrt{a} > 0$, $\sqrt{b} > 0$ and $\sqrt{a} \neq \sqrt{b}$. Therefore it follows from 2.2.5(a) that $(\sqrt{a} - \sqrt{b})^2 > 0$. Expanding this square, we obtain

$$a - 2\sqrt{ab} + b > 0,$$

whence it follows that

$$\sqrt{ab} < \tfrac{1}{2}(a + b).$$

Therefore (1) holds (with strict inequality) when $a \neq b$. Moreover, if $a = b$ (>0), then both sides of (1) equal a, so (1) becomes an equality. This proves that (1) holds for $a > 0$, $b > 0$.

On the other hand, suppose that $a > 0$, $b > 0$ and that $\sqrt{ab} = \frac{1}{2}(a + b)$. Then, squaring both sides and multiplying by 4, we obtain

$$4ab = (a + b)^2 = a^2 + 2ab + b^2,$$

whence it follows that

$$0 = a^2 - 2ab + b^2 = (a - b)^2.$$

But this equality implies that $a = b$ (why?). Thus, equality in (1) implies that $a = b$.

Remark. The general Arithmetic-Geometric Mean Inequality for the strictly positive real numbers $a_1, a_2, \ldots, a_n$ is

$$(a_1 a_2 \cdots a_n)^{1/n} \leqslant \frac{a_1 + a_2 + \cdots + a_n}{n} \tag{2}$$

with equality occurring if and only if $a_1 = a_2 = \ldots = a_n$. It is possible to prove this more general statement using mathematical induction, but the proof is somewhat intricate. A more elegant proof that uses properties of the exponential function will be indicated in Chapter 7.

(c) **Bernoulli's Inequality.** If $x > -1$, then

$$(1 + x)^n \geq 1 + nx \qquad \text{for all} \quad n \in \mathbf{N}. \tag{3}$$

The proof utilizes mathematical induction. The case $n = 1$ yields equality so that the assertion is valid in this case. Thus, we assume the validity of the inequality (3) for a positive integer n, and shall deduce it for $n + 1$. The assumption $(1 + x)^n \geq 1 + nx$ and the fact that $1 + x > 0$ imply that

$$\begin{aligned}(1 + x)^{n+1} &= (1 + x)^n (1 + x) \\ &\geq (1 + nx)(1 + x) = 1 + (n + 1)x + nx^2 \\ &\geq 1 + (n + 1)x.\end{aligned}$$

Thus, the inequality (3) for $n + 1$ follows. Hence, the inequality is true for all $n \in \mathbf{N}$.

‡(d) **Cauchy's Inequality.** If $n \in \mathbf{N}$ and $a_1, \ldots, a_n$ and $b_1, \ldots, b_n$ are real numbers, then:

$$(a_1 b_1 + \cdots + a_n b_n)^2 \leq (a_1^2 + \cdots + a_n^2)(b_1^2 + \cdots + b_n^2). \tag{4}$$

Moreover, if not all of the $b_j = 0$, then equality holds in (4) if and only if there exists a number $s \in \mathbf{R}$ such that $a_1 = sb_1, \ldots, a_n = sb_n$.

To prove this, we define a function $F : \mathbf{R} \to \mathbf{R}$ for $t \in \mathbf{R}$ by

$$F(t) := (a_1 - tb_1)^2 + \cdots + (a_n - tb_n)^2.$$

It follows from 2.2.5(a) and 2.2.1(i) that $F(t) \geq 0$ for all $t \in \mathbf{R}$. If we expand the squares, we obtain

$$F(t) = A + 2Bt + Ct^2 \geq 0,$$

where we have set

$$A := a_1^2 + \cdots + a_n^2, \qquad B := a_1 b_1 + \cdots + a_n b_n, \qquad C := b_1^2 + \cdots + b_n^2.$$

‡ The remainder of this section can be omitted on a first reading.

Since the quadratic function $F(t)$ is positive for all $t \in \boldsymbol{R}$, it cannot have two distinct real roots. Therefore its discriminant

$$\Delta := (2B)^2 - 4AC = 4(B^2 - AC)$$

must satisfy $\Delta \leqslant 0$. Consequently we must have $B^2 \leqslant AC$, which is precisely (4).

If $b_j = 0$ for all $j = 1, \ldots, n$, then equality holds in (4) for any choice of the a_j. Suppose now that not all $b_j = 0$ $(j = 1, \ldots, n)$. It is readily seen that if $a_j = sb_j$ for some $s \in \boldsymbol{R}$ and all $j = 1, \ldots, n$, then both sides of (4) equal $s^2(b_1^2 + \cdots + b_n^2)^2$. On the other hand, if equality holds in (4), then we must have $\Delta = 0$ so that there exists a unique root s of the quadratic equation $F(t) = 0$. But this implies (why?) that

$$a_1 - sb_1 = 0, \ldots, a_n - sb_n = 0$$

whence it follows that $a_j = sb_j$ for all $j = 1, \ldots, n$.

(e) **The Triangle Inequality.** If $n \in \boldsymbol{N}$ and $a_1, \ldots, a_n$ and $b_1, \ldots, b_n$ are real numbers, then

$$(5) \quad [(a_1 + b_1)^2 + \cdots + (a_n + b_n)^2]^{1/2} \leqslant [a_1^2 + \cdots + a_n^2]^{1/2} + [b_1^2 + \cdots + b_n^2]^{1/2}.$$

Moreover, if not all of the $b_j = 0$, then equality holds in (5) if and only if there exists a number s such that $a_1 = sb_1, \ldots, a_n = sb_n$.

Since $(a_j + b_j)^2 = a_j^2 + 2a_jb_j + b_j^2$ for $j = 1, \ldots, n$, it follows from Cauchy's Inequality (4) that (if A, B, and C are as in the preceding part (d)) we have

$$(a_1 + b_1)^2 + \cdots + (a_n + b_n)^2 = A + 2B + C \leqslant A + 2\sqrt{AC} + C = (\sqrt{A} + \sqrt{C})^2.$$

In view of part (a) we have

$$[(a_1 + b_1)^2 + \cdots + (a_n + b_n)^2]^{1/2} \leqslant \sqrt{A} + \sqrt{C},$$

which is (5).

If equality holds in (5), then we must have $B = \sqrt{AC}$, so that equality holds in Cauchy's Inequality.

Exercises for Section 2.2

1. (a) If $a \leqslant b$ and $c < d$, prove that $a + c < b + d$.
   (b) If $a \leqslant b$ and $c \leqslant d$, prove that $a + c \leqslant b + d$.
2. (a) If $0 < a < b$ and $0 < c < d$, prove that $0 < ac < bd$.
   (b) If $0 < a < b$ and $0 \leqslant c \leqslant d$, prove that $0 \leqslant ac \leqslant bd$. Also show by example that it does *not* follow that $ac < bd$.

3. If $a < b$ and $c < d$, prove that $ad + bc < ac + bd$.
4. Find numbers a, b, c, d in $\mathbf{R}$ satisfying $0 < a < b$ and $c < d < 0$ and either (i) $ac < bd$, or (ii) $bd < ac$.
5. If $a, b \in \mathbf{R}$ show that $a^2 + b^2 = 0$ if and only if $a = 0$ and $b = 0$.
6. If $0 \leq a < b$, prove that $a^2 \leq ab < b^2$. Also show by example that it does *not* follow that $a^2 < ab < b^2$.
7. Show that if $0 < a < b$, then $a < \sqrt{ab} < b$ and $0 < 1/b < 1/a$.
8. If $n \in \mathbf{N}$, show that $n^2 \geq n$ and hence $1/n^2 \leq 1/n$.
9. Find all real numbers x such that:
   (a) $x^2 > 3x + 4$;
   (b) $1 < x^2 < 4$;
   (c) $\dfrac{1}{x} < x$.
10. Let $a, b \in \mathbf{R}$, and suppose that for every $\varepsilon > 0$ we have $a - \varepsilon < b$.
   (a) Show that $a \leq b$.
   (b) Show that it does *not* follow that $a < b$.
11. Prove that $\left(\frac{1}{2}(a + b)\right)^2 \leq \frac{1}{2}(a^2 + b^2)$ for all $a, b \in \mathbf{R}$. Show that equality holds if and only if $a = b$.
12. (a) If $0 < c < 1$, show that $0 < c^2 < c < 1$.
   (b) If $1 < c$, show that $1 < c < c^2$.
13. If $c > 1$, show that $c^n \geq c$ for all $n \in \mathbf{N}$. (Consider Bernoulli's Inequality with $c = 1 + x$.)
14. If $c > 1$ and $m, n \in \mathbf{N}$, show that $c^m > c^n$ if and only if $m > n$.
15. If $0 < c < 1$, show that $c^n \leq c$ for all $n \in \mathbf{N}$.
16. If $0 < c < 1$ and $m, n \in \mathbf{N}$, show that $c^m < c^n$ if and only if $m > n$.
17. If $a > 0$, $b > 0$ and $n \in \mathbf{N}$, show that $a < b$ if and only if $a^n < b^n$.
18. Let $c_k > 0$ for $k = 1, \ldots, n$. Prove that

$$n^2 \leq (c_1 + c_2 + \cdots + c_n)\left(\frac{1}{c_1} + \frac{1}{c_2} + \cdots + \frac{1}{c_n}\right).$$

19. Let $c_k > 0$ for $k = 1, \ldots, n$. Show that

$$\frac{c_1 + c_2 + \cdots + c_2}{\sqrt{n}} \leq [c_1^2 + c_2^2 + \cdots + c_n^2]^{1/2}$$
$$\leq c_1 + c_2 + \cdots + c_n.$$

SECTION 2.3 Absolute Value

From the Trichotomy Property, we are assured that if $a \in \mathbf{R}$ and $a \neq 0$, then exactly one of the numbers a and $-a$ is strictly positive. The absolute value of $a \neq 0$ is defined to be the positive one of the pair $\{a, -a\}$.

2.3.1 Definition. If $a \in \mathbf{R}$, the **absolute value** of a, denoted by $|a|$, is defined by

$$|a| := a \quad \text{if} \quad a \geq 0,$$
$$:= -a \quad \text{if} \quad a < 0.$$

For example, $|3| = 3$ and $|-1| = 1$. We see from the definition that $|a| \geq 0$ for all $a \in \mathbf{R}$. Moreover, if $|a| = 0$, then $a = 0$, because if $a \neq 0$, then since also $-a \neq 0$, we would have $|a| \neq 0$.

2.3.2 Theorem. (a) $|-a| = |a|$ *for all* $a \in \mathbf{R}$.
(b) $|ab| = |a|\,|b|$ *for all* $a, b, \in \mathbf{R}$.
(c) *If* $c > 0$, *then* $|a| \leq c$ *if and only if* $-c \leq a \leq c$.
(d) $-|a| \leq a \leq |a|$ *for all* $a \in \mathbf{R}$.

Proof. (a) If $a = 0$, then $|0| = 0 = |-0|$. If $a > 0$, then $-a \leq 0$ so that $|a| = a = -(-a) = |-a|$. If $a < 0$, then $-a > 0$ so that $|-a| = -a = |a|$.

(b) If either a or b is 0, then both $|ab|$ and $|a|\,|b|$ are equal to 0. If $a > 0$ and $b > 0$, then $ab > 0$ so that $|ab| = ab = |a|\,|b|$. If $a > 0$ and $b < 0$, then $ab < 0$ so that $|ab| = -ab = a(-b) = |a|\,|b|$. The case $a < 0$ and $b > 0$ is treated similarly. Finally, if $a < 0$ and $b < 0$, then $|ab| = ab = (-a)(-b) = |a|\,|b|$.

(c) Suppose $|a| \leq c$. Then we have both $a \leq c$ and $-a \leq c$. (Why?) Since the latter inequality is equivalent to $-c \leq a$, we have $-c \leq a \leq c$. Conversely, if $-c \leq a \leq c$, then we have both $a \leq c$ and $-a \leq c$, so that $|a| \leq c$.

(d) Take $c = |a|$ in part (c). Q.E.D.

The following inequality will be used frequently in the sequel.

2.3.3 Triangle Inequality. *For any a and b in* $\mathbf{R}$, *we have*

$$|a + b| \leq |a| + |b|.$$

Proof. From 2.3.2(d), we know that $-|a| \leq a \leq |a|$ and $-|b| \leq b \leq |b|$. Then, employing 2.2.6(b), we obtain $-(|a| + |b|) \leq a + b \leq |a| + |b|$. Hence, we have $|a + b| \leq |a| + |b|$ by 2.3.2(c). Q.E.D.

There are some useful consequences of the Triangle Inequality.

2.3.4 Corollary. *For any a, b in* $\mathbf{R}$, *we have*
(a) $||a| - |b|| \leq |a - b|$,
(b) $|a - b| \leq |a| + |b|$.

Proof. (a) From $|a| = |a - b + b| \leq |a - b| + |b|$, we obtain $|a| - |b| \leq |a - b|$. Similarly, from $|b| = |b - a + a| \leq |b - a| + |a|$, we obtain $-(|a| - |b|) = |b| - |a| \leq |a - b|$. Combining these two inequalities via 2.3.2(c) yields (a).

(b) Replace b in the Triangle Inequality by $-b$ to get $|a - b| \leq |a| + |-b|$. Now apply 2.3.2(a) to get $|a - b| \leq |a| + |b|$. Q.E.D.

A straightforward argument using mathematical induction extends the Triangle Inequality to any finite number of elements of $\mathbf{R}$.

2.3.5 Corollary. *For any $a_1, a_2, \ldots, a_n$ in **R**, we have*

$$|a_1 + a_2 + \cdots + a_n| \leqslant |a_1| + |a_2| + \cdots + |a_n|.$$

The following examples illustrate how the preceding properties of absolute value can be used.

2.3.6 Examples. (a) Determine the set A of all real numbers x that satisfy $|2x + 3| < 6$.

From 2.3.2(c), we see that $x \in A$ if and only if $-6 < 2x + 3 < 6$, which is satisfied if and only if $-9 < 2x < 3$. Dividing by 2, we conclude that $A = \{x \in R : -\frac{9}{2} < x < \frac{3}{2}\}$.

(b) Determine the set $B := \{x \in \mathbf{R} : |x - 1| < |x|\}$.

One procedure is to consider cases for which the absolute value symbols can be removed. Here we take cases (i) $x > 1$, (ii) $0 \leqslant x \leqslant 1$, (iii) $x < 0$. (Why did we choose these three cases?) In case (i) the inequality becomes $x - 1 < x$, which is satisfied without any further restriction. Therefore all x satisfying $x \geqslant 1$ belong to the set B. In case (ii), the inequality becomes $-(x - 1) < x$, which imposes the further restriction that $x > \frac{1}{2}$. Thus, case (ii) contributes all x satisfying $\frac{1}{2} < x \leqslant 1$ to the set B. In case (iii), the inequality becomes $-(x - 1) < -x$ which is equivalent to $1 < 0$. Since this statement is always false, no value of x covered by case (iii) satisfies the inequality. Combining the three cases, we conclude that $B = \{x \in \mathbf{R} : x > \frac{1}{2}\}$.

There is an alternative method of determining the set B based on the fact that $a < b$ if and only if $a^2 < b^2$ when a and b are strictly positive. (See 2.2.13(a).) Thus, the inequality $|x - 1| < |x|$ is satisfied if and only if $|x - 1|^2 < |x|^2$. Since $|a|^2 = a^2$ for any a (why?), we can expand the square to obtain $x^2 - 2x + 1 < x^2$, which upon simplification becomes $x > \frac{1}{2}$. Thus, we again find that $B = \{x \in \mathbf{R} : x > \frac{1}{2}\}$. This procedure of squaring can be used to advantage under certain circumstances, but often a case analysis cannot be avoided when dealing with absolute values.

(c) Suppose the function f is defined by $f(x) := (2x^2 - 3x + 1)/(2x - 1)$ for $2 \leqslant x \leqslant 3$. Find a constant M such that $|f(x)| \leqslant M$ for all x satisfying $2 \leqslant x \leqslant 3$.

We consider separately the numerator and denominator of

$$|f(x)| = \frac{|2x^2 - 3x + 1|}{|2x - 1|}$$

From the Triangle Inequality, we obtain

$$|2x^2 - 3x + 1| \leqslant 2|x|^2 + 3|x| + 1 \leqslant 2 \cdot 3^2 + 3 \cdot 3 + 1 = 28$$

since $|x| \leqslant 3$ for the x under consideration. Also, $|2x - 1| \geqslant 2|x| - 1 \geqslant 2 \cdot 2 - 1 = 3$ since $|x| \geqslant 2$ for the x under consideration. Thus, $1/|2x - 1| \leqslant \frac{1}{3}$ for

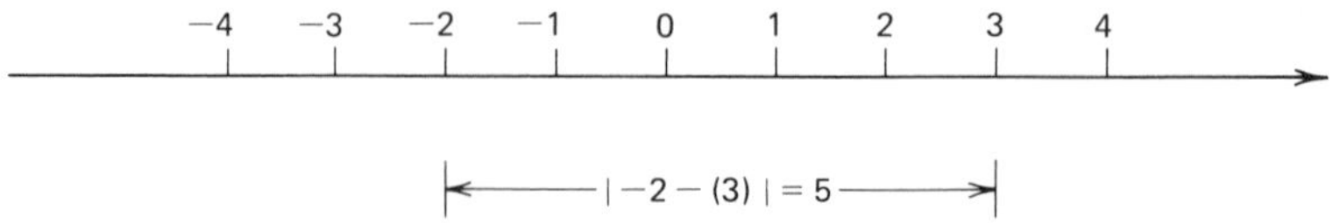

FIGURE 2.3.1 The distance between *a* = −2 and *b* = 3.

$x \geqslant 2$. (Why?) Therefore, for $2 \leqslant x \leqslant 3$ we have $|f(x)| \leqslant 28/3$. Hence we can take $M = 28/3$. (Note that we have found ***one*** such constant, M; evidently any number $M > 28/3$ will also satisfy $|f(x)| \leqslant M$. It is also possible that 28/3 is not the smallest possible choice for M.)

The Real Line

A convenient and familiar geometric interpretation of the real number system is the real line. In this interpretation, the absolute value $|a|$ of an element a in $\boldsymbol{R}$ is regarded as the distance from a to the origin 0. More generally, the **distance** between elements a and b in $\boldsymbol{R}$ is $|a - b|$. (See Figure 2.3.1.)

We shall later need precise language to discuss the notion of one real number being "close to" another. If a is a given real number, then saying that a real number x is "close to" a should mean that the distance $|x - a|$ between them is "small". A context in which this idea can be discussed is provided by the terminology of neighborhoods, which we now define.

2.3.7 Definition. Let a be an element of $\boldsymbol{R}$.

(i) For $\varepsilon > 0$, the ε-**neighborhood** of a is the set $V_\varepsilon(a) := \{x \in \boldsymbol{R} : |x - a| < \varepsilon\}$.

(ii) A **neighborhood** of a is any set that contains an ε-neighborhood of a for some $\varepsilon > 0$.

For $a \in \boldsymbol{R}$, the statement that x belongs to $V_\varepsilon(a)$ is equivalent to the statement that x satisfies the condition

$$a - \varepsilon < x < a + \varepsilon.$$

(See Figure 2.3.2) Thus the ε-neighborhood $V_\varepsilon(a)$ is a set that is "symmetric" about a. The idea of neighborhood as defined in 2.3.7(ii) relaxes this particular feature, but still serves the same purpose.

2.3.8 Theorem. *Let $a \in \boldsymbol{R}$. If $x \in \boldsymbol{R}$ is such that x belongs to every neighborhood of a, then $x = a$.*

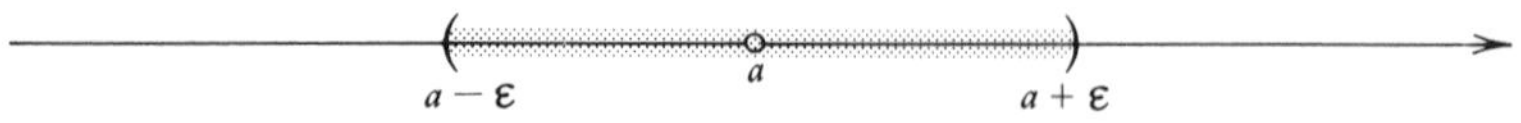

FIGURE 2.3.2 An ε-neighborhood of *a*.

Proof. The hypothesis implies that x belongs to $V_\varepsilon(a)$ for every $\varepsilon > 0$. This requires that $|x - a| < \varepsilon$ for every $\varepsilon > 0$. Then by 2.2.9 we have $|x - a| = 0$, and hence $x = a$. Q.E.D.

2.3.9 Examples. (a) Let $U := \{x : 0 < x < 1\}$. If $a \in U$, then $0 < a < 1$; hence, if ε is the smaller of the two numbers a and $1 - a$, we see that $V_\varepsilon(a)$ is an ε-neighborhood of a contained in U. Thus U is a neighborhood of each of its points.

(b) If $I := \{x : 0 \leq x \leq 1\}$, then I is not a neighborhood of 0 because for any $\varepsilon > 0$ there exists a number $x < 0$ (therefore $x \notin I$) satisfying $|x| < \varepsilon$.

(c) The idea of decimal place accuracy can be cast in terms of ε-neighborhoods. If $a \in \boldsymbol{R}$ and if $\varepsilon = 0.5 \times 10^{-k}$, then the inequality $|x - a| < \varepsilon$ entails that after rounding off to k places the decimal representations of x and a agree in the first k places. Thus, $x \in V_\varepsilon(a)$ if and only if x approximates a to at least k decimal places.

(d) If $|x - a| < \varepsilon$ and $|y - b| < \varepsilon$, then the Triangle Inequality implies that

$$\begin{aligned} |(x + y) - (a + b)| &= |(x - a) + (y - b)| \\ &\leq |x - a| + |y - b| \\ &< 2\varepsilon. \end{aligned}$$

Thus if x, y belong to the ε-neighborhoods of a, b, respectively, then $x + y$ belongs to the 2ε-neighborhood of $a + b$ (but not necessarily to the ε-neighborhood of $a + b$). In particular, addition will not preserve the order of decimal place accuracy. However, since $10^{-k} < 0.5 \times 10^{-(k-1)}$, at most one decimal place will be lost.

Exercises for Section 2.3

1. Let $a \in \boldsymbol{R}$. Show that we have:
   (a) $|a| = \sqrt{a^2}$, (b) $|a^2| = a^2$.
2. If $a, b \in \boldsymbol{R}$ and $b \neq 0$, show that $|a/b| = |a|/|b|$.
3. If $a, b \in \boldsymbol{R}$, show that $|a + b| = |a| + |b|$ if and only if $ab \geq 0$.
4. If $x, y, z \in \boldsymbol{R}$, $x \leq z$, show that $x < y < z$ if and only if $|x - y| + |y - z| = |x - z|$. Interpret this geometrically.
5. Find all $x \in \boldsymbol{R}$ that satisfy the following inequalities:
   (a) $|4x - 5| \leq 13$; (b) $|x^2 - 1| \leq 3$;
   (c) $|x - 1| > |x + 1|$; (d) $|x| + |x + 1| < 2$.
6. Show that $|x - a| < \varepsilon$ if and only if $a - \varepsilon < x < a + \varepsilon$.
7. If $a < x < b$ and $a < y < b$, show that $|x - y| < b - a$. Interpret this geometrically.
8. Determine and sketch the set of pairs (x,y) in $\boldsymbol{R} \times \boldsymbol{R}$ that satisfy:
   (a) $|x| = |y|$; (b) $|x| + |y| = 1$;
   (c) $|xy| = 2$; (d) $|x| - |y| = 2$.

9. Determine and sketch the set of pairs (x,y) in $\boldsymbol{R} \times \boldsymbol{R}$ that satisfy the inequalities:
   (a) $|x| \leqslant |y|$; (b) $|x| + |y| \leqslant 1$;
   (c) $|xy| \leqslant 2$; (d) $|x| - |y| \geqslant 2$.
10. Show that if U and V are neighborhoods of $a \in \boldsymbol{R}$, then $U \cap V$ and $U \cup V$ are neighborhoods of a.
11. Show that if $a, b \in \boldsymbol{R}$, and $a \neq b$, then there exist neighborhoods U of a and V of b such that $U \cap V = \emptyset$.
12. Let $a, c \in \boldsymbol{R}$ with $c > 0$. Show that if $|x - a| < \varepsilon$ then $|cx - ac| < c\varepsilon$. What can be concluded about the decimal place accuracy under multiplication by a fixed number c when $0 < c < 1$? When $0 < c < 1000$?

SECTION 2.4 The Completeness Property of *R*

Thus far in this chapter, we have discussed the algebraic properties and the order properties of the real number system $\boldsymbol{R}$. In this section we shall present one more property of $\boldsymbol{R}$ that is often called the "completeness property" since it guarantees the existence of elements in $\boldsymbol{R}$ under certain hypotheses. The system $\boldsymbol{Q}$ of rational numbers satisfies both the Algebraic Properties 2.1.1 and the Order Properties 2.2.1, but we have seen that $\sqrt{2}$ cannot be represented as a rational number; therefore $\sqrt{2}$ does not belong to $\boldsymbol{Q}$. This observation shows the necessity of an additional property to characterize the real number system. This additional property, the Completeness (or the Supremum) Property, is an essential feature of $\boldsymbol{R}$.

There are several different versions of the Completeness Property. We choose to give here what is probably the most efficient method by assuming that each non-empty, bounded set in $\boldsymbol{R}$ has a supremum.

Suprema and Infima

We now introduce the notion of an upper bound of a set of real numbers. This idea will be of utmost importance in later sections.

2.4.1 Definition. Let S be a subset of $\boldsymbol{R}$.
(i) An element $u \in \boldsymbol{R}$ is said to be an **upper bound** of S if $s \leqslant u$ for all $s \in S$.
(ii) An element $w \in \boldsymbol{R}$ is said to be a **lower bound** of S if $w \leqslant s$ for all $s \in S$.

We note that a subset S of $\boldsymbol{R}$ may not have an upper bound (for example, take $S = \boldsymbol{R}$). However, if S has one upper bound, then it will have infinitely many upper bounds because if u is an upper bound of S then any v such that $u < v$ is also an upper bound of S; see Figure 2.4.1. (A similar observation is valid for lower bounds.)

We also note that it is possible for a set to have lower bounds but no upper bounds (and vice versa). For example, consider the sets $S_1 := \{x \in \boldsymbol{R} : x \geqslant 0\}$ and $S_2 := \{x \in \boldsymbol{R} : x < 0\}$.

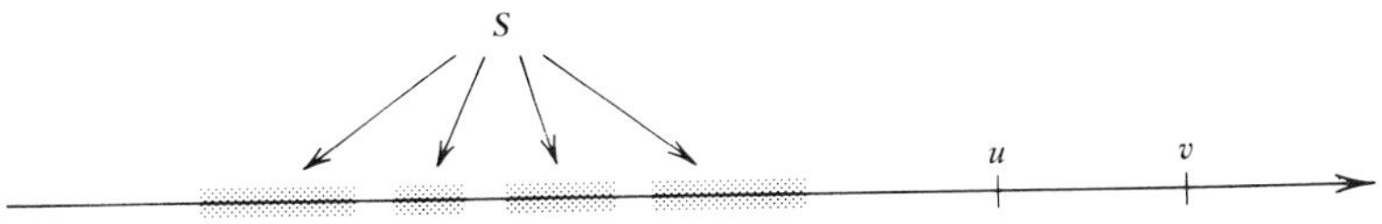

FIGURE 2.4.1 Upper bounds of S.

Remark. If we apply the definitions to the empty set $\emptyset$, we are forced to conclude that every real number is an upper bound of $\emptyset$. For, in order that a number $u \in \boldsymbol{R}$ is *not* an upper bound of a set S, an element $s \in S$ must exist such that $u < s$. If $S = \emptyset$, there is no such element. Hence every real number is an upper bound of the empty set. Similarly, every real number is a lower bound of the empty set $\emptyset$. This may seem artificial, but it is a logical consequence of the definition.

As a matter of terminology, we say that a set in $\boldsymbol{R}$ is **bounded above** if it has an upper bound. Similarly, if a set in $\boldsymbol{R}$ has a lower bound, we say it is **bounded below**. If a set in $\boldsymbol{R}$ has both an upper bound and a lower bound, we say it is **bounded**. We say that a set in $\boldsymbol{R}$ is **unbounded** if it lacks either an upper bound or a lower bound. For example, the set $\{x \in \boldsymbol{R} : x \leq 2\}$ is unbounded (even though it is bounded above) since it is not bounded below.

2.4.2 Definition. Let S be a subset of $\boldsymbol{R}$.

(i) If S is bounded above, then an upper bound is said to be a **supremum** (or a **least upper bound**) of S if it is less than every other upper bound of S.

(ii) If S is bounded below, then a lower bound is said to be an **infimum** (or a **greatest lower bound**) of S if it is greater than every other lower bound of S.

The definition of supremum can be expressed as follows. A number $u \in \boldsymbol{R}$ is a supremum of $S \subseteq \boldsymbol{R}$ if it satisfies the two conditions:

(1) $s \leq u$ for all $s \in S$;
(2) if $s \leq v$ for all $s \in S$, then $u \leq v$.

Indeed, condition (1) makes u an upper bound of S, and condition (2) shows that u is less than any other upper bound of the set S. (The definition of infimum of a set can be reformulated similarly. Do so.)

It follows from the definition that there can be at most one supremum of a given subset S of $\boldsymbol{R}$. For suppose that u_1 and u_2 are suprema of S. Then they are both upper bounds of S, and applying condition (2) to the supremum u_1 of S gives us $u_1 \leq u_2$. Since u_2 is also a supremum of S, we similarly have $u_2 \leq u_1$. Therefore, $u_1 = u_2$. (In the same way, one shows that there can be at most one infimum for a given subset of $\boldsymbol{R}$.)

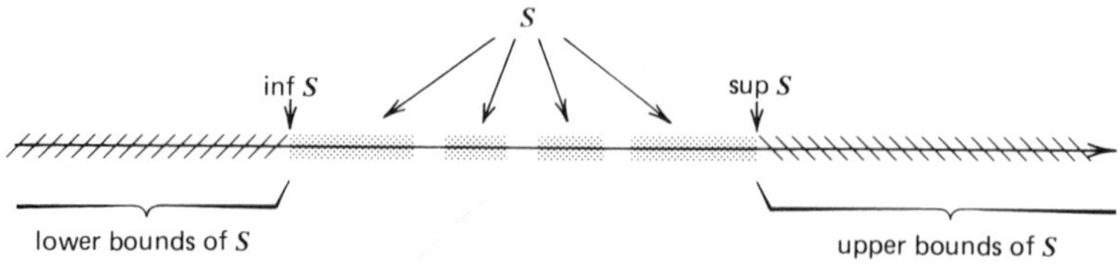

FIGURE 2.4.2 inf S and sup S.

Where the supremum or the infimum of a set S exists (see Figure 2.4.2), we shall denote them by

$$\sup S \qquad \text{and} \qquad \inf S.$$

The following criterion is often useful in establishing that a certain upper bound of a given set is actually the supremum of the set.

2.4.3 Lemma. *An upper bound u of a non-empty set S in **R** is the supremum of S if and only if for each $\varepsilon > 0$ there exists $s_\varepsilon \in S$ such that $u - \varepsilon < s_\varepsilon$.*

Proof. Suppose that u satisfies the stated condition; we shall show that $u = \sup S$. To do so, let v be any upper bound of S with $v \neq u$. If we assume that $v < u$ and take $\varepsilon := u - v > 0$, the condition implies that there exists a number $s_\varepsilon \in S$ such that $v = u - \varepsilon < s_\varepsilon$. But this contradicts the assertion that v is an upper bound of S. Since the hypothesis that $v < u$ has led to a contradiction, we conclude that $u < v$. Therefore we have shown $u = \sup S$.

Conversely, suppose $u = \sup S$ and let $\varepsilon > 0$. Since $u - \varepsilon < u$, then $u - \varepsilon$ is not an upper bound of S. Therefore, some element s_ε of S must exceed $u - \varepsilon$; that is, $u - \varepsilon < s_\varepsilon$. (See Figure 2.4.3.) Q.E.D.

It is important to realize that the supremum of a set may or may not belong to the set in question.

We consider a few elementary examples.

2.4.4 Examples. (a) If S_1 has only a finite number of elements, then it can be shown that S_1 has a largest element u and a least element w. Then u

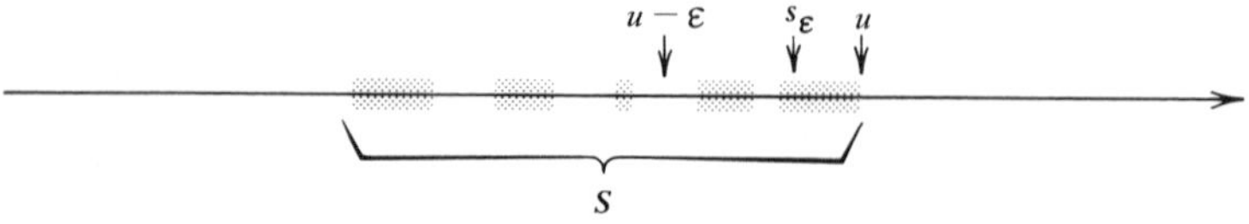

FIGURE 2.4.3 u = sup S.

$= \sup S_1$, $w = \inf S_1$, and they are members of the set S_1. (See Exercises 2.4.4 and 2.4.5.)

(b) The set $S_2 := \{x \in \boldsymbol{R} : 0 \leq x \leq 1\}$ has 1 for an upper bound. Since $1 \in S_2$, any other upper bound v of S_2 must satisfy $1 < v$. Thus, $\sup S_2 = 1$ and S_2 contains its supremum. Similarly, $\inf S_2 = 0$ is contained in S_2.

(c) The set $S_3 := \{x \in \boldsymbol{R} : 0 < x < 1\}$ again has 1 for an upper bound. To show 1 is the supremum of S_3, we use Lemma 2.4.3. If $0 < \varepsilon \leq 1$, then $s_\varepsilon := 1 - \varepsilon/2$ is an element of S_3 and satisfies $1 - \varepsilon < s_\varepsilon$. It follows that $1 = \sup S_3$. In this case, the supremum of S_3 is not an element of S_3. Similarly, $0 = \inf S_3$ is not contained in S_3.

(d) As noted above, every real number is both an upper bound and a lower bound for the empty set $\emptyset$. Thus, the set $\emptyset$ has no supremum or infimum.

In the preceding examples, the set S_2 contains its supremum, but the set S_3 does not. Thus, *when we say that a set has a supremum, we are making no statement as to whether the set contains the supremum as an element or not.* When a set contains its supremum, we sometimes call this supremum the **maximum** of the set; similarly, if a set contains its infimum, we sometimes call this infimum the **minimum** of the set.

The Supremum Property of *R*

It is not possible to prove on the basis of the assumptions we have made about $\boldsymbol{R}$ to this point that every non-empty subset of $\boldsymbol{R}$ that is bounded above has a supremum. However, it is a deep and fundamental property of the real number system that this is indeed the case. We shall make frequent and essential use of this property. The following statement (which is also referred to as the Completeness Property of $\boldsymbol{R}$) is our final assumption about $\boldsymbol{R}$.

2.4.5 The Supremum Property of *R*. *Every non-empty set of real numbers that has an upper bound has a supremum.*

The analogous property of infima can be readily established from the Supremum Property. For, suppose that S is a non-empty subset of $\boldsymbol{R}$ that is bounded below. Then the set $S_1 := \{-s : s \in S\}$ is bounded above and the Supremum Property implies that $u := \sup S_1$ exists. It then follows that $-u$ is the infimum of S, as the reader can verify. Thus, we have the Infinimum Property of $\boldsymbol{R}$: *Every non-empty set of real numbers that has a lower bound has an infimum.*

The following examples illustrate techniques of working with suprema and infima and also foreshadow ways in which these notions will be put to use.

2.4.6 Examples. (a) It is important that suprema and infima of sets are compatible with the Algebraic Properties of $\boldsymbol{R}$. We present here one example of this compatibility with addition; others are given in the exercises.

Let S be a non-empty subset of $\boldsymbol{R}$ that is bounded above and let $a \in \boldsymbol{R}$. Define the set $a + S := \{a + x : x \in S\}$. We shall show that

$$\sup (a + S) = a + \sup S.$$

If we let $u := \sup S$, then, since $x \leqslant u$ for any $x \in S$, we have $a + x \leqslant a + u$. Therefore, $a + u$ is an upper bound for $a + S$; consequently, we have $\sup (a + S) \leqslant a + u$. If v is *any* upper bound of the set $a + S$, then $a + x \leqslant v$ for all $x \in S$. Then $x \leqslant v - a$ for all $x \in S$, which implies that $u = \sup S \leqslant v - a$. Thus, $a + u \leqslant v$, and (since $a + u$ is an upper bound for $a + S$) we conclude that

$$\sup (a + S) = a + u = a + \sup S.$$

For similar relationships between the suprema and infima of sets and the operations of addition and multiplication, see the exercises.

(b) Suppose f and g are real-valued functions with common domain $D \subseteq \boldsymbol{R}$. We assume their ranges $f(D) := \{f(x) : x \in D\}$ and $g(D) := \{g(x) : x \in D\}$ are bounded sets in $\boldsymbol{R}$.

(i) If $f(x) \leqslant g(x)$ for all $x \in D$, then $\sup f(D) \leqslant \sup g(D)$.

To verify this assertion, we note that the number $\sup g(D)$ is an upper bound for the set $f(D)$ because for any $x \in D$, we have $f(x) \leqslant g(x) \leqslant \sup g(D)$. Therefore, $\sup f(D) \leqslant \sup g(D)$.

(ii) If $f(x) \leqslant g(y)$ for all $x, y \in D$, then $\sup f(D) \leqslant \inf g(D)$.

The proof is in two stages. First, for a particular $y \in D$, we see that since $f(x) \leqslant g(y)$ for all $x \in D$, then $g(y)$ is an upper bound for the set $f(D)$. Consequently, $\sup f(D) \leqslant g(y)$. Since the last inequality persists for all $y \in D$, we may now conclude that the number $\sup f(D)$ is a lower bound for the set $g(D)$. Therefore, the inequality $\sup f(D) \leqslant \inf g(D)$ is established.

It should be noted that the hypothesis $f(x) \leqslant g(x)$ for all $x \in D$ does not imply any relation between $\sup f(D)$ and $\inf g(D)$. For example, if $f(x) := x^2$ and $g(x) := x$ with $D := \{x \in \boldsymbol{R} : 0 < x < 1\}$, then $f(x) \leqslant g(x)$ for all $x \in D$ but $\sup f(D) = 1$ and $\inf g(D) = 0$. However, $\sup g(D) = 1$ so that the conclusion of (i) holds.

Further relationships between suprema and infima of sets of function values are given in the exercises.

The Archimedean Property

One important consequence of the Supremum Property is that the subset $\boldsymbol{N}$ of natural numbers is not bounded above in $\boldsymbol{R}$. This means that: given any real number x there exists a natural number n (depending on x) such that $x < n$. Because of familiarity with numbers and the customary picture of the real line, this may seem an obvious fact. However, it is significant that this property cannot be deduced from the Algebraic and Order Properties given in the previous sections of this chapter. The proof, which is now given, makes essential use of the Supremum Property of $\mathbf{R}$.

2.4.7 Archimedean Property. *If $x \in \mathbf{R}$, then there exists $n_x \in \mathbf{N}$ such that $x < n_x$.*

Proof. If the conclusion fails, then x is an upper bound of $\mathbf{N}$. Therefore, by the Supremum Property, the non-empty set $\mathbf{N}$ has a supremum $u \in \mathbf{R}$. Since $u - 1 < u$, it follows from Lemma 2.4.3 that there exists $m \in \mathbf{N}$ such that $u - 1 < m$. But then $u < m + 1$, and since $m + 1 \in \mathbf{N}$, this contradicts the assumption that u is an upper bound of $\mathbf{N}$. Q.E.D.

The Archimedean Property can be stated in a variety of ways. We present three of these variations:

2.4.8 Corollary. *Let y and z be strictly positive real numbers. Then:*
(a) *There exists $n \in \mathbf{N}$ such that $z < ny$.*
(b) *There exists $n \in \mathbf{N}$ such that $0 < 1/n < y$.*
(c) *There exists $n \in \mathbf{N}$ such that $n - 1 \leq z < n$.*

Proof. (a) Since $x := z/y > 0$, there exists $n \in \mathbf{N}$ such that $z/y = x < n$ and hence $z < ny$.

(b) Setting $z = 1$ in (a) gives us $1 < ny$, which implies $1/n < y$.

(c) The Archimedean Property assures us that the subset $\{m \in \mathbf{N} : z < m\}$ of $\mathbf{N}$ is not empty. Let n be the least element of this set (see Section 1.3). Then we have $n - 1 \leq z < n$. Q.E.D.

The Existence of $\sqrt{2}$

The importance of the Supremum Property lies in the fact that it guarantees the existence of real numbers under certain hypotheses. We shall make use of it in this way many times. At the moment, we shall illustrate this use by proving the existence of a positive real number x such that $x^2 = 2$, that is, the positive square root of 2. It was shown earlier (see Theorem 2.1.7) that such an x cannot be a rational number; thus, we will be deriving the existence of at least one irrational number.

2.4.9 Theorem. *There exists a positive real number x such that $x^2 = 2$.*

Proof. Let $S := \{s \in \mathbf{R} : 0 \leq s,\ s^2 < 2\}$. The set is not empty since $1 \in S$. Also, S is bounded above by 2, for if $t > 2$ then $t^2 > 4$ so that $t \notin S$. Thus, the Supremum Property implies that the set S has a supremum in $\mathbf{R}$ and we let $x := \sup S$.

We shall prove that $x^2 = 2$. If not, then either $x^2 < 2$ or $x^2 > 2$.

Suppose first that $x^2 < 2$. We show that this contradicts the fact that x is an upper bound of S. To this end, we use the fact that $(2 - x^2)/(2x + 1) > 0$ and the Archimedean Property (Corollary 2.4.8(b)) to obtain $n \in \mathbf{N}$ such that $1/n < (2 - x^2)/(2x + 1)$. Then $x + 1/n \in S$, because

$$\left(x + \frac{1}{n}\right)^2 = x^2 + \frac{2x}{n} + \frac{1}{n^2} = x^2 + \frac{1}{n}\left(2x + \frac{1}{n}\right)$$
$$\leqslant x^2 + \frac{1}{n}\left(2x + 1\right) < x^2 + (2 - x^2) = 2.$$

Thus, we have $x < x + 1/n \in S$, which is not possible since x is an upper bound of S. Therefore we cannot have $x^2 < 2$.

Now suppose that $x^2 > 2$. We show that this leads to an upper bound of S smaller than x, which contradicts the fact that x is the supremum of S. Since $(x^2 - 2)/2x > 0$, if we choose $m \in \mathbf{N}$ such that $1/m < (x^2 - 2)/2x$, we have

$$\left(x - \frac{1}{m}\right)^2 = x^2 - \frac{2x}{m} + \frac{1}{m^2} > x^2 - \frac{2x}{m}$$
$$> x^2 - (x^2 - 2) = 2.$$

It follows that $x - 1/m > s$ for any $s \in S$, whence $x - 1/m$ is an upper bound for S. Since $x > x - 1/m$, this contradicts the fact that $x = \sup S$. Therefore we cannot have $x^2 > 2$.

Since the possibilities $x^2 < 2$ and $x^2 > 2$ have been excluded, we must have $x^2 = 2$. Q.E.D.

By slightly modifying the above argument, the reader can show that if $a > 0$, then there is a unique $b > 0$ such that $b^2 = a$. We call b the **positive square root** of a and denote it by $b = \sqrt{a}$ or $b = a^{1/2}$. A slightly more complicated argument involving the binomial theorem can be formulated to establish the existence of a unique positive nth root of a, denoted by $\sqrt[n]{a}$ or $a^{1/n}$, for each $n \in \mathbf{N}$.

Density of Rational Numbers in *R*

We now know that there exists at least one irrational real number, namely $\sqrt{2}$. Actually there are "more" irrational numbers than rational numbers in the sense that the set of rational numbers is countable, while the set of irrational numbers is uncountable (see Section 1.4). We next show that in spite of this apparent disparity, the set of rational numbers is "dense" in $\mathbf{R}$ in the sense that a rational number (in fact, infinitely many of them) can be found between any two distinct real numbers.

2.4.10 The Density Theorem. *If x and y are real numbers with $x < y$, then there exists a rational number r such that $x < r < y$.*

Proof. It is no loss of generality to assume that $x > 0$. (Why?) By the Archimedean Property 2.4.7, there exists $n \in \mathbf{N}$ such that $n > 1/(y - x)$. For such an n, we have $ny - nx > 1$. Applying Corollary 2.4.8(c) to $nx > 0$, we obtain $m \in \mathbf{N}$ such that $m - 1 \leqslant nx < m$. This m also satisfies $m < ny$ since $m \leqslant nx + 1 < ny$. Thus, we have $nx < m < ny$ so that $r := m/n$ is a rational number satisfying $x < r < y$. Q.E.D.

To round out the discussion of the interlacing of rational and irrational numbers, we have the same betweenness property for the set of irrational numbers.

2.4.11 Corollary. *If x and y are real numbers with $x < y$, then there exists an irrational number z such that $x < z < y$.*

Proof. Applying the Density Theorem 2.4.10 to the real numbers $x/\sqrt{2}$ and $y/\sqrt{2}$, we obtain a rational number $r \neq 0$ such that

$$\frac{x}{\sqrt{2}} < r < \frac{y}{\sqrt{2}}.$$

Then $z := r\sqrt{2}$ is irrational (why?) and satisfies $x < z < y$. Q.E.D.

Exercises for Section 2.4

1. Let $S := \{1 - (-1)^n/n : n \in \mathbf{N}\}$. Find sup S and inf S.
2. Show in detail that the set $S_1 := \{x \in \mathbf{R} : x \geqslant 0\}$ has lower bounds but no upper bounds.
3. Let $S \subseteq \mathbf{R}$ and suppose that $s^* :=$ sup S belongs to S. If $u \notin S$ show that sup $(S \cup \{u\})$ is the larger of the two numbers s^*, u.
4. Show that a non-empty finite set $S \subseteq \mathbf{R}$ contains its supremum and its infimum. (*Hint*: Use induction.)
5. If a set $S \subseteq \mathbf{R}$ contains one of its upper bounds, show that this upper bound is the supremum of S.
6. Let $S \subseteq \mathbf{R}$ be non-empty. Show that $u \in \mathbf{R}$ is an upper bound of S if and only if the conditions $t \in \mathbf{R}$ and $t > u$ imply that $t \notin S$.
7. Let $S \subseteq \mathbf{R}$ be non-empty. Show that $u =$ sup S if and only if for every $n \in \mathbf{N}$ the number $u - 1/n$ is not an upper bound of S but the number $u + 1/n$ is an upper bound of S.
8. Show that if A and B are bounded subsets in $\mathbf{R}$, then $A \cup B$ is a bounded set. Show that sup $(A \cup B) =$ sup $\{$sup A, sup $B\}$.
9. Give an example of a countable collection of bounded subsets of $\mathbf{R}$ where (i) the union is bounded, and one where (ii) the union is unbounded.
10. Let S be a bounded set in $\mathbf{R}$ and let S_0 be a non-empty subset of S. Show that

$$\inf S \leqslant \inf S_0 \leqslant \sup S_0 \leqslant \sup S.$$

11. Let S be a non-empty bounded set in $\boldsymbol{R}$.
(a) Let $a > 0$, and let $aS := \{as : s \in S\}$. Prove that

$$\inf (aS) = a \inf S, \qquad \sup (aS) = a \sup S.$$

(b) Let $b < 0$ and let $bS := \{bs : s \in S\}$. Prove that

$$\inf (bS) = b \sup S, \qquad \sup (bS) = b \inf S.$$

12. Let X be a non-empty set and let $f : X \to \boldsymbol{R}$ have bounded range in $\boldsymbol{R}$. If $a \in \boldsymbol{R}$, show that Example 2.4.6(a) implies that

$$\sup \{a + f(x) : x \in X\} = a + \sup \{f(x) : x \in X\}.$$

Show that we also have

$$\inf \{a + f(x) : x \in X\} = a + \inf \{f(x) : x \in X\}.$$

13. Let A and B be subsets of $\boldsymbol{R}$ that are bounded, and let $A + B := \{a + b : a \in A, b \in B\}$. Prove that $\sup (A + B) = \sup A + \sup B$ and $\inf (A + B) = \inf A + \inf B$.
14. Let X be a non-empty set, and let f and g be defined on X and have bounded ranges in $\boldsymbol{R}$. Show that

$$\sup \{f(x) + g(x) : x \in X\} \leqslant \sup \{f(x) : x \in X\} + \sup \{g(x) : x \in X\}.$$

and that

$$\inf \{ f(x) : x \in X\} + \inf \{g(x) : x \in X\} \leqslant \inf \{ f(x) + g(x) : x \in X\}.$$

Give examples to show that each of these inequalities can be either equalities or strict inequalities.
15. Let $X = Y := \{x \in \boldsymbol{R} : 0 < x < 1\}$. Define $h : X \times Y \to \boldsymbol{R}$ by $h(x,y) := 2x + y$.
(a) For each $x \in X$, find $f(x) := \sup \{h(x,y) : y \in Y\}$; then find $\inf \{f(x) : x \in X\}$.
(b) For each $y \in Y$, find $g(y) := \inf \{h(x,y) : x \in X\}$; then find $\sup \{g(y) : y \in Y\}$. Compare with the result found in part (a).
16. Perform the computations in (a) and (b) of the preceding exercise for the function $h : X \times Y \to \boldsymbol{R}$ defined by

$$\begin{aligned} h(x,y) &:= 0 \quad \text{if} \quad x < y, \\ &:= 1 \quad \text{if} \quad x \geqslant y. \end{aligned}$$

17. Let X and Y be non-empty sets and let $h : X \times Y \to \boldsymbol{R}$ have bounded range in $\boldsymbol{R}$. Let $f : X \to \boldsymbol{R}$ and $g : Y \to \boldsymbol{R}$ be defined by

$$f(x) := \sup \{h(x,y) : y \in Y\}, \qquad g(y) := \inf \{h(x,y) : x \in X\}.$$

Prove that

$$\sup \{g(y) : y \in Y\} \leq \inf \{f(x) : x \in X\}.$$

We sometimes express this by writing

$$\sup_y \inf_x h(x,y) \leq \inf_x \sup_y h(x,y).$$

Note that Exercises 15 and 16 show that the inequality may be either equality or a strict inequality.

18. Let X and Y be non-empty sets and let $h : X \times Y \to \boldsymbol{R}$ have bounded range in $\boldsymbol{R}$. Let $F : X \to \boldsymbol{R}$ and $G : Y \to \boldsymbol{R}$ be defined by

$$F(x) := \sup \{h(x,y) : y \in Y\}, \qquad G(y) := \sup \{h(x,y) : x \in X\}.$$

Establish the Principle of the Iterated Suprema:

$$\sup \{h(x,y) : x \in X,\ y \in Y\} = \sup \{F(x) : x \in X\} = \sup \{G(y) : y \in Y\}.$$

We sometimes express this in symbols by

$$\sup_{x,y} h(x,y) = \sup_x \sup_y h(x,y) = \sup_y \sup_x h(x,y).$$

19. Given any $x \in \boldsymbol{R}$ show that there exists a *unique* $n \in \boldsymbol{Z}$ such that $n - 1 \leq x < n$.
20. If $y > 0$ show that there exists $n \in \boldsymbol{N}$ such that $1/2^n < y$.
21. Modify the argument in Theorem 2.4.9 to show that there exists a positive real number y such that $y^2 = 3$.
22. Modify the argument in Theorem 2.4.9 to show that if $a > 0$, then there exists a positive real number z such that $z^2 = a$. This number will be denoted by $\sqrt{a}$ or $a^{1/2}$ and will be called the **positive square root of** a.
23. Modify the argument in Theorem 2.4.9 to show that there exists a positive real number u such that $u^3 = 2$.
24. If $a > 0$ and $n \in \boldsymbol{N}$, modify the argument in Theorem 2.4.9 to show that there exists a positive real number v such that $v^n = a$. This number will be denoted by $\sqrt[n]{a}$ or $a^{1/n}$ and will be called the **positive nth root of** a.
25. Complete the proof of the Density Theorem 2.4.10 by removing the hypothesis that $x > 0$.
26. If $u > 0$ is any number and $x < y$, show that there exists a rational number r such that $x < ru < y$. (Hence the set $\{ru : r \in \boldsymbol{Q}\}$ is dense in $\boldsymbol{R}$.)

SECTION 2.5 Intervals, Cluster Points, and Decimals

The order relation on $\boldsymbol{R}$ determines a natural collection of subsets known as intervals. The notation and terminology for these special sets are as follows. If $a, b \in \boldsymbol{R}$ and $a \leq b$, then the **open interval** determined by a and b is

(1) $$(a, b) := \{x \in \boldsymbol{R} : a < x < b\}.$$

The points a and b are called the **end points** of the open interval (a,b), but the end points are not included. If both end points are adjoined to the open interval, we have the **closed interval**

(2) $$[a, b] := \{x \in \boldsymbol{R} : a \leq x \leq b\}.$$

The sets

(3) $$[a, b) := \{x \in \boldsymbol{R} : a \leq x < b\}$$

and

(4) $$(a, b] := \{x \in \boldsymbol{R} : a < x \leq b\}$$

are **half-open** (or **half-closed**) intervals determined by the end points a and b. (See Figure 2.5.1.) Each of the above intervals have **length** defined by $b - a$. If $a = b$, note that the corresponding open interval is the empty set

(5) $$(a, a) = \emptyset,$$

whereas the corresponding closed interval is the singleton set $[a, a] = \{a\}$.

If $a \in \boldsymbol{R}$ then the sets defined by

(6) $$(a, \infty) := \{x \in \boldsymbol{R} : x > a\},$$

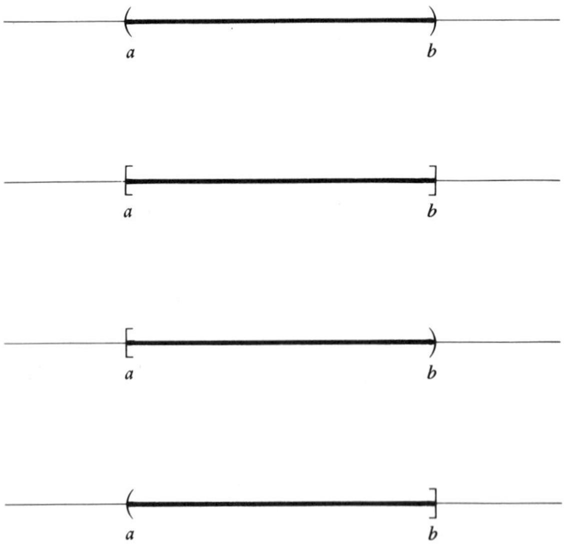

FIGURE 2.5.1 Types of intervals.

(7) $$(-\infty, a) := \{x \in \boldsymbol{R} : x < a\},$$

are called **open rays** (or **infinite open intervals**). Similarly, the sets defined by

(8) $$[a, \infty) := \{x \in \boldsymbol{R} : x \geq a\},$$

(9) $$(-\infty, a] := \{x \in \boldsymbol{R} : x \leq a\},$$

are called **closed rays** (or **infinite closed intervals**). In these cases the point a is called the **end point** of these intervals. It is often convenient to think of the entire set $\boldsymbol{R}$ as an infinite interval. In this case we write

(10) $$(-\infty, \infty) := \boldsymbol{R},$$

and we do *not* consider any point to be an end point of $(-\infty, \infty)$.

It will be noted that intervals of the forms (1), . . . , (5) are **bounded** intervals. Intervals of the forms (6), . . . , (10) are **unbounded**. In denoting these unbounded intervals we have used the symbols $-\infty$ and ∞. These symbols should be considered merely as notational conveniences; *they are not elements of* $\boldsymbol{R}$.

The **unit interval** is the closed interval $[0, 1] := \{x \in \boldsymbol{R} : 0 \leq x \leq 1\}$. It will be denoted by the standard notation $\boldsymbol{I}$. (In some books, the unit interval is referred to as the **continuum**.)

Nested Intervals

We say that a sequence of intervals I_n, $n \in \boldsymbol{N}$, is **nested** (see Figure 2.5.2) if the following chain of inclusions holds:

$$I_1 \supseteq I_2 \supseteq I_3 \supseteq \cdots \supseteq I_n \supseteq I_{n+1} \supseteq \cdots.$$

For example, if $I_n := [0, 1/n]$, $n \in \boldsymbol{N}$, then $I_n \supseteq I_{n+1}$ for each n so that the intervals are nested. In this case, the element 0 belongs to all I_n and the Archimedean Property 2.4.7 can be used to prove that 0 is the only such element. (Prove this.) In other terms, $\bigcap_{n=1}^{\infty} I_n = \{0\}$.

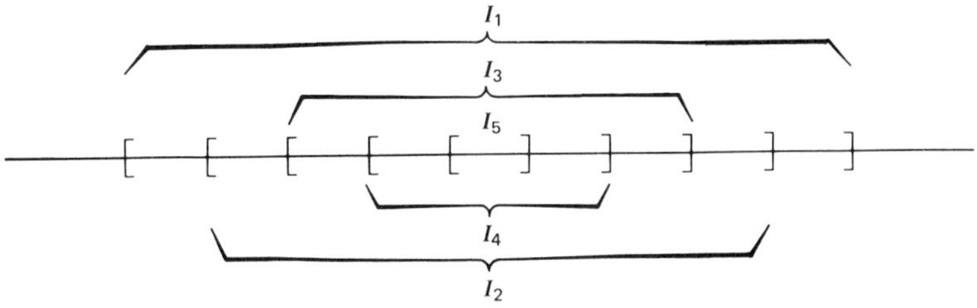

FIGURE 2.5.2 Nested intervals.

It is important to realize that, in general, a nested sequence of intervals *need not* have a common point.

For example, if $J_n := (0, 1/n)$ for $n \in N$, then this sequence of intervals is nested but has no common point because any $x \in R$ satisfying $0 < x < 1/n$ for all n would have to satisfy the contradictory inequalities $x > 0$ and $x \leq 0$. (Why?) Similarly, the nested sequence of intervals $K_n := (n, \infty)$, $n \in N$, has no common point since no $x \in R$ is greater than all natural numbers.

However, it is a very important property of R that every nested sequence of *closed bounded* intervals does have a common point. The completeness of R plays an essential role in establishing this property.

2.5.1 Nested Intervals Property. *If $I_n = [a_n, b_n]$, $n \in N$, is a nested sequence of closed bounded intervals, then there exists a number $\xi \in R$ such that $\xi \in I_n$ for all $n \in N$.*

Furthermore, if the lengths $b_n - a_n$ of I_n satisfy $\inf \{b_n - a_n : n \in N\} = 0$, then the common element ξ is unique.

Proof. Since the intervals are nested, we have $I_n \subseteq I_1$ for all $n \in N$, so that $a_n \leq b_1$ for all $n \in N$. Hence, the non-empty set $\{a_n : n \in N\}$ is bounded above and we let ξ be its supremum. Clearly $a_n \leq \xi$ for all $n \in N$.

We claim also that $\xi \leq b_n$ for all n. This is established by showing that for any particular n, the number b_n is an upper bound for the set $\{a_k : k \in N\}$. We consider two cases. (i) If $n \leq k$, then since $I_n \supseteq I_k$, we have $a_k \leq b_k \leq b_n$. (ii) If $k < n$, then since $I_k \supseteq I_n$, we have $a_k \leq a_n \leq b_n$. (See Figure 2.5.3.) Thus, we conclude that $a_k \leq b_n$ for all k so that b_n is an upper bound of the set $\{a_k : k \in N\}$. Hence, $\xi \leq b_n$ for each $n \in N$. Since $a_n \leq \xi \leq b_n$ for all n, we have $\xi \in I_n$ for all $n \in N$.

If $\eta := \inf \{b_n : n \in N\}$, then an analogous argument (which the reader should write out) can be used to show that $a_n \leq \eta$ for all n, and hence $\xi \leq \eta$. In fact, it is an exercise (see Exercise 2.5.8) to show that $x \in I_n$ for all $n \in N$ if and only if $\xi \leq x \leq \eta$.

Now suppose that $\inf \{b_n - a_n : n \in N\} = 0$. Then for any $\varepsilon > 0$, there exists an $n \in N$ such that $0 \leq \eta - \xi \leq b_n - a_n < \varepsilon$. Since this holds for all $\varepsilon > 0$, it follows from 2.2.9 that $\eta - \xi = 0$. Therefore, in this case, $\xi = \eta$ is the only point that belongs to I_n for every $n \in N$. Q.E.D.

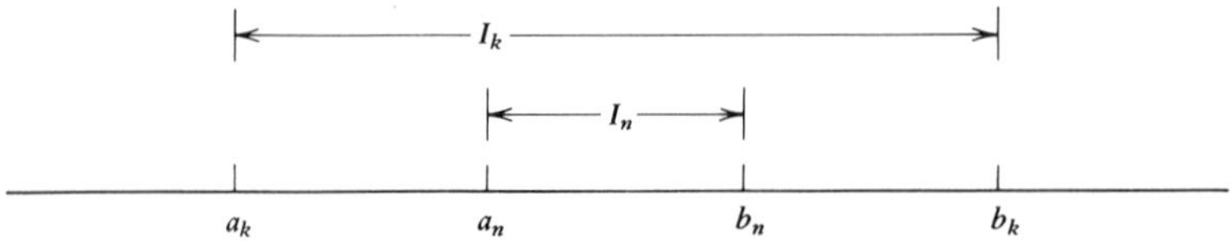

FIGURE 2.5.3 If $k < n$, then $I_n \subseteq I_k$.

Cluster Points

2.5.2 Definition. A point $x \in \boldsymbol{R}$ is a **cluster point** (or a **point of accumulation**) of a subset $S \subseteq \boldsymbol{R}$ if each ε-neighborhood $V_\varepsilon := (x - \varepsilon, x + \varepsilon)$ of x contains at least one point of S distinct from x.

This definition can be reformulated as follows. A point x is a cluster point of S if for each $n \in \boldsymbol{N}$, there exists $s_n \in S$ such that $0 < |x - s_n| < 1/n$. The reader should convince himself that this statement is equivalent to the definition.

We say that a set S "has a cluster point" if some point $x \in \boldsymbol{R}$ is a cluster point of S. However, no assertion is made as to whether the cluster point x is a member of S or not. It is possible for a set to have cluster points that are members and other clusters points that are not members of the set.

2.5.3 Examples. (a) If S_1 is the open interval $(0, 1)$, then every point of the closed interval $[0, 1]$ is a cluster point of S_1. Note that S_1 has the cluster points 0 and 1 that are not elements of S_1.

(b) A finite set has no cluster points. (Why?) The unbounded set $S_2 := \boldsymbol{N}$ has no cluster points even though it is an infinite set.

(c) The set $S_3 := \{1/n : n \in \boldsymbol{N}\}$ has 0 as its only cluster point. None of the points of S_3 are cluster points of S_3.

(d) The set $S_4 := \boldsymbol{I} \cap \boldsymbol{Q}$ consists of all the rational numbers in the unit interval. Every point of $\boldsymbol{I}$ is a cluster point of S_4, as can be shown by using the Density Theorem 2.4.10. The set $\boldsymbol{I} \setminus \boldsymbol{Q}$ of irrational numbers in $\boldsymbol{I}$ also has $\boldsymbol{I}$ as its set of cluster points.

(e) Let S be an infinite set that is bounded above and let $u := \sup S$. If $u \notin S$, then u is a cluster point of S, since for any $\varepsilon > 0$ there exists $x \in S$ such that $x \in (u - \varepsilon, u + \varepsilon)$.

The following result is one of the fundamental theorems of real analysis. As Example 2.5.3(b) shows, the conclusion may fail if either hypothesis is removed.

2.5.4 Bolzano-Weierstrass Theorem. *Every bounded infinite subset of* $\boldsymbol{R}$ *has at least one cluster point.*

Proof. Suppose S is a bounded set with infinitely many members. Because S is bounded, there is a bounded closed interval $I_1 := [a,b]$ containing S. The proof proceeds by repeated bisection to produce a sequence of nested intervals, whose common point will be shown to be a cluster point of S.

We first bisect I_1 into the subintervals $[a, \frac{1}{2}(a + b)]$ and $[\frac{1}{2}(a + b), b]$ and note that at least one of these two intervals must contain infinitely many points of S. For, if not, then S would be the union of two finite sets, and therefore be a finite set itself. Let I_2 denote one of these intervals such that $S \cap I_2$ is infinite. We now bisect I_2 and, as before, we select one of the resulting subintervals I_3 such that $S \cap I_3$ is infinite. If we continue in this manner, we obtain a sequence

$I_1 \supseteq I_2 \supseteq \cdots \supseteq I_n \supseteq \cdots$ of nested bounded closed intervals such that: the length of I_n is $l_n := (b - a)/2^{n-1}$, and $S \cap I_n$ is infinite for every $n \in \mathbf{N}$. Applying the Nested Intervals Property 2.5.1, we obtain a point $x \in \bigcap_{n=1}^{\infty} I_n$.

It remains to show that x is a cluster point of S. If $\varepsilon > 0$ is given and $V := (x - \varepsilon, x + \varepsilon)$ is the ε-neighborhood of x, we choose $n \in \mathbf{N}$ such that $(b - a)/2^{n-1} < \varepsilon$. Since $x \in I_n$ and $l_n < \varepsilon$, it follows that $I_n \subseteq V$. (Why?) Since I_n contains infinitely many points of S, the neighborhood V contains (infinitely many) points of S distinct from x. Hence, x is a cluster point of S. Q.E.D.

Binary and Decimal Representations

We shall digress briefly to discuss (in an informal manner) the notions of binary and decimal representations of real numbers from the perspective of nested intervals. It suffices to consider the situation for real numbers between 0 and 1, since representations for other real numbers can then be obtained by adjoining a positive or negative integer as appropriate.

We shall consider first the idea of a binary representation for a given x in the interval $[0, 1]$. By using the bisection procedure employed in the proof of the Nested Intervals Theorem 2.5.1, we can associate a sequence of 0's and 1's as follows. If x belongs to the left subinterval $[0, \frac{1}{2}]$, then for the first term a_1 of the sequence we take $a_1 = 0$; if x belongs to the right subinterval $[\frac{1}{2}, 1]$, then we take $a_1 = 1$. If $x = \frac{1}{2}$, the bisection point, then we can take a_1 to be either 0 or 1. In any case, we have the inequality

$$\frac{a_1}{2} \leqslant x \leqslant \frac{a_1}{2} + \frac{1}{2}.$$

When the first subinterval has been selected, it is bisected. For the second term a_2 we take $a_2 = 0$ if x belongs to the left subinterval, and we take $a_2 = 1$ if x belongs to the right subinterval. If either $x = \frac{1}{4}$ or $x = \frac{3}{4}$, then a_2 can be taken to be either 0 or 1. At this stage, we have the inequality

$$\frac{a_1}{2} + \frac{a_2}{2^2} \leqslant x \leqslant \frac{a_1}{2} + \frac{a_2}{2^2} + \frac{1}{2^2}.$$

We continue the bisection procedure, assigning at the nth stage the value $a_n = 0$ if x lies in the left subinterval and the value $a_n = 1$ if x lies in the right subinterval. In this way we obtain a sequence $a_1, a_2, \ldots, a_n, \ldots$ of 0's and 1's that corresponds to a sequence of nested intervals whose intersection is the point x. For each n, we have the inequality

$$(*) \qquad \frac{a_1}{2} + \frac{a_2}{2^2} + \cdots + \frac{a_n}{2^n} \leqslant x \leqslant \frac{a_1}{2} + \frac{a_2}{2^2} + \cdots + \frac{a_n}{2^n} + \frac{1}{2^n}.$$

If x happens to be the bisection point at the nth stage, then x is of the form $x = m/2^n$ with m odd. In this case, we may choose either the left or right subinterval so that $a_n = 0$ or $a_n = 1$; however, once this subinterval is chosen, then all subsequent subintervals in the bisection process are determined. For instance, if we choose the left subinterval so that $a_n = 0$, then x is the right end point of all subsequent subintervals and hence $a_k = 1$ for all $k \geqslant n + 1$. On the other hand, if we choose $a_n = 1$, then we must have $a_k = 0$ for all $k \geqslant n + 1$. (For example, if $x = \frac{1}{2}$, then the two possible sequences for x are $0,1,1,1, \ldots$ and $1,0,0,0, \ldots$.)

To summarize: *If $x \in [0,1]$, then there exists a sequence $a_1, a_2, \ldots, a_n, \ldots$ of 0's and 1's such that inequality* $(*)$ *holds for all n.* We shall write $x = (.a_1a_2 \cdots a_n \cdots)_2$ and shall call this a **binary representation** of x. The representation is unique except when x is of the form $x = m/2^n$, where m is odd, in which case there are two possible representations

$$x = (.a_1a_2 \cdots a_{n-1}100 \cdots)_2 = (.a_1a_2 \cdots a_{n-1}011 \cdots)_2,$$

one ending in 0's, and the other one ending in 1's.

Conversely, each sequence of 0's and 1's is the binary representation of a unique real number in $[0,1]$. Indeed, if we are given $a_1, a_2, \ldots, a_n, \ldots$ where $a_n = 0$ or $a_n = 1$ for all $n \in \mathbf{N}$, then the inequality $(*)$ determines a closed subinterval of $[0,1]$ of length $1/2^n$ for each n. It is easily verified that the sequence of intervals obtained in this way is nested; thus, by the Nested Intervals Theorem 2.5.1, there is a unique real number x satisfying $(*)$ for every $n \in \mathbf{N}$. But this means that x has the binary representation $(.a_1a_2 \cdots a_n \cdots)_2$.

Remark. The concept of binary representation is extremely important in this era of digital computers. A number is entered in a digital computer on "bits" and each bit can be put in one of two states—either it will pass current or it will not. These two states correspond to the values 1 and 0, respectively. Thus, a number represented by a sequence of 1's and 0's can be stored in a digital computer on a string of bits. Of course, in actual practice, since only finitely many bits can be used for a number, the binary representation must be truncated. If n binary digits are used for a number $x \in [0,1]$, then accuracy is at most $1/2^n$. For example, to assure four decimal place accuracy, it is necessary to use at least 15 binary digits (or 15 bits).

The geometric insight into the decimal representations of real numbers is essentially similar to that for binary representations, except that in the case of decimal representations we subdivide each interval into *ten* equal subintervals instead of two. If we are given $x \in [0,1]$ and if we subdivide $[0,1]$ into ten equal subintervals, then x lies in a subinterval $[b_1/10, (b_1 + 1)/10]$ for some integer b_1 in $\{0, 1, \ldots, 9\}$. If x is one of the subdivision points, then two values of b_1 are possible and either may be chosen. Thus we have the inequality

$$\frac{b_1}{10} \leqslant x \leqslant \frac{b_1}{10} + \frac{1}{10}$$

where $b_1 \in \{0, 1, \ldots, 9\}$. The chosen subinterval is subdivided into ten equal subintervals, and the process is continued. In this way we obtain a sequence b_1, $b_2, \ldots, b_n, \ldots$ of integers with $0 \leqslant b_n \leqslant 9$ for all $n \in \mathbf{N}$ such that x satisfies the inequality

$$(**) \qquad \frac{b_1}{10} + \frac{b_2}{10^2} + \cdots + \frac{b_n}{10^n} \leqslant x \leqslant \frac{b_1}{10} + \frac{b_2}{10^2} + \cdots + \frac{b_n}{10^n} + \frac{1}{10^n}$$

for all $n \in \mathbf{N}$. We write $x = .b_1b_2 \cdots b_n \cdots$ and we call this a **decimal representation** of x. If $x \geqslant 1$ and if $B \in \mathbf{N}$ is such that $B < x \leqslant B + 1$, then $x = B.b_1b_2 \cdots b_n \cdots$ where the decimal representation of $x - B \in [0,1]$ is as above.

The fact that each decimal determines a unique real number follows from the Nested Intervals Theorem 2.5.1. From a decimal $.b_1b_2 \cdots b_n \cdots$, we get a sequence of nested intervals with lengths $1/10^n$ via the inequalities $(**)$ and hence there is a unique real number x in the intersection. Since x satisfies the inequalities $(**)$, it is evident that $x = .b_1b_2 \cdots b_n \cdots$.

The decimal representation of $x \in [0,1]$ is unique except when x is a point of subdivision at some stage. Suppose that x is such a point, so that $x = m/10^n$ for some $m, n \in \mathbf{N}$, $1 \leqslant m \leqslant 10^n$. (We may assume that m is not divisible by 10.) Then x appears as a subdivision point at the nth stage, and two values for the nth digit are possible. One choice of b_n corresponds to choosing the left subinterval for the next stage. Since x is the right end point of this subinterval, it follows that all subsequent digits will have the value 9; that is, $b_k = 9$ for all $k \geqslant n + 1$. Thus, one decimal representation of x has the form $x = .b_1b_2 \cdots b_n 99 \cdots$. The other choice for the nth decimal place is then $b_n + 1$, which corresponds to choosing the right subinterval at the nth stage. Since x is the left end point of the nth subinterval, all subsequent values will be 0; that is, $b_k = 0$ for all $k \geqslant n + 1$. Thus, the other decimal representation of x has the form $x = .b_1b_2 \cdots (b_n + 1)00 \cdots$. (For example, if $x := \frac{1}{2}$, then $x = .499 \cdots = .500 \cdots$. Similarly, if $y := 38/100$, then $y = .3799 \cdots = .3800 \cdots$.)

We shall conclude this brief look at decimal representations of real numbers by describing the contrasting types of decimal representations that occur for rational and irrational numbers. For this we need the idea of a periodic decimal.

A decimal $B.a_1a_2 \cdots a_n \cdots$ is called **periodic** (or **repeating**) if there exist natural numbers k and m such that $a_n = a_{n+m}$ for all $n > k$. In this case, the block of digits $a_k\, a_{k+1} \cdots a_{k+m-1}$ is repeated once the kth digit is reached. The smallest number m with this property is called the **period** of the decimal. For example, $19/88 = 0.2159090 \cdots 90 \cdots$ has period $m = 2$ with repeating block 90 starting at the digit $k = 4$. A **terminating decimal** is a repeating decimal where the repeated block is simply the number 0.

The relationship between the rationality or irrationality of a real number and the nature of its decimal representations is that: *a positive real number is rational if and only if its decimal representation is periodic.*

Instead of presenting a formal proof of this statement, we shall only indicate the ideas underlying such a proof. Suppose that we have a rational number p/q where p, q are natural numbers with no common prime factors. It suffices to consider the case where $0 < p < q$ (why?). It can be shown that the familiar process of long division of q into p produces the decimal representation of p/q. Each step in the division algorithm produces a remainder that is an integer from 0 to $q - 1$. Therefore, after at most q steps, some remainder will occur a second time and at that point the digits in the quotient will begin to repeat themselves in cycles. Hence the decimal representation of a rational number is periodic.

Conversely, if a decimal is periodic, then it represents a rational number. The idea of the proof is easily illustrated by an example. Suppose that $x = 7.31414 \cdots 14 \cdots$. We first multiply by 10 to move the decimal point to the first repeating block, that is, $10x = 73.1414 \cdots$. We next multiply $10x$ by 10^2 to move one block to the left of the decimal point, that is $1000x = 7314.1414 \cdots$. Subtraction now gives us an integer: $1000x - 10x = 7314 - 73 = 7241$. Hence, $x = 7241/990$, a rational number.

Exercises for Section 2.5

1. If $I := [a,b]$ and $I' := [a',b']$ are closed intervals in $\boldsymbol{R}$, show that $I \subseteq I'$ if and only if $a' \leqslant a$ and $b \leqslant b'$.
2. If $S \subseteq \boldsymbol{R}$ is non-empty, show that S is bounded if and only if there is some closed interval $I \subseteq \boldsymbol{R}$ such that $S \subseteq I$.
3. If $S \subseteq \boldsymbol{R}$ is a non-empty bounded set, and I_S is the interval $I_S := [\inf S, \sup S]$, show that $S \subseteq I_S$. Moreover, if J is any closed bounded interval of $\boldsymbol{R}$ such that $S \subseteq J$, show that $I_S \subseteq J$.
4. Show that if $I_1 \supseteq I_2 \supseteq \cdots \supseteq I_n \supseteq \cdots$ is a nested sequence of closed intervals in $\boldsymbol{R}$, and if $I_n = [a_n,b_n]$, then $a_1 \leqslant a_2 \leqslant \cdots \leqslant a_n \leqslant \cdots$ and $b_1 \geqslant b_2 \geqslant \cdots \geqslant b_n \geqslant \cdots$.
5. Let $I_n := [0,1/n]$ for $n \in \boldsymbol{N}$. Prove that if $x > 0$, then $x \notin \bigcap_{n=1}^{\infty} I_n$.
6. Prove that if $J_n := (0,1/n)$ for $n \in \boldsymbol{N}$, then $\bigcap_{n=1}^{\infty} J_n = \emptyset$.
7. Prove that if $K_n := (n, +\infty)$ for $n \in \boldsymbol{N}$, then $\bigcap_{n=1}^{\infty} K_n = \emptyset$.
8. Using the notation in the proof of the Nested Intervals Theorem 2.5.1, show that $\eta \in \bigcap_{n=1}^{\infty} I_n$. Also show that $[\xi,\eta] = \bigcap_{n=1}^{\infty} I_n$.
9. Prove that every point of $[0,1]$ is a cluster point of $S := (0,1)$.
10. Write out the proof of the assertion in Example 2.5.3(b) that a finite set in $\boldsymbol{R}$ has no cluster point.
11. If $x \in \boldsymbol{R}$, $x > 0$, and if $0 < \varepsilon < x$, show that there are at most a finite number of $n \in \boldsymbol{N}$ such that $1/n$ belongs to the interval $(x - \varepsilon, x + \varepsilon)$.
12. Prove the assertion in Example 2.5.3(d) that every point of $\boldsymbol{I} = [0,1]$ is a cluster point of $\boldsymbol{I} \cap \boldsymbol{Q}$ and $\boldsymbol{I} \setminus \boldsymbol{Q}$.
13. Show that if

$$\frac{a_1}{10} + \frac{a_2}{10^2} + \cdots + \frac{a_n}{10^n} = \frac{b_1}{10} + \frac{b_2}{10^2} + \cdots + \frac{b_m}{10^m} \neq 0,$$

where the a_k and b_k belong to $\{0,1, \ldots, 9\}$, then $n = m$ and $a_k = b_k$ for $k = 1, \ldots, n$.

14. Express 1/7 and 2/19 as periodic decimals.
15. Find the rational number represented by the periodic decimals $1.25137 \cdots 137 \cdots$ and $37.14653 \cdots 653 \cdots$.

SECTION 2.6 Open and Closed Sets in *R*

There are special types of subsets of $\boldsymbol{R}$ that play an important role in analysis. These are the open sets and the closed sets in $\boldsymbol{R}$. Recall that a neighborhood of a point $x \in \boldsymbol{R}$ is any set V containing some ε-neighborhood $(x - \varepsilon, x + \varepsilon)$ of x.

2.6.1 Definition. (i) A subset G of $\boldsymbol{R}$ is **open** in $\boldsymbol{R}$ if for each $x \in G$ there exists a neighborhood V of x such that $V \subseteq G$.

(ii) A subset F of $\boldsymbol{R}$ is **closed** in $\boldsymbol{R}$ if the complement $\mathscr{C}(F) = \boldsymbol{R} \setminus F$ is open in $\boldsymbol{R}$.

To show that a set $G \subseteq \boldsymbol{R}$ is open, it suffices to show that each point in G has an ε-neighborhood contained in G. In fact, G is open if and only if for each $x \in G$, there exists $\varepsilon_x > 0$ such that $(x - \varepsilon_x, x + \varepsilon_x)$ is contained in G.

To show that a set $F \subseteq \boldsymbol{R}$ is closed, it suffices to show that each point $y \notin F$ has an ε-neighborhood disjoint from F. In fact, F is closed if and only if for each $y \notin F$ there exists $\varepsilon_y > 0$ such that $F \cap (y - \varepsilon_y, y + \varepsilon_y) = \emptyset$.

2.6.2 Examples. (a) The entire set $\boldsymbol{R} = (-\infty, \infty)$ is open.

For any $x \in \boldsymbol{R}$, we may take $\varepsilon := 1$.

(b) The set $G := \{x \in \boldsymbol{R} : 0 < x < 1\}$ is open.

For any $x \in G$ we may take ε_x to be the smaller of the numbers x, $1 - x$. We leave it to the reader to show that if $|u - x| < \varepsilon_x$ then $u \in G$.

(c) Any open interval $I := (a,b)$ is an open set.

In fact, if $x \in I$, we can take ε_x to be the smaller of the numbers $x - a$, $b - x$. The reader can then show that $(x - \varepsilon_x, x + \varepsilon_x) \subseteq I$. Similarly, $(-\infty, b)$ and (a, ∞) are open sets.

(d) The set $\boldsymbol{I} := [0,1]$ is not open.

This follows since every neighborhood of $0 \in \boldsymbol{I}$ contains points not in $\boldsymbol{I}$.

(e) The set $\boldsymbol{I}$ is closed.

To see this let $y \notin \boldsymbol{I}$; then either $y < 0$ or $y > 1$. If $y < 0$, we take $\varepsilon_y := |y|$, and if $y > 1$ we take $\varepsilon_y := y - 1$. We leave it to the reader to show that in either case, we have $\boldsymbol{I} \cap (y - \varepsilon_y, y + \varepsilon_y) = \emptyset$.

(f) The set $H := \{x : 0 \leqslant x < 1\}$ is neither open nor closed. (Why?)

(g) The empty set $\emptyset$ is open in $\boldsymbol{R}$.

In fact, the empty set contains no points at all, so the requirement in Definition 2.6.1(i) is vacuously satisfied. The empty set is also closed since its complement $\boldsymbol{R}$ is open, as was seen in part (a).

In ordinary parlance, when applied to doors, windows, and minds, the words "open" and "closed" are antonyms. However, when applied to subsets of $\boldsymbol{R}$, these words are not antonyms. For example, we noted above that the sets $\emptyset$, $\boldsymbol{R}$ are *both* open and closed in $\boldsymbol{R}$. (The reader will probably be relieved to learn that there are no other subsets of $\boldsymbol{R}$ which have both properties.) In addition, there are many subsets of $\boldsymbol{R}$ that are *neither* open nor closed; in fact, most subsets of $\boldsymbol{R}$ have this neutral character.

The following basic result describes the manner in which open sets relate to the operations of the union and intersection of sets in $\boldsymbol{R}$.

2.6.3 Open Set Properties. (a) *The union of an arbitrary collection of open subsets of* $\boldsymbol{R}$ *is open.*

(b) *The intersection of any finite collection of open sets is open.*

Proof. (a) Let $\{G_\lambda : \lambda \in \Lambda\}$ be a family of sets in $\boldsymbol{R}$ that are open, and let G be their union. Consider an element $x \in G$. By the definition of union, x must belong to G_{λ_0} for some $\lambda_0 \in \Lambda$. Since G_{λ_0} is open, there exists a neighborhood V of x such that $V \subseteq G_{\lambda_0}$. But $G_{\lambda_0} \subseteq G$, so that $V \subseteq G$. Since x is an arbitrary element of G, we conclude that G is open in $\boldsymbol{R}$.

(b) Suppose G_1 and G_2 are open and let $G := G_1 \cap G_2$. To show that G is open, we consider any $x \in G$; then $x \in G_1$ and $x \in G_2$. Since G_1 is open, there exists $\varepsilon_1 > 0$ such that $(x - \varepsilon_1, x + \varepsilon_1)$ is contained in G_1. Similarly, since G_2 is open, there exists $\varepsilon_2 > 0$ such that $(x - \varepsilon_2, x + \varepsilon_2)$ is contained in G_2. If we now take ε to be the smaller of ε_1 and ε_2, then the ε-neighborhood $U := (x - \varepsilon, x + \varepsilon)$ satisfies both $U \subseteq G_1$ and $U \subseteq G_2$. Thus, $x \in U \subseteq G$. Since x is an arbitrary element of G, we conclude that G is open in $\boldsymbol{R}$.

It now follows by an induction argument (which we leave to the reader to write out) that the intersection of any finite collection of open sets is open. Q.E.D.

The corresponding properties for closed sets will be established by using the De Morgan identities for sets and their complements. (See Theorem 1.1.7.)

2.6.4 Corollary. (a) *The intersection of an arbitrary collection of closed sets in* $\boldsymbol{R}$ *is closed.*

(b) *The union of any finite collection of closed sets in* $\boldsymbol{R}$ *is closed.*

Proof. (a) If $\{F_\lambda : \lambda \in \Lambda\}$ is a family of closed sets in $\boldsymbol{R}$ and $F := \bigcap_{\lambda \in \Lambda} F_\lambda$, then $\mathscr{C}(F) = \bigcup_{\lambda \in \Lambda} \mathscr{C}(F_\lambda)$ is the union of open sets. Hence, $\mathscr{C}(F)$ is open by Theorem 2.6.3(a), and consequently, F is closed.

(b) Suppose $F_1, F_2, \cdots, F_n$ are closed in $\boldsymbol{R}$ and let $F := F_1 \cup F_2 \cup \cdots \cup F_n$. By the De Morgan identity the complement of F is given by

$$\mathscr{C}(F) = \mathscr{C}(F_1) \cap \cdots \cap \mathscr{C}(F_n).$$

Since each $\mathscr{C}(F_i)$ is open, it follows from Theorem 2.6.3(b) that $\mathscr{C}(F)$ is open. Hence F is closed. Q.E.D.

The finiteness restrictions in 2.6.3(b) and 2.6.4(b) *cannot* be removed. Consider the following examples:

2.6.5 Examples. (a) Let $G_n := (0, 1 + 1/n)$ for $n \in \boldsymbol{N}$. Then G_n is open for each $n \in \boldsymbol{N}$, by Example 2.6.2(c). However, the intersection $G := \bigcap_{n=1}^{\infty} G_n$ is the interval $(0,1]$ which is not open. Thus, *the intersection of infinitely many open sets in* $\boldsymbol{R}$ *need not be open*.

(b) Let $F_n := [1/n, 1]$, $n \in \boldsymbol{N}$. Each F_n is closed, but the union $F := \bigcup_{n=1}^{\infty} F_n$ is the set $(0, 1]$ which is not closed. Thus, *the union of infinitely many closed sets in* $\boldsymbol{R}$ *need not be closed*.

The characterization of open sets

The idea of an open set in $\boldsymbol{R}$ is a generalization of the notion of an open interval. That this generalization does not lead to extremely exotic sets that are open is revealed by the next result, whose proof may be omitted on a first reading.

2.6.6 Theorem. *A subset of* $\boldsymbol{R}$ *is open if and only if it is the union of countably many disjoint open intervals in* $\boldsymbol{R}$.

Proof. Suppose that $G \neq \emptyset$ is an open set in $\boldsymbol{R}$. For each $x \in G$, let $A_x := \{a \in \boldsymbol{R} : (a,x] \subseteq G\}$ and let $B_x := \{b \in \boldsymbol{R} : [x,b) \subseteq G\}$. Since G is open, it follows that A_x and B_x are not empty. (Why?) If the set A_x is bounded below, we set $a_x := \inf A_x$; if A_x is not bounded below, we set $a_x := -\infty$. Note that, in either case, $a_x \notin G$. If the set B_x is bounded above, we set $b_x := \sup B_x$; if B_x is not bounded above, we set $b_x := \infty$. Note that, in either case, $b_x \notin G$.

We now define $I_x := (a_x, b_x)$; clearly I_x is an open interval containing x. We claim that $I_x \subseteq G$. To see this, let $y \in I_x$ and suppose that $y < x$. It follows from the definition of a_x that there exists $a' \in A_x$ with $a' < y$, whence it follows that $y \in (a', x] \subseteq G$. Similarly, if $y \in I_x$ and $x < y$, there exists $b' \in B_x$ with $y < b'$, whence it follows that $y \in [x, b') \subseteq G$. Since $y \in I_x$ is arbitrary, we have that $I_x \subseteq G$.

Since $x \in G$ is arbitrary, we conclude that

$$\bigcup_{x \in G} I_x \subseteq G.$$

On the other hand, since for each $x \in G$ there is an open interval I_x with $x \in I_x \subseteq G$, we also have

$$G \subseteq \bigcup_{x \in G} I_x.$$

Therefore we conclude that $G = \bigcup_{x \in G} I_x$.

We claim that if $x, y \in G$ and $x \neq y$, then either $I_x = I_y$ or $I_x \cap I_y = \emptyset$. To prove this suppose that $z \in I_x \cap I_y$, whence it follows that $a_x < z < b_y$ and $a_y < z < b_x$. (Why?) We shall show that $a_x = a_y$. If not, it follows from the Trichotomy Property that either (i) $a_x < a_y$, or (ii) $a_y < a_x$. In case (i), then $a_y \in I_x = (a_x, b_x) \subseteq G$, which contradicts the fact that $a_y \notin G$. Similarly, in case (ii), then $a_x \in I_y = (a_y, b_y) \subseteq G$, which contradicts the fact that $a_x \notin G$. Therefore we must have $a_x = a_y$ and a similar argument implies that $b_x = b_y$. Therefore, we conclude that if $I_x \cap I_y \neq \emptyset$, then $I_x = I_y$.

It remains to show that the collection of distinct intervals $\{I_x : x \in G\}$ is countable. To do this, we enumerate the set $\boldsymbol{Q}$ of rational numbers $\boldsymbol{Q} = \{r_1, r_2, \ldots, r_n, \ldots\}$ (see Section 1.4). It follows from the Density Theorem 2.4.9 that each interval I_x contains rational numbers; we select that rational number in I_x that has the smallest index n in this enumeration of $\boldsymbol{Q}$. That is, we chose $r_{n(x)} \in \boldsymbol{Q}$ such that $I_{r_{n(x)}} = I_x$ and $n(x)$ is the smallest index n such that $I_{r_n} = I_x$. Thus the set of distinct intervals I_x, $x \in G$, is put into correspondence with a subset of $\boldsymbol{N}$. Hence this set of distinct intervals is countable. Q.E.D.

It is left as an exercise to show that the representation of G as a disjoint union of open intervals is uniquely determined.

Note. It does *not* follow from the preceding theorem that a subset of $\boldsymbol{R}$ is closed if and only if it is the intersection of a countable collection of closed *intervals* (Why not?). In fact, there are closed sets in $\boldsymbol{R}$ that cannot be expressed as the intersection of a countable collection of closed intervals in $\boldsymbol{R}$. A set consisting of two points is one example. (Why?) For the construction of a much more interesting example called the **Cantor set**, see *The Elements of Real Analysis*, pp. 48–49.

A characterization of closed sets

There is a characterization of closed sets in terms of cluster points that is sometimes useful. First, recall from Example 2.5.3(a) that a cluster point of a set need not belong to the set. We now show that closed sets are precisely those sets that contain all of their cluster points.

2.6.7 Theorem. *A subset of $\boldsymbol{R}$ is closed if and only if it contains all of its cluster points.*

Proof. Let F be a closed set in $\boldsymbol{R}$ and let x be a cluster point of F; we shall show that $x \in F$. If not, then x belongs to the open set $\mathscr{C}(F)$. Therefore there exists a neighborhood V of x such that $V \subseteq \mathscr{C}(F)$. Consequently $V \cap F = \emptyset$, which contradicts the assumption that x is a cluster point of F.

Conversely, let F be a subset of $\boldsymbol{R}$ that contains all of its cluster points; we shall show that $\mathscr{C}(F)$ is open. For if $y \in \mathscr{C}(F)$, then y is not a cluster point of F. It follows from Definition 2.5.2 that there exists an ε-neighborhood V_ε of y that does not contain a point of F (except possibly y). But since $y \in \mathscr{C}(F)$ it follows that $V_\varepsilon \subseteq \mathscr{C}(F)$. Since y is an arbitrary element of $\mathscr{C}(F)$, we deduce that for every point in y there is a neighborhood that is entirely contained in $\mathscr{C}(F)$. But this means that $\mathscr{C}(F)$ is open in $\boldsymbol{R}$. Therefore F is closed in $\boldsymbol{R}$. Q.E.D.

Exercises for Section 2.6

1. If $x \in (0,1)$, let ε_x be as in Example 2.6.2(b). Show that if $|u - x| < \varepsilon_x$ then $u \in (0,1)$.
2. Write out the induction argument in the proof of part (b) of the Open Set Properties 2.6.3.
3. Prove that $(0,1] = \bigcap_{n=1}^{\infty} (0, 1 + 1/n)$, as asserted in Example 2.6.5(a).
4. If G is open and $x \in G$, show that the sets A_x and B_x in the proof of Theorem 2.6.6 are not empty.
5. If the set A_x in the proof of Theorem 2.6.6 is bounded below, show that $a_x := \inf A_x$ does not belong to G.
6. If in the notation used in the proof of Theorem 2.2.6, we have $a_x < y < x$, show that $y \in G$.
7. If in the notation used in the proof of Theorem 2.2.6, we have $I_x \cap I_y \neq \emptyset$, show that $b_x = b_y$.
8. A point $x \in \boldsymbol{R}$ is said to be an **interior point** of $A \subseteq \boldsymbol{R}$ in case there is a neighborhood V of x such that $V \subseteq A$. Show that a set $A \subseteq \boldsymbol{R}$ is open if and only if every point of A is an interior point of A.
9. A point $x \in \boldsymbol{R}$ is said to be a **boundary point** of $A \subseteq \boldsymbol{R}$ in case every neighborhood V of x contains points in A and points in $\mathscr{C}(A)$. Show that a set A and its complement $\mathscr{C}(A)$ have exactly the same boundary points.
10. Show that a set $G \subseteq \boldsymbol{R}$ is open if and only if it does not contain any of its boundary points.
11. Show that a set $F \subseteq \boldsymbol{R}$ is closed if and only if it contains all of its boundary points.
12. If $A \subseteq \boldsymbol{R}$, let A° be the union of all open sets that are contained in A; the set A° is called the **interior** of A. Show that A° is an open set, that it is the largest open set contained in A, and that a point z belongs to A° if and only if z is an interior point of A.
13. Using the notation of the preceding exercise, let A, B be sets in $\boldsymbol{R}$. Show that $A^\circ \subseteq A$, $(A^\circ)^\circ = A^\circ$, and that $(A \cap B)^\circ = A^\circ \cap B^\circ$. Show also that $A^\circ \cup B^\circ \subseteq (A \cup B)^\circ$ and give an example to show that the inclusion may be proper.
14. If $A \subseteq \boldsymbol{R}$, let A^- be the intersection of all closed sets containing A; the set A^- is called the **closure** of A. Show that A^- is a closed set, that it is the smallest closed set containing A, and that a point w belongs to A^- if and only if w is either an interior point or a boundary point of A.
15. Using the notation of the preceding exercise, let A, B be sets in $\boldsymbol{R}$. Show that we have $A \subseteq A^-$, $(A^-)^- = A^-$, and that $(A \cup B)^- = A^- \cup B^-$. Show that $(A \cap B)^- \subseteq A^- \cap B^-$ and give an example to show that the inclusion may be proper.
16. Give an example of a set $A \subseteq \boldsymbol{R}$ such that $A^\circ = \emptyset$ and $A^- = \boldsymbol{R}$.
17. Show that if $F \subseteq \boldsymbol{R}$ is a closed non-empty set that is bounded above, then $\sup F$ belongs to F.

CHAPTER THREE

SEQUENCES

The material in the preceding two chapters should provide an adequate understanding of the real number system $\boldsymbol{R}$. Now that these foundations have been laid we are prepared to pursue questions of a more analytic nature, and we shall begin with a study of the convergence of sequences. Some of these results may be familiar to the reader from calculus, but the presentation here is intended to be rigorous and will give certain more profound results than are usually discussed in earlier courses.

We shall first introduce the meaning of the convergence of a sequence of real numbers and establish some elementary (but useful) results about convergent sequences. We then present some important criteria for the convergence of sequences. It is important for the reader to learn both the theorems and how the theorems apply to special sequences.

Because of the linear limitations inherent in a book it is necessary to decide whether to follow this chapter with a discussion of series, or whether to defer this discussion until after continuity, differentiation, and integration have been studied. While we have decided to defer the discussion of series until later, the instructor may wish to give a brief introduction to series along with this chapter.

SECTION 3.1 Sequences and Their Limits

The reader will recall that a sequence in a set S is a function on the set $\boldsymbol{N} = \{1, 2, \ldots\}$ of natural numbers whose range is contained in the set S. In this chapter we will be concerned with sequences in $\boldsymbol{R}$.

3.1.1 Definition. A **sequence of real numbers** (or a **sequence in $\boldsymbol{R}$**) is a function on the set $\boldsymbol{N}$ of natural numbers whose range is contained in the set $\boldsymbol{R}$ of real numbers.

In other words, a sequence in $\boldsymbol{R}$ assigns to each natural number $n = 1, 2, \ldots$ a uniquely determined real number. The real numbers so obtained are called the **elements** of the sequence, or the **values** of the sequence, or the **terms** in the sequence. It is customary to denote the element of $\boldsymbol{R}$ assigned to $n \in \boldsymbol{N}$ by a symbol such as x_n (or a_n, or z_n), and we shall generally use this traditional notation. Thus, if $X : \mathbf{N} \to \mathbf{R}$ is a sequence, we shall ordinarily denote the value of X at n by x_n, rather than by $X(n)$. We will denote this sequence by the notations

$$X, \quad (x_n), \quad (x_n : n \in \mathbf{N}).$$

We use the parentheses to indicate that the ordering induced by that in $\mathbf{N}$ is a matter of importance. Thus, we distinguish notationally between the sequence $X = (x_n : n \in \mathbf{N})$, whose terms have an ordering, and the set $\{x_n : n \in \mathbf{N}\}$ of values of this sequence which are not considered to be ordered. For example, the sequence $X := ((-1)^n : n \in \mathbf{N})$ alternates between -1 and 1, whereas the set of values $\{(-1)^n : n \in \mathbf{N}\}$ is equal to the set $\{-1, 1\}$.

In defining sequences it is often convenient to list in order the terms in the sequence, stopping when the rule of formation seems evident. Thus we may write

$$X := (2, 4, 6, 8, \ldots)$$

for the sequence of even natural numbers, or

$$Y := \left(\frac{1}{1}, \frac{1}{2}, \frac{1}{3}, \frac{1}{4}, \ldots\right)$$

for the sequence of reciprocals of the natural numbers, or

$$Z := \left(\frac{1}{1^2}, \frac{1}{2^2}, \frac{1}{3^2}, \frac{1}{4^2}, \ldots\right)$$

for the sequence of reciprocals of the squares of the natural numbers. A more satisfactory method is to specify a formula for the general term of the sequence, such as

$$X := (2n : n \in \mathbf{N}), \qquad Y := \left(\frac{1}{m} : m \in \mathbf{N}\right), \qquad Z := \left(\frac{1}{s^2} : s \in \mathbf{N}\right).$$

In practice, it is often convenient to specify the value x_1 and a formula for obtaining x_{n+1} $(n \geq 1)$ when x_n is known. Still more generally we may specify x_1 and a formula for obtaining x_{n+1} $(n \geq 1)$ from $x_1, x_2, \ldots, x_n$. We refer to either of these methods as an **inductive** or **recursive** definition of the sequence. In this way, the sequence X of even natural numbers can be defined by

$$x_1 := 2, \qquad x_{n+1} := x_n + 2 \quad (n \geq 1);$$

or by the definition

$$x_1 := 2, \qquad x_{n+1} := x_1 + x_n \quad (n \geq 1).$$

Remark. Sequences that are given by an inductive process often arise in computer science. In particular, sequences defined by an inductive process of the form $x_1 :=$ given, $x_{n+1} := f(x_n)$ for $n \in \boldsymbol{N}$ are especially amenable to study using computers. Sequences defined by the process: $y_1 :=$ given, $y_{n+1} := g_n(y_1, y_2, \ldots, y_n)$ for $n \in \boldsymbol{N}$, can also be treated. However, the calculation of the terms of such a sequence becomes unwieldy for large n, since one must store each of the values $y_1, \ldots, y_n$ in order to calculate y_{n+1}.

3.1.2 Examples (a) If $b \in \boldsymbol{R}$, the sequence $B := (b, b, b, \ldots)$, all of whose terms equal b, is called the **constant sequence** b. Thus the constant sequence 1 is the sequence $(1, 1, 1, \ldots)$, all of whose terms equal 1, and the constant sequence 0 is the sequence $(0, 0, 0, \ldots)$.

(b) The sequence of squares of the natural numbers is the sequence $S := (1^2, 2^2, 3^2, \ldots) = (n^2 : n \in \boldsymbol{N})$, which, of course, is the same as the sequence $(1, 4, 9, \ldots, n^2, \ldots)$.

(c) If $a \in \boldsymbol{R}$, then the sequence $A := (a^n : n \in \boldsymbol{N})$ is the sequence $A = (a, a^2, a^3, \ldots, a^n, \ldots)$. In particular, if $a = \frac{1}{2}$, then we obtain the sequence

$$\left(\frac{1}{2^n} : n \in \boldsymbol{N}\right) = \left(\frac{1}{2}, \frac{1}{4}, \frac{1}{8}, \ldots, \frac{1}{2^n}, \ldots\right).$$

(d) The **Fibonacci sequence** $F := (f_n : n \in \boldsymbol{N})$ is given by the inductive definition

$$f_1 := 1, \qquad f_2 := 1, \qquad f_{n+1} := f_{n-1} + f_n \quad (n \geq 2).$$

The first ten terms of the Fibonacci sequence are seen to be $F = (1, 1, 2, 3, 5, 8, 13, 21, 34, 55, \ldots)$.

We now introduce some important ways of constructing new sequences from given ones.

3.1.3 Definition. If $X = (x_n)$ and $Y = (y_n)$ are sequences of real numbers, then we define their **sum** to be the sequence $X + Y := (x_n + y_n : n \in \boldsymbol{N})$, their **difference** to be the sequence $X - Y := (x_n - y_n : n \in \boldsymbol{N})$, and their **product** to be the sequence $X \cdot Y := (x_n y_n : n \in \boldsymbol{N})$. If $c \in \boldsymbol{R}$ we define the **multiple** of X by c to be the sequence $cX := (cx_n : n \in \boldsymbol{N})$. Finally, if $Z = (z_n)$ is a sequence of real numbers with $z_n \neq 0$ for all $n \in \boldsymbol{N}$, then we define the **quotient** of X and Z to be the sequence $X/Z := (x_n/z_n : n \in \boldsymbol{N})$.

For example, if X and Y are the sequences

$$X := (2, 4, 6, \ldots, 2n, \ldots), \qquad Y := \left(\frac{1}{1}, \frac{1}{2}, \frac{1}{3}, \ldots, \frac{1}{n}, \ldots\right),$$

then we have

$$X + Y = \left(\frac{3}{1}, \frac{9}{2}, \frac{19}{3}, \ldots, \frac{2n^2+1}{n}, \ldots\right),$$

$$X - Y = \left(\frac{1}{1}, \frac{7}{2}, \frac{17}{3}, \ldots, \frac{2n^2-1}{n}, \ldots\right),$$

$$X \cdot Y = (2, 2, 2, \ldots, 2, \ldots),$$

$$3X = (6, 12, 18, \ldots, 6n, \ldots),$$

$$X/Y = (2, 8, 18, \ldots, 2n^2, \ldots).$$

We note that if Z denotes the sequence

$$Z := (0, 2, 0, \ldots, 1+(-1)^n, \ldots),$$

then we have defined $X + Z$, $X - Z$ and $X \cdot Z$; but X/Z is not defined since some of the terms of Z are equal to 0.

We now come to the notion of the limit of a sequence.

3.1.4 Definition. Let $X = (x_n)$ be a sequence of real numbers. A real number x is said to be a **limit of** X if, for each neighborhood V of x there is a natural number $K(V)$ such that for all $n \geqslant K(V)$, the terms x_n belong to V. If x is a limit of the sequence X, we also say that X **converges to** x (or **has a limit** x). If a sequence has a limit, we say that the sequence is **convergent**; if it has no limit, we say that the sequence is **divergent**.

The notation $K(V)$ is used to suggest that the choice of K may depend on the choice of the neighborhood V. It is clear that a "small" neighborhood V will usually require a "large value" of $K(V)$ in order to guarantee that $x_n \in V$ for all $n \geqslant K(V)$.

When a sequence $X = (x_n)$ of real numbers has a limit $x \in \mathbf{R}$, we often write

$$x = \lim X, \qquad x = \lim (x_n), \qquad \text{or} \qquad x = \lim_n (x_n);$$

sometimes we use the symbolism $x_n \to x$.

One can describe the definition of convergence of $X = (x_n)$ to x by saying: *for every neighborhood V of x, all but a finite number of the terms of X belong to V.*

3.1.5 Uniqueness of Limits. *A sequence of real numbers can have at most one limit.*

Proof. Suppose, on the contrary, that x' and x'' are limits of $X = (x_n)$, and that $x' \neq x''$. Let V' and V'' be disjoint neighborhoods of x' and x'', respectively.

Moreover, let K' and K'' be natural numbers such that if $n \geqslant K'$ then $x_n \in V'$, and if $n \geqslant K''$ then $x_n \in V''$. Now let $K := \sup\{K', K''\}$ so that both $x_K \in V'$ and $x_K \in V''$. Then $x_K \in V' \cap V''$, which contradicts the supposition that V' and V'' are disjoint. Consequently, we must have $x' = x''$. Q.E.D.

We have defined the limit of a sequence $X = (x_n)$ of real numbers in terms of neighborhoods. We now give some equivalent formulations of this definition.

3.1.6 Theorem. *Let $X = (x_n)$ be a sequence of real numbers, and let $x \in \mathbf{R}$. The following assertions are equivalent:*

(a) *X converges to x.*

(b) *For every ε-neighborhood V_ε of x there is a natural number $K(\varepsilon)$ such that for all $n \geqslant K(\varepsilon)$, the terms x_n belong to V_ε.*

(c) *For every $\varepsilon > 0$ there is a natural number $K(\varepsilon)$ such that for all $n \geqslant K(\varepsilon)$, then $x - \varepsilon < x_n < x + \varepsilon$.*

(d) *For every $\varepsilon > 0$ there is a natural number $K(\varepsilon)$ such that for all $n \geqslant K(\varepsilon)$, then $|x_n - x| < \varepsilon$.*

Proof. (a) $\Rightarrow$ (b). If X converges to x in the sense of Definition 3.1.4, then since an ε-neighborhood $V_\varepsilon = (x - \varepsilon, x + \varepsilon)$ of x is a neighborhood of x, there is a natural number $K(\varepsilon) := K(V_\varepsilon)$ such that for all $n \geqslant K(\varepsilon)$, the terms x_n belong to V_ε.

(b) $\Rightarrow$ (c) We note that if $x_n \in V_\varepsilon$, then $x - \varepsilon < x_n < x + \varepsilon$.

(c) $\Rightarrow$ (d) We note that if $x - \varepsilon < x_n < x + \varepsilon$, then $|x_n - x| < \varepsilon$.

(d) $\Rightarrow$ (a) Suppose that (d) holds for arbitrary choice of $\varepsilon > 0$; we now show that $X = (x_n)$ converges to x in the sense of Definition 3.1.4. To do this, let V be an arbitrary neighborhood of x. By the definition of a neighborhood of a point, there exists an $\varepsilon > 0$ such that $V \supseteq V_\varepsilon = (x - \varepsilon, x + \varepsilon)$. Let $K(\varepsilon)$ be such that for all $n \geqslant K(\varepsilon)$ then $|x_n - x| < \varepsilon$; it follows that for $n \geqslant K(\varepsilon)$ we have

$$x - \varepsilon < x_n < x + \varepsilon.$$

Hence, if $n \geqslant K(\varepsilon)$, then $x_n \in V$; consequently, we can take the $K(V)$ of Definition 3.1.4 to be equal to $K(\varepsilon)$ for the above choice of ε. Q.E.D.

Remark. The definition of limit of a sequence of real numbers is used to verify that a proposed value x is indeed the limit. It does *not* provide a means for initially determining what that value should be. Later results will contribute to this end, but quite often it is necessary in practice to arrive at a conjectured value of the limit by direct calculation of a number of terms of the sequence. Computers can be very helpful in this respect, but since they can calculate only a finite number of terms of a sequence, such computations do not in any way constitute a proof, but only a conjecture, of the value of the limit.

The K(ε) Game

Theorem 3.1.6 forms the basis for what we will refer to as "The $K(\varepsilon)$ Game". In this game, Player A asserts that a certain number x is the limit of a sequence of real numbers. In making this assertion Player A is issuing a challenge to Player B to give a specific value for $\varepsilon > 0$, after which Player A will provide a value $K(\varepsilon)$ such that if $n \in \mathbf{N}$ and $n \geq K(\varepsilon)$ it will be true that $|x_n - x| < \varepsilon$. If Player A can always come up with a value of $K(\varepsilon)$ and show that this value will work, he wins. However, if Player B can give a value of $\varepsilon > 0$ for which Player A cannot give an adequate response, then Player B wins and Player A is publicly humiliated.

To be sure of success, Player A should have a formula for $K(\varepsilon)$, or at least an organized procedure (i.e., an "algorithm") for finding a value for $K(\varepsilon)$ once $\varepsilon > 0$ has been specified.

Note that Player A does *not* have to produce the smallest number $K(\varepsilon)$ such that $|x_n - x| < \varepsilon$ for all $n \geq K(\varepsilon)$, although he may be able to do so. All he wagers is that he will be able to give *some* value for $K(\varepsilon)$ that works, no matter what value of $\varepsilon > 0$ that Player B might specify.

We recall that Player B wins when he can specify a value of $\varepsilon > 0$ for which Player A's response is not adequate in the sense that no matter what value Player A gives for $K(\varepsilon)$, Player B can produce a natural number (say $n_{K(\varepsilon)}$) with $n_{K(\varepsilon)} \geq K(\varepsilon)$ such that $|x_{K(\varepsilon)} - x| \geq \varepsilon$.

3.1.7 Examples. (a) $\lim (1/n) = 0$.

To show this, Player A notes that, given $\varepsilon > 0$, then $1/\varepsilon > 0$. Hence by the Archimedean Property 2.4.7 there is a natural number exceeding $1/\varepsilon$. Now if $K(\varepsilon)$ is a natural number with $K(\varepsilon) > 1/\varepsilon$, then for any $n \in \mathbf{N}$ such that $n \geq K(\varepsilon)$, we will have $n > 1/\varepsilon$ so that $1/n < \varepsilon$. That is, if $n \geq K(\varepsilon)$, then

$$\left|\frac{1}{n} - 0\right| = \frac{1}{n} < \varepsilon.$$

Therefore Player A can confidently assert that: "The sequence $(1/n)$ converges to 0", since no matter what value of $\varepsilon > 0$ Player B names, A merely needs to produce a natural number $K(\varepsilon) > 1/\varepsilon$.

(b) $\lim (1/n^2) = 0$.

Given $\varepsilon > 0$, Player A wants to get

$$\left|\frac{1}{n^2} - 0\right| = \frac{1}{n^2} < \varepsilon$$

for n sufficiently large. If we extract positive square roots, the inequality $1/n^2 < \varepsilon$ leads us to

$$\frac{1}{n} < \sqrt{\varepsilon} \qquad \text{or} \qquad \frac{1}{\sqrt{\varepsilon}} < n.$$

Thus, Player A is tempted to take $K(\varepsilon)$ to be a natural number such that $K(\varepsilon) > 1/\sqrt{\varepsilon}$ (the Archimedean Property 2.4.7 assures that there always is one). Before Player A makes his claim, he must assure himself that if $n \geqslant K(\varepsilon)$ then $n > 1/\sqrt{\varepsilon}$ so that $n^2 > 1/\varepsilon$ (why?), whence it follows that $\varepsilon > 1/n^2$ (why?). Having answered both of these questions affirmatively, Player A knows that if $n \geqslant K(\varepsilon)$, then

$$\left|\frac{1}{n^2} - 0\right| = \frac{1}{n^2} < \varepsilon ;$$

consequently, he makes the assertion that "the sequence $(1/n^2)$ converges to 0".

(c) The sequence $(0, 2, 0, 2, \ldots, 0, 2, \ldots)$ does *not* converge to 0.

If Player A asserts that 0 is the limit of this sequence, he will lose the $K(\varepsilon)$ game when Player B picks a value of $\varepsilon < 2$. To be definite, let Player B choose $\varepsilon = 1$. Then, no matter what value Player A picks for $K(1)$, his response will not be adequate, for Player B can always select an even number $n > K(1)$ for which the corresponding value $x_n = 2$ and for which $|x_n - 0| = |2 - 0| = 2 > 1$. Thus, the number 0 is not the limit of the sequence (x_n).

Tails of Sequences

It is important to realize that the convergence (or the divergence) of a sequence $X = (x_n)$ depends only on the "ultimate behavior" of its terms. By this we mean that if, for any fixed natural number M, we drop the first M terms in the sequence, then the resulting sequence X_M converges if and only if the original sequence X converges, and in this case $\lim X_M = \lim X$. We shall state this formally; however, first it is convenient to define a new term.

3.1.8 Definition. Let $X = (x_1, x_2, \ldots, x_n, \ldots,)$ be a sequence of real numbers and let M be a natural number. Then the M**-tail of** X is the sequence

$$X_M := (x_{M+n} : n \in \mathbf{N}) = (x_{M+1}, x_{M+2}, x_{M+3}, \ldots)$$

Thus the 3-tail of the sequence $(2, 4, 6, 8, \ldots, 2n, \ldots)$ is the sequence $(8, 10, 12, \ldots, 2n + 6 \ldots)$.

3.1.9 Theorem. *Let $X = (x_n : n \in \mathbf{N})$ be a sequence of real numbers and let $M \in \mathbf{N}$. Then the M-tail $X_M = (x_{M+n} : n \in \mathbf{N})$ of X converges if and only if X converges. In this case* $\lim X_M = \lim X$.

Proof. It is evident that if $p \in \mathbf{N}$, the the pth term of X_M is the $(p + M)$-th term of X. Similarly, if $q \in \mathbf{N}$ and $q > M$, then the qth term of X is the $(q - M)$-th term of X_M.

Hence, given any $\varepsilon > 0$, if the terms for $n \geqslant K(\varepsilon)$ of X satisfy $|x_n - x| < \varepsilon$, then the terms for $m \geqslant K(\varepsilon) - M$ of X_M satisfy $|x_m - x| < \varepsilon$. Conversely, if the terms for $m \geqslant K_M(\varepsilon)$ of X_M satisfy $|x_m - x| < \varepsilon$, then the terms for $n \geqslant K_M(\varepsilon) + M$ of X satisfy $|x_n - x| < \varepsilon$. Thus, the sequence X converges to x if and only if the sequence X_M converges to x. Q.E.D.

We shall sometimes say that a sequence X *ultimately* has a certain property, if some tail of X has this property. For example, we say that the sequence $(3, 4, 5, 5, 5, \ldots, 5, \ldots)$ is "ultimately constant". On the other hand, the sequence $(3, 5, 3, 5, \ldots, 3, 5, \ldots)$ is not ultimately constant. A number of other instances of this "ultimate terminology" will be noted below.

Some Examples

We now present some examples, establishing the convergence of certain sequences. If we use Definition 3.1.4 or Theorem 3.1.6, we must, in effect, play the $K(\varepsilon)$ game. Instead, we shall often find it convenient to use the following result.

3.1.10 Theorem. *Let* $A = (a_n)$ *and* $X = (x_n)$ *be sequences of real numbers and let* $x \in \mathbf{R}$. *If for some* $C > 0$ *we have*

$$|x_n - x| \leqslant C\,|a_n| \qquad \textit{for all} \quad n \in \mathbf{N},$$

and if $\lim (a_n) = 0$, *then it follows that* $\lim (x_n) = x$.

Proof. If $\varepsilon > 0$ is given, then since $\lim (a_n) = 0$, it follows that there exists a natural number $K_A(\varepsilon/C)$ such that if $n \geqslant K_A(\varepsilon/C)$ then

$$|a_n| = |a_n - 0| < \varepsilon/C.$$

Therefore it follows that if $n \geqslant K_A(\varepsilon/C)$ then

$$|x_n - x| \leqslant C|a_n| < C(\varepsilon/C) = \varepsilon.$$

Since $\varepsilon > 0$ is arbitrary, we conclude that $x = \lim (x_n)$. Q.E.D.

Note that we have shown that the number $K_A(\varepsilon/C)$ serves as a suitable natural number $K_X(\varepsilon)$ for the sequence X.

3.1.11 Examples. (a) If $a > 0$, then $\lim \left(\dfrac{1}{1+na}\right) = 0$.

Since $a > 0$, it follows that $0 < na < 1 + na$. Therefore we conclude that $0 < 1/(1+na) < 1/(na)$, which evidently implies that

$$\left|\frac{1}{1+na} - 0\right| \leqslant \left(\frac{1}{a}\right)\frac{1}{n} \qquad \text{for all} \quad n \in \mathbf{N}.$$

Since $\lim (1/n) = 0$, we may invoke Theorem 3.1.10 with $C = 1/a$ to infer that $\lim \big(1/(1 + na)\big) = 0$.

(b) $\lim (1/2^n) = 0$.

Since $0 < n < 2^n$ for all $n \in \mathbf{N}$ (see 2.2.13(d)), we have $0 < 1/2^n < 1/n$ from which it follows that

$$\left|\frac{1}{2^n} - 0\right| \leqslant \frac{1}{n} \qquad \text{for all} \quad n \in \mathbf{N}.$$

We invoke Theorem 3.1.10 to infer that $\lim (1/2^n) = 0$.

(c) If $0 < b < 1$, then $\lim (b^n) = 0$.

Since $0 < b < 1$, we can write $b = 1/(1+a)$, where $a := (1/b) - 1$ so that $a > 0$. By Bernoulli's Inequality 2.2.13(c) we have $(1 + a)^n \geqslant 1 + na$. Hence

$$0 < b^n = \frac{1}{(1+a)^n} \leqslant \frac{1}{1+na} < \frac{1}{na}$$

so that from Theorem 3.1.10, we get $\lim (b^n) = 0$.

(d) If $c > 0$, then $\lim (c^{1/n}) = 1$.

The case $c = 1$ is trivial since then $(c^{1/n})$ is the constant sequence $(1, 1, 1, \ldots)$ which evidently converges to 1.

If $c > 1$, then $c^{1/n} = 1 + d_n$ with $d_n > 0$. Hence by Bernoulli's Inequality 2.2.13(c),

$$c = (1 + d_n)^n \geqslant 1 + nd_n \qquad \text{for} \quad n \in \mathbf{N}.$$

Therefore we have $c - 1 \geqslant nd_n$, so that $d_n \leqslant (c-1)/n$. Consequently we have

$$|c^{1/n} - 1| = d_n \leqslant (c-1)\frac{1}{n} \qquad \text{for} \quad n \in \mathbf{N}.$$

We now invoke Theorem 3.1.10 to infer that $\lim (c^{1/n}) = 1$ when $c > 1$.

Now suppose that $0 < c < 1$; then $c^{1/n} = 1/(1+h_n)$ with $h_n > 0$. Hence by Bernoulli's Inequality it follows that

$$c = \frac{1}{(1+h_n)^n} \leqslant \frac{1}{1+nh_n} < \frac{1}{nh_n},$$

from which it follows that $0 < h_n < 1/nc$ for $n \in \mathbf{N}$. Therefore we have

$$0 < 1 - c^{1/n} = \frac{h_n}{1+h_n} < h_n < \frac{1}{nc}$$

so that

$$|c^{1/n} - 1| < \left(\frac{1}{c}\right)\frac{1}{n} \qquad \text{for} \quad n \in \mathbf{N}.$$

We now invoke Theorem 3.1.10 to infer that $\lim (c^{1/n}) = 1$ when $0 < c < 1$.

(e) $\lim (n^{1/n}) = 1$.

Since $n^{1/n} > 1$ for $n > 1$, we can write $n^{1/n} = 1 + k_n$ with $k_n > 0$ when $n > 1$. Hence $n = (1 + k_n)^n$ for $n > 1$. By the Binomial Theorem, if $n > 1$ we have

$$n = 1 + nk_n + \tfrac{1}{2}n(n-1)k_n^2 + \cdots \geqslant 1 + \tfrac{1}{2}n(n-1)k_n^2,$$

whence it follows that

$$n - 1 \geqslant \tfrac{1}{2}n(n-1)k_n^2.$$

Hence $k_n^2 \leqslant 2/n$ for $n > 1$. Now if $\varepsilon > 0$ is given, it follows from the Archimedean Property [see Corollary 2.4.8(b)] that there exists a natural number N_ε such that $2/N_\varepsilon < \varepsilon^2$. It follows that if $n \geqslant \sup\{2, N_\varepsilon\}$ then $2/n < \varepsilon^2$, whence it follows that

$$0 < n^{1/n} - 1 = k_n \leqslant (2/n)^{1/2} < \varepsilon.$$

Since $\varepsilon > 0$ is arbitrary, we deduce that $\lim (n^{1/n}) = 1$.

Exercises for Section 3.1

1. The sequence (x_n) is defined by the following formulas for the nth term. Write the first five terms in each case:
   (a) $x_n := 1 + (-1)^n$,
   (b) $x_n := (-1)^n/n$,
   (c) $x_n := \dfrac{1}{n(n+1)}$,
   (d) $x_n := \dfrac{1}{n^2+2}$.
2. The first few terms of a sequence (x_n) are given below. Assuming that the "natural pattern" indicated by these terms persists, write down a formula for the nth term x_n.
   (a) $5, 7, 9, 11, \ldots$;
   (b) $1/2, -1/4, 1/8, -1/16, \ldots$;
   (c) $\dfrac{1}{2}, \dfrac{2}{3}, \dfrac{3}{4}, \dfrac{4}{5}, \ldots$;
   (d) $1, 4, 9, 16, \ldots$.
3. List the first five terms of the following inductively defined sequences.
   (a) $x_1 := 1, \quad x_{n+1} := 3x_n + 1$;
   (b) $y_1 := 2, \quad y_{n+1} := \frac{1}{2}(y_n + 2/y_n)$;
   (c) $z_1 := 1, \quad z_2 := 2, \quad z_{n+2} := (z_{n+1} + z_n)/(z_{n+1} - z_n)$;
   (d) $s_1 := 3, \quad s_2 := 5, \quad s_{n+2} := s_n + s_{n+1}$.
4. For any $b \in \boldsymbol{R}$, prove that $\lim (b/n) = 0$.

5. Use the ε-K formulation of limit of a sequence to establish the following limits.

(a) $\lim\left(\dfrac{1}{n^2+1}\right) = 0$; (b) $\lim\left(\dfrac{2n}{n+1}\right) = 2$;

(c) $\lim\left(\dfrac{3n+1}{2n+5}\right) = \dfrac{3}{2}$; (d) $\lim\left(\dfrac{n^2-1}{2n^2+3}\right) = \dfrac{1}{2}$.

6. Show that

(a) $\lim\left(\dfrac{1}{\sqrt{n+7}}\right) = 0$; (b) $\lim\left(\dfrac{2n}{n+2}\right) = 2$;

(c) $\lim\left(\dfrac{\sqrt{n}}{n+1}\right) = 0$; (d) $\lim\left(\dfrac{(-1)^n n}{n^2+1}\right) = 0$.

7. Prove that $\lim(x_n) = 0$ if and only if $\lim(|x_n|) = 0$. Give an example to show that the convergence of $(|x_n|)$ need not imply the convergence of (x_n).
8. Show that if $x_n \geqslant 0$ for all $n \in \boldsymbol{N}$ and $\lim(x_n) = 0$, then $\lim(\sqrt{x_n}) = 0$.
9. Prove that if $\lim(x_n) = x$ and if $x > 0$, then there exists a natural number M such that $x_n > 0$ for all $n \geqslant M$.
10. Show that $\lim\left(\dfrac{1}{n} - \dfrac{1}{n+1}\right) = 0$.
11. Show that $\lim(1/3^n) = 0$.
12. Let $b \in \boldsymbol{R}$ satisfy $0 < b < 1$. Show that $\lim(nb^n) = 0$. [*Hint:* Use the Binomial Theorem as in Example 3.1.11(e).]
13. Show that $\lim((2n)^{1/n}) = 1$.
14. Show that $\lim(n^2/n!) = 0$.
15. Show that $\lim(2^n/n!) = 0$. [*Hint:* If $n \geqslant 3$, then $0 < 2^n/n! \leqslant 2(\frac{2}{3})^{n-2}$.]

SECTION 3.2 Limit Theorems

In this section we shall obtain some results that often enable us to evaluate the limits of certain sequences of real numbers. These results will enable us to expand our collection of convergent sequences rather extensively.

3.2.1 Definition. A sequence $X = (x_n)$ of real numbers is said to be **bounded** if there exists a real number $M > 0$ such that $|x_n| \leqslant M$ for all $n \in \boldsymbol{N}$.

Thus a sequence $X = (x_n)$ is bounded if and only if the set $\{x_n : n \in \boldsymbol{N}\}$ of its values is bounded in $\boldsymbol{R}$.

3.2.2 Theorem. *A convergent sequence of real numbers is bounded.*

Proof. Suppose that $\lim(x_n) = x$ and let $\varepsilon := 1$. By Theorem 3.1.6(d), there is a natural number $K := K(1)$ such that if $n \geqslant K$ then $|x_n - x| < 1$. Hence, by the Triangle Inequality 2.2.3, we infer that if $n \geqslant K$, then $|x_n| < |x| + 1$. If we set

$$M := \sup \{|x_1|, |x_2|, \ldots, |x_{K-1}|, |x| + 1\},$$

then it follows that $|x_n| \leq M$ for all $n \in \mathbf{N}$. Q.E.D.

In Definition 3.1.3 we defined the sum, difference, product, multiple, and (sometimes) the quotient of sequences of real numbers. We now show that sequences obtained in these ways from convergent sequences give rise to new sequences whose limits can be predicted.

3.2.3 Theorem. (a) *Let $X = (x_n)$ and $Y = (y_n)$ be sequences of real numbers that converge to x and y, respectively, and let $c \in \mathbf{R}$. Then the sequences $X + Y$, $X - Y$, $X \cdot Y$, and cX converge to $x + y$, $x - y$, xy, and cx, respectively.*

(b) *If $X = (x_n)$ converges to x and $Z = (z_n)$ is a sequence of non-zero real numbers that converges to z and if $z \neq 0$, then the quotient sequence X/Z converges to x/z.*

Proof. (a) To show that $\lim(x_n + y_n) = x + y$, we need to estimate the magnitude of $|(x_n + y_n) - (x + y)|$. To do this we use the Triangle Inequality 2.2.3 to obtain

$$\begin{aligned} |(x_n + y_n) - (x + y)| &= |(x_n - x) + (y_n - y)| \\ &\leq |x_n - x| + |y_n - y|. \end{aligned}$$

By hypothesis, if $\varepsilon > 0$ there exists a natural number K_1 such that if $n \geq K_1$, then $|x_n - x| < \varepsilon/2$; also there exists a natural number K_2 such that if $n \geq K_2$, then $|y_n - y| < \varepsilon/2$. Hence if $K(\varepsilon) := \sup \{K_1, K_2\}$, it follows that if $n \geq K(\varepsilon)$ then

$$\begin{aligned} |(x_n + y_n) - (x + y)| &\leq |x_n - x| + |y_n - y| \\ &< \tfrac{1}{2}\varepsilon + \tfrac{1}{2}\varepsilon = \varepsilon. \end{aligned}$$

Since $\varepsilon > 0$ is arbitrary, we infer that $X + Y = (x_n + y_n)$ converges to $x + y$.

Precisely the same argument can be used to show that $X - Y = (x_n - y_n)$ converges to $x - y$.

To show that $X \cdot Y = (x_n y_n)$ converges to xy, we make the estimate

$$\begin{aligned} |x_n y_n - xy| &= |(x_n y_n - x_n y) + (x_n y - xy)| \\ &\leq |x_n(y_n - y)| + |(x_n - x)y| \\ &= |x_n|\,|y_n - y| + |x_n - x|\,|y|. \end{aligned}$$

According to Theorem 3.2.2 there exists a real number $M_1 > 0$ such that $|x_n| \leqslant M_1$ for all $n \in \mathbf{N}$ and we set $M := \sup\{M_1, |y|\}$. Hence we have the estimate

$$|x_n y_n - xy| \leqslant M|y_n - y| + M|x_n - x|.$$

From the convergence of X and Y we conclude that if $\varepsilon > 0$ is given, then there exist natural numbers K_1 and K_2 such that if $n \geqslant K_1$ then $|x_n - x| < \varepsilon/2M$, and if $n \geqslant K_2$ then $|y_n - y| < \varepsilon/2M$. Now let $K(\varepsilon) := \sup\{K_1, K_2\}$; then, if $n \geqslant K(\varepsilon)$ we infer that

$$\begin{aligned} |x_n y_n - xy| &\leqslant M|y_n - y| + M|x_n - x| \\ &\leqslant M(\varepsilon/2M) + M(\varepsilon/2M) = \varepsilon. \end{aligned}$$

Since $\varepsilon > 0$ is arbitrary, this proves that the sequence $X \cdot Y = (x_n y_n)$ converges to xy.

The fact that $cX = (cx_n)$ converges to cx can be proved in the same way; it can also be deduced by taking Y to be the constant sequence $(c, c, c, \ldots)$. We leave the details to the reader.

(b) We next show that if $Z = (z_n)$ is a sequence of non-zero real numbers that converges to a non-zero limit z, then the sequence $(1/z_n)$ of reciprocals converges to $1/z$. First let $\alpha := \frac{1}{2}|z|$ so that $\alpha > 0$. Since $\lim (z_n) = z$, there exists a natural number K_1 such that if $n \geqslant K_1$ then $|z_n - z| < \alpha$. It follows from the Triangle Inequality that $-\alpha \leqslant -|z_n - z| \leqslant |z_n| - |z|$ for $n \geqslant K_1$, whence it follows that $\frac{1}{2}|z| = |z| - \alpha \leqslant |z_n|$ for $n \geqslant K_1$. Therefore $1/|z_n| \leqslant 2/|z|$ for $n \geqslant K_1$ so we have the estimate

$$\begin{aligned} \left|\frac{1}{z_n} - \frac{1}{z}\right| &= \left|\frac{z - z_n}{z_n z}\right| = \frac{1}{|z_n z|}|z - z_n| \\ &\leqslant \frac{2}{|z|^2}|z - z_n|. \end{aligned}$$

Now, if $\varepsilon > 0$ is given, there exists a natural number K_2 such that if $n \geqslant K_2$ then $|z_n - z| \leqslant \frac{1}{2}\varepsilon|z|^2$. Therefore, it follows that if $K(\varepsilon) := \sup\{K_1, K_2\}$, then

$$\left|\frac{1}{z_n} - \frac{1}{z}\right| \leqslant \varepsilon \qquad \text{for all} \quad n > K(\varepsilon).$$

Since $\varepsilon > 0$ is arbitrary, it follows that

$$\lim\left(\frac{1}{z_n}\right) = \frac{1}{z}.$$

The proof of (b) is now completed by taking Y to be the sequence $(1/z_n)$ and using the fact that $X \cdot Y = (x_n/z_n)$ converges to $x(1/z) = x/z$. Q.E.D.

Some of the results of Theorem 3.2.3 can be extended, by mathematical induction, to a finite number of convergent sequences. For example, if $A = (a_n)$, $B = (b_n), \ldots, Z = (z_n)$ are convergent sequences of real numbers, then their sum $A + B + \cdots + Z := (a_n + b_n + \cdots + z_n)$ is a convergent sequence and

(1) $$\lim (a_n + b_n + \cdots + z_n) = \lim (a_n) + \lim (b_n) + \cdots + \lim (z_n).$$

Also their product $A \cdot B \cdots Z := (a_n b_n \cdots z_n)$ is a convergent sequence and

(2) $$\lim (a_n b_n \cdots z_n) = (\lim (a_n))(\lim (b_n)) \cdots (\lim (z_n)).$$

Hence, if $k \in \mathbf{N}$ and if $A = (a_n)$ is a convergent sequence, then

(3) $$\lim (a_n^k) = (\lim (a_n))^k.$$

We leave the proof of these assertions to the reader.

3.2.4 Theorem. *If $X = (x_n)$ is a convergent sequence of real numbers and if $x_n \geq 0$ for all $n \in \mathbf{N}$, then $x := \lim (x_n) \geq 0$.*

Proof. Suppose the conclusion is not true and that $x < 0$; then $\varepsilon := -x$ is strictly positive. Since X converges to x, there is a natural number K such that

$$x - \varepsilon < x_n < x + \varepsilon \qquad \text{for all} \quad n \geq K.$$

In particular, we have $x_K < x + \varepsilon = x + (-x) = 0$. But this contradicts the hypothesis that $x_n \geq 0$ for all $n \in \mathbf{N}$. Therefore, this contradiction implies that $x \geq 0$. Q.E.D.

We now give a result that is formally stronger than Theorem 3.2.4.

3.2.5 Theorem. *If $X = (x_n)$ and $Y = (y_n)$ are convergent sequences of real numbers and if $x_n \leq y_n$ for all $n \in \mathbf{N}$, then $\lim (x_n) \leq \lim (y_n)$.*

Proof. Let $z_n := y_n - x_n$ so that $Z := (z_n) = Y - X$ and $z_n \geq 0$ for all $n \in \mathbf{N}$. It follows from Theorems 3.2.4 and 3.2.3 that

$$0 \leq \lim Z = \lim (y_n) - \lim (x_n),$$

so that $\lim (x_n) \leq \lim (y_n)$. Q.E.D.

The next result asserts that if all the terms of a convergent sequence satisfy an inequality of the form $a \leq x_n \leq b$, then the limit of the sequence satisfies the same inequality. Thus if the sequence is convergent, one may "pass to the limit" in an inequality of this type.

3.2.6 Theorem. *If $X = (x_n)$ is a convergent sequence and if $a \leqslant x_n \leqslant b$ for all $n \in \mathbf{N}$, then $a \leqslant \lim (x_n) \leqslant b$.*

Proof. Let Y be the constant sequence $(b, b, b, \ldots)$. It follows from Theorem 3.2.5 that $\lim X \leqslant \lim Y = b$. Similarly one shows that we have $a \leqslant \lim X$. Q.E.D.

The next result asserts that a sequence Y, squeezed between two sequences that converge to the *same limit*, must also converge to this limit.

3.2.7 Squeeze Theorem. *Suppose that $X = (x_n)$, $Y = (y_n)$, and $Z = (z_n)$ are sequences of real numbers such that*

$$x_n \leqslant y_n \leqslant z_n \qquad \textit{for all} \quad n \in \mathbf{N},$$

and that $\lim (x_n) = \lim (z_n)$. *Then $Y = (y_n)$ is convergent and*

$$\lim (x_n) = \lim (y_n) = \lim (z_n).$$

Proof. Let $w := \lim (x_n) = \lim (z_n)$. If $\varepsilon > 0$ is given, then it follows from the convergence of X and Z to w that there exists a natural number K such that if $n \geqslant K$ then

$$|x_n - w| < \varepsilon \qquad \text{and} \qquad |z_n - w| < \varepsilon.$$

Since the hypothesis implies that

$$x_n - w \leqslant y_n - w \leqslant z_n - w \qquad \text{for all} \quad n \in \mathbf{N},$$

it follows that

$$|y_n - w| \leqslant \sup \{|x_n - w|, |z_n - w|\} < \varepsilon$$

for all $n \geqslant K$. Since $\varepsilon > 0$ is arbitrary, this implies that $\lim (y_n) = w$. Q.E.D.

Remark. Since any tail of a convergent sequence has the same limit, the hypotheses of Theorems 3.2.4, 3.2.5, 3.2.6, and 3.2.7 can be weakened to apply to the tail of a sequence. For example, in Theorem 3.2.4, if $X = (x_n)$ is "ultimately positive" in the sense that there exists $M \in \mathbf{N}$ such that $x_n \geqslant 0$ for all $n \geqslant M$, then the same conclusion that $x \geqslant 0$ will hold. Similar modifications are valid for the other theorems, as the reader should verify.

3.2.8 Examples. (a) The sequence (n) is divergent.

It follows from Theorem 3.2.2 that if the sequence $X := (n)$ is convergent then there exists a real number $M > 0$ such that $n = |n| < M$ for all $n \in \mathbf{N}$. But this violates the Archimedean Property 2.4.7.

(b) The sequence $((-1)^n)$ is divergent.

This sequence $X := ((-1)^n)$ is bounded (take $M := 1$), so we cannot invoke Theorem 3.2.2. However, assume that $a := \lim X$ exists. Let $\varepsilon := 1$ so that there exists a natural number K_1 such that

$$|(-1)^n - a| < 1 \qquad \text{for all} \quad n \geqslant K_1.$$

If n is an odd natural number with $n \geqslant K_1$, this gives $|-1-a| < 1$, so that $-2 < a < 0$. On the other hand, if n is an even natural number with $n \geqslant K_1$, this inequality gives $|1 - a| < 1$ so that $0 < a < 2$. Since a cannot satisfy both of these inequalities, the hypothesis that X is convergent leads to a contradiction. Therefore the sequence X is divergent.

(c) $\lim \left(\dfrac{2n+1}{n}\right) = 2.$

If we let $X := (2)$ and $Y := (1/n)$, then $((2n + 1)/n) = X + Y$. Hence it follows from Theorem 3.2.3(a) that $\lim (X + Y) = \lim X + \lim Y = 2 + 0 = 2.$

(d) $\lim \left(\dfrac{2n+1}{n+5}\right) = 2.$

Since the sequences $(2n + 1)$ and $(n + 5)$ are not convergent (why?), it is not possible to use Theorem 3.2.3(b) directly. However, if we write

$$\frac{2n + 1}{n + 5} = \frac{2 + 1/n}{1 + 5/n},$$

we can obtain the given sequence as one to which Theorem 3.2.3(b) applies when we take $X := (2 + 1/n)$ and $Z := (1 + 5/n)$. (Check that all hypotheses are satisfied.) Since $\lim X = 2$ and $\lim Z = 1 \neq 0$, we deduce that we have $\lim ((2n+1)/(n+5)) = 2/1 = 2.$

(e) $\lim \left(\dfrac{2n}{n^2+1}\right) = 0.$

Theorem 3.2.3(b) does not apply directly. (Why?) We note that

$$\frac{2n}{n^2+1} = \frac{2}{n+1/n},$$

but Theorem 3.2.3(b) does not apply here either, because $(n + 1/n)$ is not a convergent sequence. (Why not?) However, if we write

$$\frac{2n}{n^2+1} = \frac{2/n}{1+1/n^2},$$

then we can apply Theorem 3.2.3(b), since $\lim (2/n) = 0$ and $\lim (1 + 1/n^2) = 1 \neq 0$. Therefore $\lim \big(2n/(n^2+1)\big) = 0/1 = 0$.

(f) $\lim \left(\dfrac{\sin n}{n}\right) = 0$.

We cannot apply Theorem 3.2.2(b) directly, at least, since the sequence (n) is not convergent [neither is the sequence $(\sin n)$]. It does not appear that a simple algebraic manipulation will enable us to reduce the sequence into one to which Theorem 3.2.3 will apply. However, if we note that $-1 \leqslant \sin n \leqslant 1$, then it follows that

$$-\frac{1}{n} \leqslant \frac{\sin n}{n} \leqslant \frac{1}{n} \qquad \text{for all} \quad n \in \mathbf{N}.$$

Hence we can apply the Squeeze Theorem 3.2.7 to infer that $\lim (n^{-1} \sin n) = 0$. (We note that Theorem 3.1.10 could also be applied to this sequence.)

(g) Let $X = (x_n)$ be a sequence of real numbers that converges to $x \in \boldsymbol{R}$. Let p be a polynomial; for example, let

$$p(t) := a_k t^k + a_{k-1} t^{k-1} + \cdots + a_1 t + a_0,$$

where $k \in \mathbf{N}$ and $a_j \in \boldsymbol{R}$ for $j = 0, 1, \ldots, k$. It follows from Theorem 3.2.3 that the sequence $\big(p(x_n)\big)$ converges to $p(x)$. We leave the details to the reader as an exercise.

(h) Let $X = (x_n)$ be a sequence of real numbers that converges to $x \in \boldsymbol{R}$. Let r be a rational function (that is, $r(t) := p(t)/q(t)$, where p and q are polynomials). Suppose that $q(x_n) \neq 0$ for all $n \in \mathbf{N}$ and that $q(x) \neq 0$. Then the sequence $\big(r(x_n)\big)$ converges to $r(x) = p(x)/q(x)$. We leave the details to the reader as an exercise.

We conclude this section with several results that will be useful in the work that follows.

3.2.9 Theorem. *Let the sequence $X = (x_n)$ converge to x. Then the sequence $(|x_n|)$ of absolute values converges to $|x|$. That is, if $x = \lim (x_n)$, then $|x| = \lim (|x_n|)$.*

Proof. It follows from the Triangle Inequality (see Corollary 2.3.4) that

$$\big||x_n| - |x|\big| \leqslant |x_n - x| \qquad \text{for all} \quad n \in \mathbf{N}.$$

The convergence of $(|x_n|)$ to $|x|$ is then an immediate consequence of the convergence of (x_n) to x. Q.E.D.

3.2.10 Theorem. *Let $X = (x_n)$ be a sequence of real numbers that converges to x and suppose that $x_n \geqslant 0$. Then the sequence $(\sqrt{x_n})$ of positive square roots converges and $\lim (\sqrt{x_n}) = \sqrt{x}$.*

Proof. It follows from Theorem 3.2.4 that $x = \lim (x_n) \geq 0$ so the assertion makes sense. We now consider the two cases: (i) $x = 0$ and (ii) $x > 0$.

(i) If $x = 0$, let $\varepsilon > 0$ be given. Since $x_n \to 0$ there exists a natural number K such that if $n \geq K$ then

$$0 \leq x_n = x_n - 0 < \varepsilon^2.$$

Therefore [see Example 2.2.13(a)], $0 \leq \sqrt{x_n} < \varepsilon$ for $n \geq K$. Since $\varepsilon > 0$ is arbitrary, this implies that $\sqrt{x_n} \to 0$.

(ii) If $x > 0$, then $\sqrt{x} > 0$ and we note that

$$\sqrt{x_n} - \sqrt{x} = \frac{(\sqrt{x_n} - \sqrt{x})(\sqrt{x_n} + \sqrt{x})}{\sqrt{x_n} + \sqrt{x}} = \frac{x_n - x}{\sqrt{x_n} + \sqrt{x}}.$$

Since $\sqrt{x_n} + \sqrt{x} \geq \sqrt{x} > 0$, it follows that

$$|\sqrt{x_n} - \sqrt{x}| \leq \left(\frac{1}{\sqrt{x}}\right) |x_n - x|.$$

The convergence of $\sqrt{x_n} \to \sqrt{x}$ is now an easy consequence of the fact that $x_n \to x$. Q.E.D.

For certain types of sequences, the following result provides a quick and easy "ratio test" for convergence. Related results can be found in the exercises.

3.2.11 Theorem. *Let* (x_n) *be a sequence of strictly positive real numbers such that* $L := \lim (x_{n+1}/x_n)$ *exists. If* $L < 1$, *then* (x_n) *converges and* $\lim (x_n) = 0$.

Proof. We first choose a number r such that $L < r < 1$, and let $\varepsilon := r - L > 0$. There exists a number $K \in \mathbf{N}$ such that if $n \geq K$ then

$$\left|\frac{x_{n+1}}{x_n} - L\right| < \varepsilon.$$

It follows from this that if $n \geq K$, then

$$\frac{x_{n+1}}{x_n} < L + \varepsilon = L + (r-L) = r.$$

Therefore, if $n \geq K$, we obtain

$$0 < x_{n+1} < x_n r < x_{n-1} r^2 < \cdots < x_K r^{n-K+1}.$$

If we set $C := x_K/r^K$, we see that $0 < x_{n+1} < Cr^{n+1}$ for all $n \geq K$. Since $0 < r < 1$, it follows from 3.1.11(c) that $\lim (r^n) = 0$ and therefore from Theorem 3.1.10 that $\lim (x_n) = 0$. Q.E.D.

As an illustration of the utility of the preceding theorem, consider the sequence (x_n) given by $x_n := n/2^n$. We have

$$\frac{x_{n+1}}{x_n} = \frac{n+1}{2^{n+1}} \cdot \frac{2^n}{n} = \frac{1}{2}\left(1 + \frac{1}{n}\right),$$

so that $\lim (x_{n+1}/x_n) = \frac{1}{2}$. Since $\frac{1}{2} < 1$, it follows from Theorem 3.2.11 that $\lim (n/2^n) = 0$.

Exercises for Section 3.2

1. For x_n given by the following formulas, establish either the convergence or divergence of the sequence $X = (x_n)$:

   (a) $x_n := \dfrac{n}{n+1}$, (b) $x_n := \dfrac{(-1)^n n}{n+1}$,

   (c) $x_n := \dfrac{n^2}{n+1}$, (d) $x_n := \dfrac{2n^2+3}{n^2+1}$.

2. Give an example of two divergent sequences X, Y such that their sum $X + Y$ converges.
3. Give an example of two divergent sequences X, Y such that their product XY converges.
4. Show that if X and Y are sequences such that X and $X + Y$ are convergent, then Y is convergent.
5. Show that if X and Y are sequences such that X converges to $x \neq 0$ and XY converges, then Y converges.
6. Show that the sequence (2^n) is not convergent.
7. Show that the sequence $((-1)^n n^2)$ is not convergent.
8. Find the limits of the following sequences:

   (a) $\lim ((2 + 1/n)^2)$, (b) $\lim ((-1)^n/(n+2))$,

   (c) $\lim \left(\dfrac{\sqrt{n} - 1}{\sqrt{n} + 1}\right)$, (d) $\lim \left(\dfrac{n+1}{n\sqrt{n}}\right)$,

9. Explain why the result in equation (3) before Theorem 3.2.4 *cannot* be used to evaluate the limit of the sequence $((1+1/n)^n)$.
10. Let $y_n := \sqrt{n+1} - \sqrt{n}$ for $n \in \mathbf{N}$. Show that (y_n) and $(\sqrt{n}\, y_n)$ converge.
11. Show that if $z_n := (a^n + b^n)^{1/n}$ where $0 < a < b$, then $\lim (z_n) = b$.
12. Apply Theorem 3.2.11 to the following sequences, where a, b satisfy $0 < a < 1$, $b > 1$.

    (a) (a^n), (b) $(b^n/2^n)$,

    (c) (n/b^n), (d) $(2^{3n}/3^{2n})$.

13. (a) Give an example of a convergent sequence (x_n) of strictly positive numbers such that $\lim (x_{n+1}/x_n) = 1$.
    (b) Give an example of a divergent sequence with this property. (Thus, this property cannot be used as a test for convergence.)
14. Let $X = (x_n)$ be a sequence of strictly positive real numbers such that $\lim (x_{n+1}/x_n) = L > 1$. Show that X is not a bounded sequence and hence is not convergent.
15. Discuss the convergence of the following sequences, where a, b satisfy $0 < a < 1$, $b > 1$.
    (a) (n^2a^n), (b) (b^n/n^2),
    (c) $(b^n/n!)$, (d) $(n!/n^n)$.
16. Let (x_n) be a sequence of strictly positive real numbers such that $\lim (x_n^{1/n}) = L < 1$. Show that there exists a number r with $0 < r < 1$ such that $0 < x_n < r^n$ for all sufficiently large $n \in \mathbf{N}$. Use this to show that $\lim (x_n) = 0$.
17. (a) Give an example of a convergent sequence (x_n) of strictly positive real numbers such that $\lim (x_n^{1/n}) = 1$.
    (b) Give an example of a divergent sequence (x_n) of strictly positive real numbers such that $\lim (x_n^{1/n}) = 1$. (Thus, this property cannot be used as a test for convergence.)

SECTION 3.3 Monotone Sequences

Until now, we have obtained several methods of showing that a sequence $X = (x_n)$ of real numbers is convergent:

(i) We can use Definition 3.1.4 or Theorem 3.1.6 directly. This is often (but not always) difficult to do.

(ii) We can dominate $|x_n - x|$ by a multiple of the terms in a sequence (a_n) that is known to converge to 0, and employ Theorem 3.1.10.

(iii) We can identify X as a sequence obtained from sequences that are known to be convergent by taking tails, algebraic combinations, absolute values, or square roots, and employ Theorems 3.1.9, 3.2.3, 3.2.9, or 3.2.10.

(iv) We can "squeeze" X between two sequences that converge to the same limit and use Theorem 3.2.7.

(v) We can use the "ratio test" of Theorem 3.2.11.

Except for (iii), all of these methods require that we must already know (or at least suspect) the correct value of the limit, and we then verify that our suspicion is correct.

There are many instances, however, in which there is no obvious candidate for the limit of a sequence, even though a preliminary analysis may suggest that convergence probably does take place. In this and the next two sections, we shall give results that are deeper than those in the preceding sections and which can be used to establish the convergence of a sequence when there is no obvious candidate for the limit. The method we introduce in this section is somewhat more restricted in scope than the method we present in the next two, but it is much easier to employ. It applies to sequences that are monotone in the following sense.

3.3.1 Definition. Let $X = (x_n)$ be a sequence of real numbers. We say that X is **increasing** if it satisfies the inequalities

$$x_1 \leqslant x_2 \leqslant \cdots \leqslant x_n \leqslant x_{n+1} \leqslant \cdots .$$

We say that X is **decreasing** if it satisfies the inequalities

$$x_1 \geqslant x_2 \geqslant \cdots \geqslant x_n \geqslant x_{n+1} \geqslant \cdots .$$

We say that X is **monotone** if it is either increasing, or it is decreasing.

The following sequences are increasing:

$$(1, 2, 3, 4, \ldots, n, \ldots), \qquad (1, 2, 2, 3, 3, 3, \ldots),$$

$$(a, a^2, a^3, \ldots, a^n, \ldots) \quad \text{if} \quad a > 1.$$

The following sequences are decreasing:

$$(1, 1/2, 1/3, \ldots, 1/n, \ldots), \qquad (1, 1/2, 1/2^2, \ldots, 1/2^{n-1}, \ldots),$$

$$(b, b^2, b^3, \ldots, b^n, \ldots) \quad \text{if} \quad 0 < b < 1.$$

The following sequences are not monotone:

$$(+1, -1, +1, \ldots, (-1)^{n+1}, \ldots), \qquad (-1, +2, -3, \ldots, (-1)^n n, \ldots).$$

The following sequences are not monotone, but they are "ultimately" monotone:

$$(7, 6, 2, 1, 2, 3, 4, \ldots), \qquad (-2, 0, 1, 1/2, 1/3, 1/4, \ldots).$$

3.3.2 Monotone Convergence Theorem. *A monotone sequence of real numbers is convergent if and only if it is bounded. Further:*

(a) *If* $X = (x_n)$ *is a bounded increasing sequence, then* $\lim (x_n) = \sup \{x_n\}$.

(b) *If* $Y = (y_n)$ *is a bounded decreasing sequence, then* $\lim (y_n) = \inf \{y_n\}$.

Proof. It was seen in Theorem 3.2.2 that a convergent sequence must be bounded.

Conversely, let X be a bounded monotone sequence. Then X is either increasing or decreasing.

(a) We first treat the case where X is a bounded increasing sequence. By hypothesis, there exists a real number M such that $x_n \leqslant M$ for all $n \in \mathbf{N}$. According to the Supremum Principle 2.4.5, the supremum $x^* := \sup \{x_n : n \in \mathbf{N}\}$ exists and we shall show that $x^* = \lim (x_n)$. If $\varepsilon > 0$ is given, then $x^* - \varepsilon$ is not an upper bound for the set $\{x_n : n \in \mathbf{N}\}$; hence there exists a natural number $K := K(\varepsilon)$ such that $x^* - \varepsilon < x_K$. But since (x_n) is an increasing sequence it follows that

$$x^* - \varepsilon < x_K \leqslant x_n \leqslant x^* \qquad \text{for all} \quad n \geqslant K.$$

Therefore it follows that

$$|x_n - x^*| < \varepsilon \qquad \text{for all} \quad n \geq K.$$

Since $\varepsilon > 0$ is arbitrary, we infer that (x_n) converges to x^*.

(b) If $Y = (y_n)$ is a bounded decreasing sequence, then it is clear that $X := -Y = (-y_n)$ is a bounded increasing sequence. We have seen in part (a) that $\lim X = \sup \{-y_n : n \in \mathbf{N}\}$. On the one hand, by Theorem 3.2.3(b), $\lim X = -\lim Y$; on the other hand, by Exercise 2.4.11(b), we have

$$\sup \{-y_n : n \in \mathbf{N}\} = -\inf \{y_n : n \in \mathbf{N}\}.$$

Therefore $\lim Y = -\lim X = \inf \{y_n : n \in \mathbf{N}\}$. Q.E.D.

The Monotone Convergence Theorem establishes the existence of the limit of a bounded monotone sequence. It also gives us a way of calculating the limit of the sequence *provided* we can evaluate the supremum in case (a), or the infimum in case (b). Sometimes it is difficult to evaluate this supremum (or infimum), but once we know that it exists, it is often possible to evaluate the limit by other methods.

3.3.3 Examples. (a) $\lim (1/\sqrt{n}) = 0$.

It is possible to handle this sequence by using Theorem 3.2.10; however, we shall use the Monotone Convergence Theorem. Clearly 0 is a lower bound for the set $\{1/\sqrt{n};\ n \in \mathbf{N}\}$, and it is not difficult to show that 0 is the infimum of $\{1/\sqrt{n}\colon n \in \mathbf{N}\}$; hence $0 = \lim (1/\sqrt{n})$.

On the other hand, once we know that $X := (1/\sqrt{n})$ is bounded and decreasing, we know that it converges to some real number x. Since $X = (1/\sqrt{n})$ converges to x, it follows from Theorem 3.2.3 that $X \cdot X = (1/n)$ converges to x^2. Therefore $x^2 = 0$, whence $x = 0$.

(b) Let $x_n := 1 + \dfrac{1}{2} + \dfrac{1}{3} + \cdots + \dfrac{1}{n}$ for $n \in \mathbf{N}$.

Since $x_{n+1} = x_n + 1/(n+1) > x_n$, we see that (x_n) is an increasing sequence. By the Monotone Convergence Theorem 3.3.2, the question of whether the sequence is convergent or not is reduced to the question of whether the sequence is bounded or not. Attempts to use direct numerical calculations to arrive at a conjecture concerning the possible boundedness of the sequence (x_n) lead to inconclusive frustration. A computer run will reveal the approximate values $x_n \approx 11.4$ for $n = 50{,}000$, and $x_n \approx 12.1$ for $n = 100{,}000$. Such numerical facts may lead the casual observer to conclude that the sequence is bounded. However, the sequence is in fact divergent, which is established by noting that

$$\begin{aligned} x_{2^n} &= 1 + \frac{1}{2} + \left(\frac{1}{3} + \frac{1}{4}\right) + \cdots + \left(\frac{1}{2^{n-1}+1} + \cdots + \frac{1}{2^n}\right) \\ &> 1 + \frac{1}{2} + \left(\frac{1}{4} + \frac{1}{4}\right) + \cdots + \left(\frac{1}{2^n} + \cdots + \frac{1}{2^n}\right) \\ &= 1 + \frac{1}{2} + \frac{1}{2} + \cdots + \frac{1}{2} \\ &= 1 + \frac{n}{2}. \end{aligned}$$

Hence the sequence (x_n) is unbounded, and therefore divergent.

(c) Let $Y = (y_n)$ be defined inductively by $y_1 := 1$, $y_{n+1} := \frac{1}{4}(2y_n + 3)$ for $n \geqslant 1$. We shall show that $\lim Y = 3/2$.

Direct calculation shows that $y_2 = 5/4$. Hence we have $y_1 < y_2 < 2$. We show, by induction, that $y_n < 2$ for all $n \in \boldsymbol{N}$. Indeed, this is true for $n = 1, 2$. If $y_k < 2$ holds for some $k \in \boldsymbol{N}$, then

$$y_{k+1} = \tfrac{1}{4}(2y_k + 3) < \tfrac{1}{4}(4 + 3) = \tfrac{7}{4} < 2,$$

so that $y_{k+1} < 2$. Therefore $y_n < 2$ for all $n \in \boldsymbol{N}$.

We now show, by induction, that $y_n < y_{n+1}$ for all $n \in \boldsymbol{N}$. The truth of this assertion has been verified for $n = 1$. Now suppose that $y_k < y_{k+1}$ for some k; then $2y_k + 3 < 2y_{k+1} + 3$, whence it follows that

$$y_{k+1} = \tfrac{1}{4}(2y_k + 3) < \tfrac{1}{4}(2y_{k+1} + 3) = y_{k+2}.$$

Thus $y_k < y_{k+1}$ implies that $y_{k+1} < y_{k+2}$. Therefore $y_n < y_{n+1}$ for all $n \in \boldsymbol{N}$.

We have shown that the sequence $Y = (y_n)$ is increasing and bounded above by 2. It follows from the Monotone Convergence Theorem that Y converges to a limit that is at most 2. In this case it is not so easy to evaluate $\lim (y_n)$ by calculating $\sup \{y_n : n \in \boldsymbol{N}\}$. However, there is another way to evaluate its limit. Since $y_{n+1} = \frac{1}{4}(2y_n + 3)$ for all $n \in \boldsymbol{N}$, the nth term in the 1-tail Y_1 of Y has a simple algebraic relation to the nth term of Y. Since, by Theorem 3.1.9, we have $y := \lim Y_1 = \lim Y$, it follows from Theorem 3.2.3 that

$$y = \tfrac{1}{4}(2y + 3),$$

from which it follows that $y = 3/2$.

(d) Let $Z = (z_n)$ be the sequence of real numbers defined by $z_1 := 1$, $z_{n+1} := \sqrt{2z_n}$ for $n \in \boldsymbol{N}$. We shall show that $\lim (z_n) = 2$.

Note that $z_1 = 1$ and $z_2 = \sqrt{2}$; hence $1 \leqslant z_1 < z_2 < 2$. We claim that the sequence Z is increasing and bounded above by 2. To show this we will show, by induction, that $1 \leqslant z_n < z_{n+1} < 2$ for all $n \in \mathbf{N}$. This fact has been verified for $n = 1$. Suppose that it is true for $n = k$; then $2 \leqslant 2z_k < 2z_{k+1} < 4$, whence it follows that

$$1 < \sqrt{2} \leqslant z_{k+1} = \sqrt{2z_k} < z_{k+2} = \sqrt{2z_{k+1}} < \sqrt{4} = 2.$$

[In this last step we have used Example 2.2.13(a).] Hence the validity of the inequality $1 \leqslant z_k < z_{k+1} < 2$, implies the validity of $1 \leqslant z_{k+1} < z_{k+2} < 2$. Therefore $1 \leqslant z_n < z_{n+1} < 2$ for all $n \in \mathbf{N}$.

Since $Z = (z_n)$ is a bounded increasing sequence, it follows from the Monotone Convergence Theorem that it converges to a number $z := \sup \{z_n\}$. It may be shown directly that $\sup \{z_n\} = 2$, so that $z = 2$. Alternatively we may use the method employed in part (c). The relation $z_{n+1} = \sqrt{2z_n}$ gives a relation between the nth term of the 1-tail Z_1 of Z and the nth term of Z. By Theorem 3.1.9, we have $\lim Z_1 = z = \lim Z$. Moreover, by Theorems 3.2.3 and 3.2.10, it follows that the limit z must satisfy the relation

$$z = \sqrt{2z}.$$

Hence z must satisfy the equation $z^2 = 2z$ which has the roots $z = 0, 2$. Since the terms of $z = (z_n)$ all satisfy $1 \leqslant z_n \leqslant 2$, it follows from Theorem 3.2.6 that we must have $1 \leqslant z \leqslant 2$. Therefore $z = 2$.

(e) Let $a > 0$; we will introduce a sequence of real numbers that converges to $\sqrt{a}$.

Let $s_1 > 0$ be arbitrary and define $s_{n+1} := \frac{1}{2}(s_n + a/s_n)$ for $n \in \mathbf{N}$. We shall now show that the sequence (s_n) converges to $\sqrt{a}$. (This process for calculating square roots was known in Mesopotamia before 1500 B.C.)

We first show that $s_n^2 \geqslant a$ for $n \geqslant 2$. Since s_n satisfies the quadratic equation $s_n^2 - 2s_{n+1}\, s_n + a = 0$, this equation has a real root. Hence the discriminant $4s_{n+1}^2 - 4a$ must be positive; that is, $s_{n+1}^2 \geqslant a$ for $n \geqslant 1$.

To see that (s_n) is ultimately decreasing, we note that for $n \geqslant 2$ we have

$$\begin{aligned} s_n - s_{n+1} &= s_n - \frac{1}{2}\left(s_n + \frac{a}{s_n}\right) \\ &= \frac{1}{2} \cdot \frac{(s_n^2 - a)}{s_n} \geqslant 0. \end{aligned}$$

Hence, $s_{n+1} \leqslant s_n$ for all $n \geqslant 2$. It follows from the Monotone Convergence Theorem that $s := \lim (s_n)$ exists. Moreover, by Theorem 3.2.3 it follows that the limit s must satisfy the relation

$$s = \frac{1}{2}\left(s + \frac{a}{s}\right),$$

whence it follows that $s = a/s$ or $s^2 = a$. Thus $s = \sqrt{a}$.

For the purposes of calculation, it is often important to have an estimate of *how rapidly* the sequence S converges to $\sqrt{a}$. As above, we have $\sqrt{a} \leq s_n$ for all $n \in \mathbf{N}$, whence it follows that $a/s_n \leq \sqrt{a} \leq s_n$. Thus we have

$$0 \leq s_n - \sqrt{a} \leq s_n - a/s_n = (s_n^2 - a)/s_n.$$

Using this inequality we can calculate $\sqrt{a}$ to any desired degree of accuracy. (How?)

(f) $\lim((1 + 1/n)^n) = e$.

In this example we introduce a sequence $E = (e_n)$, where $e_n := (1 + 1/n)^n$ for $n \in \mathbf{N}$. We shall show that this sequence is bounded and increasing; hence it is convergent. The limit of this sequence is the famous *Euler number* e, whose approximate value is 2.718281828459045 . . . , and which is taken as the base of the natural logarithms. This number is, with π, one of the two most important "transcendental" numbers in mathematics.

If we apply the Binomial Theorem, we have

$$\begin{aligned} e_n = \left(1 + \frac{1}{n}\right)^n &= 1 + \frac{n}{1}\cdot\frac{1}{n} + \frac{n(n-1)}{2!}\cdot\frac{1}{n^2} \\ &+ \frac{n(n-1)(n-2)}{3!}\cdot\frac{1}{n^3} + \cdots + \frac{n(n-1)\cdots 2\cdot 1}{n!}\cdot\frac{1}{n^n}. \end{aligned}$$

If we divide the powers of n into the terms in the numerators of the binomial coefficients, we get

$$\begin{aligned} e_n = 1 + 1 &+ \frac{1}{2!}\left(1-\frac{1}{n}\right) + \frac{1}{3!}\left(1-\frac{1}{n}\right)\left(1-\frac{2}{n}\right) \\ &+ \cdots + \frac{1}{n!}\left(1-\frac{1}{n}\right)\left(1-\frac{2}{n}\right)\cdots\left(1-\frac{n-1}{n}\right). \end{aligned}$$

Similarly we have

$$\begin{aligned} e_{n+1} = 1 + 1 &+ \frac{1}{2!}\left(1-\frac{1}{n+1}\right) + \frac{1}{3!}\left(1-\frac{1}{n+1}\right)\left(1-\frac{2}{n+1}\right) \\ &+ \cdots + \frac{1}{n!}\left(1-\frac{1}{n+1}\right)\left(1-\frac{2}{n+1}\right)\cdots\left(1-\frac{n-1}{n+1}\right) \\ &+ \frac{1}{(n+1)!}\left(1-\frac{1}{n+1}\right)\left(1-\frac{2}{n+1}\right)\cdots\left(1-\frac{n}{n+1}\right). \end{aligned}$$

Note that the expression for e_n contains $n + 1$ terms, while that for e_{n+1} contains $n + 2$ terms. Moreover, each term appearing in e_n is less than or equal to the corresponding term in e_{n+1}, and e_{n+1} has one more positive term. Therefore we have $2 \leq e_1 < e_2 < \cdots < e_n < e_{n+1} < \cdots$, so that the terms of E are increasing.

To show that the terms of E are bounded above, we note that if $p = 1, 2, \ldots, n$, then $(1 - p/n) < 1$. Moreover $2^{p-1} \leq p!$ [see 1.3.3(d)] so that $1/p! \leq 1/2^{p-1}$. Therefore, if $n > 2$, then we have

$$2 < e_n < 1 + 1 + \frac{1}{2} + \frac{1}{2^2} + \cdots + \frac{1}{2^{n-1}}.$$

Since it can be verified that [see 1.3.3(e)]

$$\frac{1}{2} + \frac{1}{2^2} + \cdots + \frac{1}{2^{n-1}} = 1 - \frac{1}{2^{n-1}} < 1,$$

we deduce that $2 \leq e_n < 3$ for all $n \in \boldsymbol{N}$. The Monotone Convergence Theorem implies that the sequence E converges to a real number that is between 2 and 3. We define the number e to be the limit of this sequence.

By refining our estimates we can find closer rational approximations to e, but we cannot evaluate it *exactly*, since e is an irrational number. However, it is possible to calculate e to as many decimal places as desired. The reader should use his calculator to evaluate e_n for "large" values of n.

Exercises for Section 3.3

1. Let $x_1 > 1$ and $x_{n+1} := 2 - 1/x_n$ for $n \geq 2$. Show that (x_n) is bounded and monotone. Find the limit.
2. Let $y_1 := 1$ and $y_{n+1} := \sqrt{2 + y_n}$. Show that (y_n) is convergent and find the limit.
3. Let $a > 0$ and let $z_1 > 0$. Define $z_{n+1} := (a + z_n)^{1/2}$ for $n \in \boldsymbol{N}$. Show that (z_n) converges and find the limit.
4. Let $x_1 := a > 0$ and $x_{n+1} := x_n + 1/x_n$. Determine if (x_n) converges or diverges.
5. Let (x_n) be a bounded sequence, and for each $n \in \boldsymbol{N}$ let $s_n := \sup \{x_k : k \geq n\}$ and $t_n := \inf \{x_k : k \geq n\}$. Prove that (s_n) and (t_n) are convergent. Also prove that if $\lim (s_n) = \lim (t_n)$, then (x_n) is convergent.
6. Let (a_n) be an increasing sequence, (b_n) a decreasing sequence, and assume that $a_n \leq b_n$ for all $n \in \boldsymbol{N}$. Show that $\lim (a_n) \leq \lim (b_n)$, and thereby deduce the Nested Intervals Theorem 2.5.1 from the Monotone Convergence Theorem 3.3.2.
7. Let A be an infinite subset of $\boldsymbol{R}$ that is bounded above and let $u := \sup A$. Show there exists an increasing sequence (x_n) with $x_n \in A$ for all $n \in \boldsymbol{N}$ such that $u = \lim (x_n)$.
8. Establish the convergence or the divergence of the sequence (y_n), where

$$y_n := \frac{1}{n+1} + \frac{1}{n+2} + \cdots + \frac{1}{2n} \quad \text{for} \quad n \in \boldsymbol{N}.$$

9. Let $x_n := 1/1^2 + 1/2^2 + \cdots + 1/n^2$ for each $n \in \mathbf{N}$. Prove that (x_n) is increasing and bounded, and hence converges. [*Hint:* Note that if $k \geqslant 2$, then $1/k^2 \leqslant 1/k(k-1) = 1/(k-1) - 1/k$.]
10. Establish the convergence and find the limits of the following sequences.
 (a) $((1 + 1/n)^{n+1})$, (b) $((1 + 1/n)^{2n})$,
 (c) $\left((1 + \frac{1}{n+1})^n\right)$, (d) $((1 - 1/n)^n)$.
11. Use the method in Example 3.3.3(e) to calculate $\sqrt{2}$, correct to within 4 decimals.
12. Use the method in Example 3.3.3(e) to calculate $\sqrt{5}$, correct to within 5 decimals.
13. Calculate the number e in Example 3.3.3(f) for $n = 2,4,8,16$.

SECTION 3.4 Subsequences and the Bolzano-Weierstrass Theorem

In this section we shall introduce the notion of a subsequence of a sequence of real numbers. This notion is somewhat more general than that of a tail of a sequence (introduced in 3.1.8) and is often useful in establishing the divergence of a sequence.

We shall also prove a sequential form of the important Bolzano-Weierstrass Theorem 2.4.5, which will be used in establishing a number of results in the sequel.

3.4.1 Definition. Let $X = (x_n)$ be a sequence of real numbers and let $r_1 < r_2 < \cdots < r_n < \cdots$ be a strictly increasing sequence of natural numbers. Then the sequence X' in $\mathbf{R}$ given by

$$(x_{r_1}, x_{r_2}, x_{r_3}, \ldots, x_{r_n}, \ldots)$$

is called a **subsequence** of X.

For example, the following are subsequences of $X := \left(\frac{1}{1}, \frac{1}{2}, \frac{1}{3}, \ldots\right)$:

$$\left(\frac{1}{3}, \frac{1}{4}, \frac{1}{5}, \ldots, \frac{1}{n+2}, \ldots\right), \quad \left(\frac{1}{1}, \frac{1}{3}, \frac{1}{5}, \ldots, \frac{1}{2n-1}, \ldots\right),$$

$$\left(\frac{1}{2!}, \frac{1}{4!}, \ldots, \frac{1}{(2n)!}, \ldots\right).$$

The following sequences are *not* subsequences of $X := (1/n)$:

$$\left(\frac{1}{2}, \frac{1}{1}, \frac{1}{4}, \frac{1}{3}, \frac{1}{6}, \frac{1}{5}, \ldots\right), \quad \left(\frac{1}{1}, 0, \frac{1}{3}, 0, \frac{1}{5}, 0, \ldots\right).$$

Of course, any tail of a sequence is a subsequence; in fact, the M-tail corresponds to the sequence of indices:

$$r_1 = M + 1,\ r_2 = M + 2,\ \ldots,\ r_n = M + n,\ \ldots .$$

However, not every subsequence of a given sequence need be a tail of the sequence.

Subsequences of convergent sequences also converge to the same limit, as we now show.

3.4.2 Theorem. *If a sequence $X = (x_n)$ of real numbers converges to a real number x, then any subsequence of X also converges to x.*

Proof. Let $\varepsilon > 0$ be given and let $K(\varepsilon)$ be such that if $n \geqslant K(\varepsilon)$, then $|x_n - x| < \varepsilon$. Since $r_1 < r_2 < \cdots < r_n < \cdots$ is an increasing sequence of natural numbers, it is easily seen that $r_n \geqslant n$. Hence, if $n \geqslant K(\varepsilon)$ we also have $r_n \geqslant n \geqslant K(\varepsilon)$ so that $|x_{r_n} - x| < \varepsilon$. Therefore the subsequence (x_{r_n}) also converges to x, as claimed. Q.E.D.

3.4.3 Examples. (a) $\lim (b^n) = 0$ if $0 < b < 1$.

We have already seen, in Example 3.1.11(c), that if $0 < b < 1$ and if $x_n := b^n$, then it follows from Bernoulli's Inequality that $\lim (x_n) = 0$. Alternatively we see that since $0 < b < 1$, then $x_{n+1} = b^{n+1} < b^n = x_n$ so that the sequence (x_n) is decreasing. It is also clear that $0 \leqslant x_n \leqslant 1$, so it follows from the Monotone Convergence Theorem 3.3.2 that the sequence is convergent. Let $x := \lim x_n$. Since (x_{2n}) is a subsequence of (x_n) it follows from Theorem 3.4.2 that $x = \lim (x_{2n})$. On the other hand it follows from the relation $x_{2n} = b^{2n} = (b^n)^2 = (x_n)^2$ and Theorem 3.2.3 that

$$x = \lim (x_{2n}) = \big(\lim (x_n)\big)^2 = x^2.$$

Therefore we must either have $x = 0$ or $x = 1$. Since the sequence (x_n) is decreasing and bounded above by 1, we deduce that $x = 0$.

(b) $\lim (c^{1/n}) = 1$ for $c > 1$.

This limit has been obtained in Example 3.1.11(d) for $c > 0$, using a rather ingenious argument. We give here an alternative approach for the case $c > 1$. Note that if $z_n := c^{1/n}$, then $z_n > 1$ and $z_{n+1} < z_n$ for all $n \in \mathbf{N}$. (Why?) Thus by the Monotone Convergence Theorem, the limit $z := \lim (z_n)$ exists. By Theorem 3.4.2, it follows that $z = \lim (z_{2n})$. On the other hand it follows from the relation

$$z_{2n} = c^{1/2n} = (c^{1/n})^{1/2} = z_n^{\ 1/2}$$

and Theorem 3.2.10 that

$$z = \lim (z_{2n}) = \big(\lim (z_n)\big)^{1/2} = z^{1/2}.$$

Therefore we have $z^2 = z$ whence it follows that either $z = 0$ or $z = 1$. Since $z_n > 1$ for all $n \in \mathbf{N}$, we deduce that $z = 1$.

We leave it as an exercise to the reader to consider the case $0 < c < 1$.

The use of subsequences makes it easy to provide a test for the divergence of a sequence.

3.4.4 Divergence Criterion. *Let $X = (x_n)$ be a sequence of real numbers. Then the following statements are equivalent:*

(i) *The sequence $X = (x_n)$ does not converge to $x \in \mathbf{R}$.*

(ii) *There exists an $\varepsilon_0 > 0$ such that for any $k \in \mathbf{N}$, there exists $r_k \in \mathbf{N}$ such that $r_k \geqslant k$ and $|x_{r_k} - x| \geqslant \varepsilon_0$.*

(iii) *There exists an $\varepsilon_0 > 0$ and a subsequence $X' = (x_{r_n})$ of X such that $|x_{r_n} - x| \geqslant \varepsilon_0$ for all $n \in \mathbf{N}$.*

Proof. (i) $\Rightarrow$ (ii) If $X = (x_n)$ does not converge to x, then for some $\varepsilon_0 > 0$ it is impossible to find a natural number $K(\varepsilon)$ such that 3.1.6(d) holds. That is, for any $k \in \mathbf{N}$ it is not true that for *all* $n \geqslant k$ the inequality $|x_n - x| < \varepsilon_0$ holds. In other words, for every $k \in \mathbf{N}$ there is a natural number $r_k \geqslant k$ such that $|x_{r_k} - x| \geqslant \varepsilon_0$.

(ii) $\Rightarrow$ (iii) Let ε_0 be as in (ii) and let $r_1 \in \mathbf{N}$ be such that $r_1 \geqslant 1$ and $|x_{r_1} - x| \geqslant \varepsilon_0$. Now let $r_2 \in \mathbf{N}$ be such that $r_2 \geqslant r_1 + 1$ and $|x_{r_2} - x| \geqslant \varepsilon_0$; let $r_3 \in \mathbf{N}$ be such that $r_3 \geqslant r_2 + 1$ and $|x_{r_3} - x| \geqslant \varepsilon_0$. Continue in this way to obtain a subsequence $X' := (x_{r_n})$ of X such that $|x_{r_n} - x| \geqslant \varepsilon_0$.

(iii) $\Rightarrow$ (i) Suppose $X = (x_n)$ has a subsequence $X' = (x_{r_n})$ satisfying the condition in (iii); then X cannot converge to x. For if it did, then, by Theorem 3.4.2, the subsequence X' would also converge to x. But this is impossible, since none of the terms of X' belongs to the ε_0-neighborhood of x. Q.E.D.

3.4.5 Examples. (a) The sequence $((-1)^n)$ is divergent.

If the sequence $X := ((-1)^n)$ were convergent to a number x, then (by Theorem 3.4.2) every subsequence of X must converge to x. Since there is a subsequence converging to $+1$ and another subsequence converging to -1, we conclude that X must be divergent.

(b) The sequence $(1, \frac{1}{2}, 3, \frac{1}{4}, \ldots)$ is divergent.

[We can define this sequence by $Y = (y_n)$, where $y_n := n$ if n is odd and $y_n := 1/n$ if n is even.] It can easily be seen that this sequence is not bounded; hence, by Theorem 3.2.2, it cannot be convergent. Alternatively, we notice that although a subsequence $(1/2, 1/4, 1/6, \ldots)$ of Y converges to 0, the entire sequence Y does not converge to 0. Indeed there is a subsequence $(3, 5, 7, \ldots)$ of Y that remains outside the 1-neighborhood of 0; hence Y does not converge to 0.

(c) The sequence $S := (\sin n)$ is divergent.

This sequence is not so easy to handle. In discussing it we must, of course, make use of elementary properties of the sine function. We recall that $\sin(\pi/6) = 1/2 = \sin(5\pi/6)$ and that $\sin x > 1/2$ for x in the interval $I_1 := (\pi/6, 5\pi/6)$.

Since the length of I_1 is $5\pi/6 - \pi/6 = 2\pi/3 > 2$, there are at least two natural numbers lying inside I_1; we let n_1 be the first such number. Similarly, for each $k \in N$, $\sin x > \frac{1}{2}$ for x in the interval $I_k := (\pi/6 + 2\pi(k-1), 5\pi/6 + 2\pi(k-1))$. Since the length of I_k is greater than 2, there are at least two natural numbers lying inside I_k; we let n_k be the first one. The subsequence $S' := (\sin n_k)$ of S obtained in this way has the property that all of its values lie in the interval $[\frac{1}{2}, 1]$.

Similarly, if $k \in N$ and J_k is the interval

$$J_k := (7\pi/6 + 2\pi(k-1), 11\pi/6 + 2\pi(k-1)),$$

then it is seen that $\sin x < -\frac{1}{2}$ for all $x \in J_k$ and the length of J_k is greater than 2. Let m_k be the first natural number lying in J_k. Then the subsequence $S'' := (\sin m_k)$ of S has the property that all of its values lie in the interval $[-1, -\frac{1}{2}]$.

Given any real number c, it is readily seen that at least one of the subsequences S' and S'' lies entirely outside of the $\frac{1}{2}$-neighborhood of c. Therefore c cannot be a limit of S. Since $c \in R$ is arbitrary, we deduce that S is divergent.

The Bolzano-Weierstrass Theorem

We have seen in Theorem 3.2.2 that every convergent sequence of real numbers must be bounded; but, of course, not every bounded sequence is convergent [for example, the sequence $((-1)^n)$ is not convergent]. It is true, however, that every bounded sequence has a convergent subsequence. This fact is an important result that will be used in a number of crucial arguments to be given later.

Earlier, in Section 2.5, we presented a result derived from the Nested Intervals Property that is called the Bolzano-Weierstrass Theorem 2.5.4. It states that a bounded set $S \subseteq R$ with infinitely many elements has a cluster point x (that is, a point x such that each neighborhood of x contains points of S other than x). The result that follows is quite similar in concept and is also known as the Bolzano-Weierstrass Theorem. The two theorems are closely related and each can be used to prove the other, but the technical differences are significant and should be clearly understood.

3.4.6 The Bolzano-Weierstrass Theorem for Sequences. *A bounded sequence of real numbers has a convergent subsequence.*

Proof. Suppose that $X = (x_n)$ is a bounded sequence and let $S := \{x_n : n \in N\}$ be its range. Two cases arise: (i) S is a finite set, or (ii) S is an infinite set.

In case (i), there must be at least one element in S that is the value of x_n for infinitely many values of n. That is, there is $s \in S$ and $n_1 < n_2 < \cdots < n_k < \cdots$ such that $x_{n_k} = s$ for all $k \in N$. In this case, the subsequence (x_{n_k}) evidently converges to s.

In case (ii), we invoke the earlier Bolzano-Weierstrass Theorem 2.5.4 to obtain a cluster point x of the bounded infinite set S. We shall now find a subsequence of $X = (x_n)$ that converges to x. For each $k \in \boldsymbol{N}$, let $U_k := (x - 1/k, x + 1/k)$ be the $(1/k)$-neighborhood of x. Because x is a cluster point of S, the set U_k contains infinitely many elements of S; that is, for each k, there are infinitely many values of n such that $x_n \in U_k$. Therefore, we first choose $x_{n_1} \in U_1$; then we choose $x_{n_2} \in U_2$ such that $n_2 > n_1$; next we choose $x_{n_3} \in U_3$ such that $n_3 > n_2$, and so on. Thus, we obtain a subsequence (x_{nk}) of $X = (x_n)$ that satisfies $|x_{n_k} - x| < 1/k$ for $k \in \boldsymbol{N}$. Hence, $\lim (x_{nk}) = x$. Q.E.D.

Note. It is readily seen that a bounded sequence can have various subsequences converging to different limits; for example, the sequence $((-1)^n)$ has subsequences that converge to -1 and subsequences that converge to $+1$. It can be shown, however, that if X is a bounded sequence such that *every* convergent subsequence of X converges to the same element x, then the entire sequence converges to x. (See Theorem 3.4.9 below.)

Characterization of Closed Sets

We shall now give a characterization of closed subsets of $\boldsymbol{R}$ in terms of sequences. As we shall see, closed sets are precisely those sets F that contain the limits of all convergent sequences whose elements are taken from F. The reader should compare the next result with Theorem 2.6.7.

3.4.7 Characterization of Closed Sets. *Let F be a set of real numbers. Then the following assertions are equivalent.*

(i) *F is a closed subset of $\boldsymbol{R}$.*

(ii) *If X is any convergent sequence whose elements belong to F, then* $\lim X$ *belongs to F.*

Proof. (i) $\Rightarrow$ (ii) Let $X = (x_n)$ be a sequence of elements in F and let $x := \lim X$. We wish to show that $x \in F$. Suppose, on the contrary, that $x \notin F$; that is, that $x \in \mathscr{C}(F)$ the complement of F. Since $\mathscr{C}(F)$ is open and $x \in \mathscr{C}(F)$, it follows that there exists an ε-neighborhood V_ε of x such that V_ε is contained in $\mathscr{C}(F)$. Since $x = \lim (x_n)$, it follows that there exists a natural number $K = K(\varepsilon)$ such that $x_K \in V_\varepsilon$. Therefore we must have $x_K \in \mathscr{C}(F)$; but this contradicts the assumption that $x_n \in F$ for all $n \in \boldsymbol{N}$. Therefore, we conclude that $x \in F$.

(ii) $\Rightarrow$ (i) Suppose, on the contrary, that F is not closed, so that $G := \mathscr{C}(F)$ is not open. Then there exists a point $y_0 \in G$ such that for each $n \in \boldsymbol{N}$, there is a number $y_n \in \mathscr{C}(G) = F$ such that $|y_n - y_0| < 1/n$. It follows that $y_0 := \lim (y_n)$, and since $y_n \in F$ for all $n \in \boldsymbol{N}$, the hypothesis (ii) implies that $y_0 \in F$, contrary to the assumption $y_0 \in G = \mathscr{C}(F)$. Thus the hypothesis that F is not closed implies that (ii) is not true. Consequently (ii) implies (i), as asserted. Q.E.D.

The next result, an elementary consequence of the Bolzano-Weierstrass Theorem 3.4.6 and the Characterization Theorem 3.4.7, is sometimes useful.

3.4.8 Theorem. *Let K be a closed and bounded set of real numbers. Then every sequence $X = (x_n)$, with $x_n \in K$ for all $n \in \mathbf{N}$, has a convergent subsequence X' and* $\lim X'$ *belongs to K.*

Proof. Since K is bounded, there exists $M > 0$ such that $|x| \leqslant M$ for all $x \in K$. Since $x_n \in K$ for all $n \in \mathbf{N}$, we conclude that the sequence $X = (x_n)$ is bounded. Therefore, by the Bolzano-Weierstrass Theorem 3.4.6, X has a convergent subsequence X'. But since K is closed, we infer that $\lim X'$ belongs to K.
Q.E.D.

3.4.9 Theorem. *Let X be a bounded sequence of real numbers and let $x \in \mathbf{R}$ have the property that every convergent subsequence of X converges to x. Then the sequence X converges to x.*

Proof. Suppose $M > 0$ is a bound for the sequence $X = (x_n)$; that is, that $|x_n| \leqslant M$ for all $n \in \mathbf{N}$. Suppose also that the sequence $X = (x_n)$ does not converge to x. It follows from the Divergence Criterion 3.4.4 that there exists an $\varepsilon_0 > 0$ and a subsequence $X' := (x_{r_n})$ of X such that

$$|x_{r_n} - x| \geqslant \varepsilon_0 \qquad \text{for all} \quad n \in \mathbf{N}.$$

Since X' is a subsequence of X, it follows that M is also a bound for X'. Hence the elements of X' belong to the closed and bounded set

$$K := [-M, x - \varepsilon_0] \cup [x + \varepsilon_0, M].$$

If we apply Theorem 3.4.8 to the sequence X' and the closed bounded set K, we deduce that X' has a convergent subsequence X'' and that $\lim X''$ belongs to K. Now it is clear that since X'' is a subsequence of X', then X'' is also a subsequence of X. It follows from the hypothesis of the theorem that we must have $x = \lim X''$. But this contradicts the fact that $\lim X''$ belongs to K. Thus the supposition that the bounded sequence X does not converge to x has led to a contradiction.
Q.E.D.

Exercises for Section 3.4

1. Given an example of an unbounded sequence that has a convergent subsequence.
2. Use the method of Example 3.4.2(b) to show that if $0 < c < 1$, then $\lim (c^{1/n}) = 1$.
3. Let $X = (x_n)$ and $Y = (y_n)$ be given sequences and let the "shuffled" sequence $Z = (z_n)$ be defined by $z_1 := x_1, z_2 := y_1, \ldots, z_{2n-1} := x_n, z_{2n} := y_n, \ldots$. Show that Z is convergent if and only if X and Y are convergent and $\lim X = \lim Y$.

4. Let $x_n := n^{1/n}$ for $n \in \boldsymbol{N}$.
   (a) Show that the inequality $x_{n+1} < x_n$ is equivalent to the inequality $(1 + 1/n)^n < n$, and infer that the inequality is valid for $n \geq 3$. [See Example 3.3.3(f).] Conclude that (x_n) is ultimately decreasing and that $x := \lim (x_n)$ exists.
   (b) Use the fact that the subsequence (x_{2n}) also converges to x to show that $x = \sqrt{x}$. Conclude that $x = 1$.
5. Suppose that every subsequence of $X = (x_n)$ has a subsequence that converges to 0. Show that $\lim X = 0$.
6. Suppose that (x_n) is a bounded sequence of distinct real numbers such that its range $\{x_n : n \in \boldsymbol{N}\}$ has exactly one cluster point. Prove that (x_n) is convergent.
7. Establish the convergence and find the limits of the following sequences:
   (a) $((1 + 1/2n)^2)$, (b) $((1 + 1/2n)^n)$,
   (c) $((1 + 1/n^2)^{n^2})$, (d) $((1 + 2/n)^n)$.
8. Let (x_n) be a bounded sequence and for each $n \in \boldsymbol{N}$ let $s_n := \sup \{x_k : k \geq n\}$ and $S := \inf \{s_n\}$. Show that there exists a subsequence of (x_n) that converges to S.
9. Suppose that $x_n \geq 0$ for all $n \in \boldsymbol{N}$ and that $\lim ((-1)^n x_n)$ exists. Show that (x_n) converges.
10. Use the characterization in 3.4.7 to show that the union and the intersection of two closed sets in $\boldsymbol{R}$ is a closed set.
11. Let F be a closed set in $\boldsymbol{R}$ such that $0 \notin F$. Show that there exists a point $x_0 \in F$ such that $|x_0| = \inf \{|x| : x \in F\}$.
12. Let (I_n) be a nested sequence of intervals. For each $n \in \boldsymbol{N}$, let $x_n \in I_n$. Use the Bolzano-Weierstrass Theorem to give a proof of the Nested Intervals Theorem.

SECTION 3.5 The Cauchy Criterion

The Monotone Convergence Theorem 3.3.2 is extraordinarily useful and important, but it has the significant drawback that it applies only to sequences that are monotone. It is important for us to have a condition implying the convergence of a sequence of real numbers that does not require the knowledge of the value of the limit, and is not restricted to monotone sequences. The Cauchy Criterion, which will be established in this section, is such a condition.

3.5.1 Definition. A sequence $X = (x_n)$ of real numbers is said to be a **Cauchy sequence** if for every $\varepsilon > 0$ there is a natural number $H(\varepsilon)$ such that for all natural numbers $n, m \geq H(\varepsilon)$, we have $|x_n - x_m| < \varepsilon$.

The reader should compare this definition closely with Definition 3.1.4 and Theorem 3.1.6 pertaining to the convergence of the sequence X. It will be seen below that the Cauchy sequences are precisely the convergent sequences. In proving this we first show that a convergent sequence is a Cauchy sequence.

3.5.2 Lemma. *If $X = (x_n)$ is a convergent sequence of real numbers, then X is a Cauchy sequence.*

Proof. Let $x := \lim X$, then given $\varepsilon > 0$ there is a natural number $K(\varepsilon/2)$ such that if $n \geqslant K(\varepsilon/2)$ then $|x_n - x| < \varepsilon/2$. Thus, if $H(\varepsilon) := K(\varepsilon/2)$ and if $n, m \geqslant H(\varepsilon)$, then we have

$$|x_n - x_m| = |(x_n - x) + (x - x_m)|$$
$$\leqslant |x_n - x| + |x_m - x| < \varepsilon/2 + \varepsilon/2 = \varepsilon.$$

Since $\varepsilon > 0$ is arbitrary, it follows that (x_n) is a Cauchy sequence. Q.E.D.

In order to establish that a Cauchy sequence is convergent, we shall need the following result.

3.5.3 Lemma. *A Cauchy sequence of real numbers is bounded.*

Proof. Let $X := (x_n)$ be a Cauchy sequence and let $\varepsilon := 1$. If $H := H(1)$ and $n \geqslant H$, then $|x_n - x_H| \leqslant 1$. Hence, by the Triangle Inequality we have that $|x_n| \leqslant |x_H| + 1$ for $n \geqslant H$. If we set

$$M := \sup \{|x_1|, |x_2|, \ldots, |x_{H-1}|, |x_H| + 1\},$$

then it follows that $|x_n| \leqslant M$ for all $n \in \mathbf{N}$. Q.E.D.

We now present the important Cauchy Convergence Criterion.

3.5.4 Cauchy Convergence Criterion. *A sequence of real numbers is convergent if and only if it is a Cauchy sequence.*

Proof. We have seen, in Lemma 3.5.2, that a convergent sequence is a Cauchy sequence.

Conversely, let $X = (x_n)$ be a Cauchy sequence. We shall show that X is convergent to some real number. First we observe from Lemma 3.5.3 that the sequence X is bounded. Therefore, by the Bolzano-Weierstrass Theorem 3.4.6, there is a subsequence $X' = (x_{n_k})$ of X that converges to some real number x^*. We shall complete the proof by showing that X converges to x^*.

Since $X = (x_n)$ is a Cauchy sequence, given $\varepsilon > 0$ there is a natural number $H(\varepsilon/2)$ such that if $n, m \geqslant H(\varepsilon/2)$ then

$$(*) \qquad |x_n - x_m| < \varepsilon/2.$$

Since the subsequence $X' = (x_{n_k})$ converges to x^*, there is a natural number $K \geqslant H(\varepsilon/2)$ belonging to the set $\{n_1, n_2, \ldots\}$ such that

$$|x_K - x^*| < \varepsilon/2.$$

Since $K \geqslant H(\varepsilon/2)$, it follows from $(*)$ with $m = K$ that

$$|x_n - x_K| < \varepsilon/2 \qquad \text{for} \quad n \geqslant H(\varepsilon/2).$$

Therefore, if $n \geqslant H(\varepsilon/2)$, we have

$$\begin{aligned} |x_n - x^*| &= |(x_n - x_K) + (x_K - x^*)| \\ &\leqslant |x_n - x_K| + |x_K - x^*| \\ &< \varepsilon/2 + \varepsilon/2 = \varepsilon. \end{aligned}$$

Since $\varepsilon > 0$ is arbitrary, we infer that $\lim (x_n) = x^*$. Therefore the sequence X is convergent. Q.E.D.

We shall now give some examples of applications of the Cauchy Criterion.

3.5.5 Examples. (a) The sequence $(1/n)$ is convergent.

Of course, we have already seen that this sequence converges to 0. To show that $(1/n)$ is a Cauchy sequence we note that if $\varepsilon > 0$ is given, then there is a natural number $H := H(\varepsilon)$ such that $H > 2/\varepsilon$. Hence, if $n, m \geqslant H$, then we have $1/n \leqslant 1/H$ and $1/m \leqslant 1/H$. Therefore it follows that if $n, m \geqslant H$, then

$$\left| \frac{1}{n} - \frac{1}{m} \right| \leqslant \frac{1}{n} + \frac{1}{m} \leqslant \frac{2}{H} < \varepsilon.$$

Since $\varepsilon > 0$ is arbitrary, we infer that $(1/n)$ is a Cauchy sequence; therefore, it is a convergent sequence.

(b) Let $X = (x_n)$ be defined by

$$x_1 := 1, \qquad x_2 := 2, \qquad \text{and} \qquad x_n := \tfrac{1}{2}(x_{n-2} + x_{n-1}) \qquad \text{for} \quad n > 2.$$

It can be shown by induction that $1 \leqslant x_n \leqslant 2$ for all $n \in \mathbf{N}$. (Do so.) Some calculation shows that the sequence X is not monotone. Since the terms are formed by averaging, it is readily seen that

$$|x_n - x_{n+1}| = \frac{1}{2^{n-1}} \qquad \text{for} \quad n \in \mathbf{N}.$$

(Prove this by induction.) Thus, if $m > n$, we may employ the Triangle Inequality to obtain

$$\begin{aligned} |x_n - x_m| &\leqslant |x_n - x_{n+1}| + |x_{n+1} - x_{n+2}| + \cdots + |x_{m-1} - x_m| \\ &= \frac{1}{2^{n-1}} + \frac{1}{2^n} + \cdots + \frac{1}{2^{m-2}} \\ &= \frac{1}{2^{n-1}} \left(1 + \frac{1}{2} + \cdots + \frac{1}{2^{m-n-1}} \right) < \frac{1}{2^{n-2}}. \end{aligned}$$

Therefore, given $\varepsilon > 0$, if n is chosen so large that $1/2^n < \varepsilon/4$ and if $m \geqslant n$, then it follows that $|x_n - x_m| < \varepsilon$. therefore, X is a Cauchy sequence in $\boldsymbol{R}$. By the Cauchy Criterion 3.5.4 we infer that the sequence X converges to a number x.

To evaluate the limit x, we might first "pass to the limit" in the rule of definition $x_n = \frac{1}{2}(x_{n-1} + x_{n-2})$ to conclude that x must satisfy the relation $x = \frac{1}{2}(x + x)$, which is true, but not informative. Hence we must try something else.

Since X converges to x, so does the subsequence X' with odd indices. By induction, the reader can establish that [see 1.3.3(e)]

$$\begin{aligned} x_{2n+1} &= 1 + \frac{1}{2} + \frac{1}{2^3} + \cdots + \frac{1}{2^{2n-1}} \\ &= 1 + \frac{2}{3}\left(1 - \frac{1}{4^n}\right). \end{aligned}$$

It follows from this (how?), that

$$x = \lim X = \lim X' = 1 + \tfrac{2}{3} = \tfrac{5}{3}.$$

(c) Let $Y = (y_n)$ be the sequence of real numbers given by

$$y_1 := \frac{1}{1!},\ y_2 := \frac{1}{1!} - \frac{1}{2!},\ \ldots,\ y_n := \frac{1}{1!} - \frac{1}{2!} + \cdots + \frac{(-1)^{n+1}}{n!},\ \ldots$$

Clearly, Y is not a monotone sequence. However, if $m > n$, then

$$y_m - y_n = \frac{(-1)^{n+2}}{(n+1)!} + \frac{(-1)^{n+3}}{(n+2)!} + \cdots + \frac{(-1)^{m+1}}{m!}.$$

Since $2^{r-1} \leqslant r!$ [see 1.3.3(d)], it follows that if $m > n$, then

$$\begin{aligned} |y_m - y_n| &\leqslant \frac{1}{(n+1)!} + \frac{1}{(n+2)!} + \cdots + \frac{1}{m!} \\ &\leqslant \frac{1}{2^n} + \frac{1}{2^{n+1}} + \cdots + \frac{1}{2^{m-1}} < \frac{1}{2^{n-1}}. \end{aligned}$$

Therefore, it follows that (y_n) is a Cauchy sequence. Hence it converges to a limit y. At the present moment we cannot evalute y directly; however, passing to the limit (with respect to m) in the above inequality, the reader may show that

$$|y_n - y| \leqslant 1/2^{n-1}.$$

Hence we can calculate y to any desired accuracy by calculating the terms y_n for sufficiently large n. The reader should do this and show that y is approximately equal to 0.632120559. (The exact value of y is $1 - 1/e$.)

(d) The sequence $\left(\frac{1}{1} + \frac{1}{2} + \cdots + \frac{1}{n}\right)$ diverges.

Let $H := (h_n)$ be the sequence defined by

$$h_n := \frac{1}{1} + \frac{1}{2} + \cdots + \frac{1}{n} \quad \text{for} \quad n \in \mathbf{N},$$

which was considered in 3.3.3(b). If $m > n$, then

$$h_m - h_n = \frac{1}{n+1} + \cdots + \frac{1}{m}.$$

Since each of these $m - n$ terms exceeds $1/m$, then $h_m - h_n > (m - n)/m = 1 - n/m$. In particular, if $m = 2n$ we have $h_{2n} - h_n > \frac{1}{2}$. This shows that H is not a Cauchy sequence; therefore H is *not* a convergent sequence. (In terms that will be introduced in Chapter 8, we have just proved that the "harmonic series" $\sum_{n=1}^{\infty} \frac{1}{n}$ is divergent.)

3.5.6 Definition. We say that a sequence $X = (x_n)$ is **contractive** if there exists a constant C, $0 < C < 1$, such that

$$|x_{n+2} - x_{n+1}| \leqslant C\,|x_{n+1} - x_n|$$

for all $n \in \mathbf{N}$. The number C is called **the constant** of the contractive sequence.

3.5.7 Theorem. *Every contractive sequence is a Cauchy sequence, and therefore is convergent.*

Proof. If we successively apply the defining condition for a contractive sequence, we can work our way back to the beginning of the sequence as follows:

$$\begin{aligned}|x_{n+2} - x_{n+1}| &\leqslant C\,|x_{n+1} - x_n| \leqslant C^2\,|x_n - x_{n-1}| \\ &\leqslant C^3\,|x_{n-1} - x_{n-2}| \leqslant \cdots \leqslant C^n\,|x_2 - x_1|.\end{aligned}$$

For $m > n$, we estimate $|x_m - x_n|$ by first applying the Triangle Inequality and then using the formula for the sum of a geometric progression (see 1.3.3(e)). This gives

$$\begin{aligned}
|x_m - x_n| &\leq |x_m - x_{m-1}| + |x_{m-1} - x_{m-2}| + \cdots + |x_{n+1} - x_n| \\
&\leq (C^{m-2} + C^{m-3} + \cdots + C^{n-1})\,|x_2 - x_1| \\
&= C^{n-1}\,(C^{m-n-1} + C^{m-n-2} + \cdots + 1)\,|x_2 - x_1| \\
&= C^{n-1}\left(\frac{1-C^{m-n}}{1-C}\right)|x_2 - x_1| \\
&\leq C^{n-1}\left(\frac{1}{1-C}\right)|x_2 - x_1| .
\end{aligned}$$

Since $0 < C < 1$, we know $\lim\,(C^n) = 0$ [see 3.1.11(c)]. Therefore, we infer that (x_n) is a Cauchy sequence. It now follows from the Cauchy Convergence Criterion 3.5.4. that (x_n) is a convergent sequence. Q.E.D.

In the process of calculating the limit of a contractive sequence, it is very important to have an estimate of the error at the nth stage. In the next result we give two such estimates; the first one involves the first two terms in the sequence and n. The second one involves the difference $x_n - x_{n-1}$.

3.5.8. Corollary. *If $X := (x_n)$ is a contractive sequence with constant C, $0 < C < 1$, and if $x^* := \lim X$, then:*

(i) $|x^* - x_n| \leq \dfrac{C^{n-1}}{1-C}|x_2 - x_1|$,

(ii) $|x^* - x_n| \leq \dfrac{C}{1-C}|x_n - x_{n-1}|$.

Proof. We have seen in the preceding proof that if $m > n$, then $|x_m - x_n| \leq (C^{n-1}/(1-C))|x_2 - x_1|$. If we pass to the limit in this inequality (with respect to m), we obtain (i).

To prove (ii), recall that if $m > n$, then

$$|x_m - x_n| \leq |x_m - x_{m-1}| + \cdots + |x_{n+1} - x_n| .$$

Since it is readily established, using induction, that

$$|x_{n+k} - x_{n+k-1}| \leq C^k\,|x_n - x_{n-1}| ,$$

we infer that

$$\begin{aligned}
|x_m - x_n| &\leq (C^{m-n} + \cdots + C^2 + C)\,|x_n - x_{n-1}| \\
&\leq \frac{C}{1-C}|x_n - x_{n-1}| .
\end{aligned}$$

We now pass to the limit in this inequality (with respect to m) to obtain assertion (ii). Q.E.D.

3.5.9 Example. We are told that the cubic equation $x^3 - 7x + 2 = 0$ has a solution between 0 and 1 and we wish to approximate this solution. This can be accomplished by means of an iteration procedure as follows. We first rewrite the equation as $x = (x^3 + 2)/7$ and use this to define a sequence. We assign x_1 an arbitrary value between 0 and 1, and then define

$$x_{n+1} := \tfrac{1}{7}(x_n^3 + 2), \qquad n \in \mathbf{N}.$$

Because $0 < x_1 < 1$, it follows that $0 < x_n < 1$ for all $n \in \mathbf{N}$. (Why?) Moreover, we have

$$\begin{aligned} |x_{n+2} - x_{n+1}| &= |\tfrac{1}{7}(x_{n+1}^3 + 2) - \tfrac{1}{7}(x_{n+1}^3 + 2)| \\ &= \tfrac{1}{7}|x_{n+1}^3 - x_n^3| \\ &= \tfrac{1}{7}|x_{n+1}^2 + x_{n+1}x_n + x_n^2|\,|x_{n+1} - x_n| \\ &\leqslant \tfrac{3}{7}|x_{n+1} - x_n|. \end{aligned}$$

Therefore, (x_n) is a contractive sequence and hence there exists r such that $\lim (x_n) = r$. If we pass to the limit on both sides of the equality $x_{n+1} = (x_n^3 + 2)/7$, we obtain $r = (r^3 + 2)/7$ and hence $r^3 - 7r + 2 = 0$. Thus r is a solution of the equation.

We can approximate r by choosing x_1 and calculating $x_2, x_3, \ldots$ successively. For example, if we take $x_1 = 0.5$, we obtain $x_2 = 0.3035714$, $x_3 = 0.2897108$, $x_4 = 0.2891880$, $x_5 = 0.2891692$, $x_6 = 0.2891686$, etc. To estimate the accuracy, we note that $|x_2 - x_1| < 0.2$. Thus, after n steps it follows from Corollary 3.5.8(i) that we are sure that $|x^* - x_n| \leqslant 3^{n-1}/(7^{n-2} \cdot 20)$. Thus, when $n = 6$, we are sure that $|x^* - x_6| \leqslant 3^5/(7^4 \cdot 20) = 243/48020 < 0.0051$. Actually the approximation is substantially better than this. In fact, since $|x_6 - x_5| < 0.0000007$, it follows from 3.5.8(ii) that $|x^* - x_6| \leqslant \tfrac{3}{4}|x_6 - x_5| < 0.0000006$. Hence x_6 is accurate to six decimal places.

Exercises for Section 3.5

1. Give an example of a bounded sequence that is not a Cauchy sequence.
2. Show directly that the following are Cauchy sequences.

   (a) $\left(\dfrac{n+1}{n}\right)$, (b) $\left(1 + \dfrac{1}{2!} + \cdots + \dfrac{1}{n!}\right)$.
3. Show directly that the following are not Cauchy sequences.

   (a) $((-1)^n)$, (b) $\left(n + \dfrac{(-1)^n}{n}\right)$.

4. Show directly that if (x_n) and (y_n) are Cauchy sequences, then $(x_n + y_n)$ and $(x_n y_n)$ are Cauchy sequences.
5. Let (x_n) be a Cauchy sequence such that x_n is an integer for all $n \in \boldsymbol{N}$. Show that (x_n) is ultimately constant.
6. Show directly that a bounded monotone increasing sequence is a Cauchy sequence.
7. If $x_1 < x_2$ are arbitrary real numbers and $x_n := \frac{1}{2}(x_{n-2} + x_{n+1})$ for $n > 2$, show that (x_n) is convergent. What is its limit?
8. If $y_1 < y_2$ are arbitrary real numbers and $y_n := \frac{1}{3}y_{n-1} + \frac{2}{3}y_{n-2}$ for $n > 2$, show that (y_n) is convergent. What is its limit?
9. If $x_1 > 0$ and $x_{n+1} := (2 + x_n)^{-1}$ for $n \geq 1$, show that (x_n) is a contractive sequence. Find the limit.
10. The polynomial equation $x^3 - 5x + 1 = 0$ has a root r with $0 < r < 1$. Use an appropriate contractive sequence to calculate r within 10^{-4}.

SECTION 3.6 Properly Divergent Sequences

For certain purposes it is convenient to define what is meant for a sequence (x_n) of real numbers to "tend to $\pm\infty$".

3.6.1 Definition. Let (x_n) be a sequence of real numbers.

(i) We say that (x_n) **tends to** ∞, and write $\lim (x_n) = \infty$, if for every $\alpha \in \boldsymbol{R}$ there exists a natural number $K(\alpha)$ such that if $n \geq K(\alpha)$, then $x_n > \alpha$.

(ii) We say that (x_n) **tends to** $-\infty$, and write $\lim (x_n) = -\infty$, if for every $\beta \in \boldsymbol{R}$ there exists a natural number $K(\beta)$ such that if $n \geq K(\beta)$, then $x_n < \beta$.

We say that (x_n) is **properly divergent** in case we have either $\lim (x_n) = \infty$ or $\lim (x_n) = -\infty$.

The reader should realize that we are using the symbols ∞ and $-\infty$ purely as a convenient *notation* in the above expressions. Results that have been proved in earlier sections for conventional limits $\lim (x_n) = L$ (for $L \in \boldsymbol{R}$) *may not* remain true when $\lim (x_n) = \pm\infty$.

3.6.2 Examples. (a) $\lim (n) = \infty$.

In fact, if $\alpha \in \boldsymbol{R}$ is given, let $K(\alpha)$ be any natural number such that $K(\alpha) > \alpha$.

(b) $\lim (n^2) = \infty$.

If $K(\alpha)$ is a natural number such that $K(\alpha) > \alpha$, and if $n \geq K(\alpha)$ then we have $n^2 \geq n > \alpha$.

(c) If $c > 1$, then $\lim (c^n) = \infty$.

Let $c := 1 + b$, where $b > 0$. If $\alpha \in \boldsymbol{R}$ is given, let $K(\alpha)$ be a natural number such that $K(\alpha) > \alpha/b$. If $n \geq K(\alpha)$ it follows from Bernoulli's Inequality that

$$c^n = (1 + b)^n \geq 1 + nb > 1 + \alpha > \alpha.$$

Therefore $\lim (c^n) = \infty$.

Monotone sequences are particularly simple in regard to their convergence. We have seen in the Monotone Convergence Theorem 3.3.2 that a monotone sequence is convergent if and only if it is bounded. The next result is a reformulation of that result.

3.6.3 Theorem *A monotone sequence of real numbers is properly divergent if and only if it is unbounded.*

(a) *If* (x_n) *is an unbounded increasing sequence, then* $\lim (x_n) = \infty$.

(b) *If* (x_n) *is an unbounded decreasing sequence, then* $\lim (x_n) = -\infty$.

Proof. (a) Suppose that (x_n) is an increasing sequence. We know that if (x_n) is bounded, then it is convergent. If (x_n) is unbounded, then for any $\alpha \in \boldsymbol{R}$ there exists $n(\alpha) \in \boldsymbol{N}$ such that $\alpha < x_{n(\alpha)}$. But since (x_n) is increasing, we have $\alpha < x_n$ for all $n \geqslant n(\alpha)$. Therefore $\lim (x_n) = \infty$.

Part (b) is proved in a similar fashion. Q.E.D.

The following "comparison theorem" is frequently used in showing that a sequence is properly divergent. [In fact, we implicitly used it in Example 3.6.2(c).]

3.6.4 Theorem. *Let* (x_n) *and* (y_n) *be two sequences of real numbers and suppose that*

$$(*) \qquad x_n \leqslant y_n \quad \textit{for all } n \in \boldsymbol{N}.$$

(a) *If* $\lim (x_n) = \infty$, *then* $\lim (y_n) = \infty$.

(b) *If* $\lim (y_n) = -\infty$, *then* $\lim (x_n) = -\infty$.

Proof. (a) If $\lim (x_n) = \infty$, and if $\alpha \in \boldsymbol{R}$ is given, then there exists a natural number $K(\alpha)$ such that if $n \geqslant K(\alpha)$, then $\alpha < x_n$. In view of $(*)$, it follows that $\alpha < y_n$ for all $n \geqslant K(\alpha)$. Therefore $\lim (y_n) = \infty$.

The proof of (b) is similar. Q.E.D.

Remarks. (a) Theorem 3.6.4 remains true if condition $(*)$ is ultimately true; that is, if there exists $M \in \boldsymbol{N}$ such that $x_n \leqslant y_n$ for all $n \geqslant M$.

(b) If condition $(*)$ of Theorem 3.6.4 holds and if $\lim (y_n) = \infty$, it does *not* follow that $\lim (x_n) = \infty$. Similarly, if $(*)$ holds and if $\lim (x_n) = -\infty$, it does *not* follow that $\lim (y_n) = -\infty$. In using Theorem 3.6.4 to show that a sequence tends to ∞ [respectively, $-\infty$] we need to show that the terms of the sequence are ultimately greater [respectively, less] than or equal to the corresponding terms of a sequence that is known to tend to ∞ [respectively, $-\infty$].

Since it is sometimes difficult to establish an inequality such as $(*)$, the following "limit comparison theorem" is often more convenient to use than Theorem 3.6.4.

3.6.5 Theorem. *Let (x_n) and (y_n) be two sequences of strictly positive real numbers and suppose that for some $L \in \mathbf{R}$, $L > 0$, we have*

$$(\dagger) \qquad \lim \left(\frac{x_n}{y_n}\right) = L.$$

Then $\lim (x_n) = \infty$ *if and only if* $\lim (y_n) = \infty$.

Proof. If ($\dagger$) holds, there exists $K \in \mathbf{N}$ such that

$$\tfrac{1}{2} L < \frac{x_n}{y_n} < 2L \qquad \text{for all } n \geq K.$$

Hence we have $(\frac{1}{2}L)y_n < x_n < (2L)y_n$ for all $n \geq K$. The conclusion now follows from a slight modification of Theorem 3.6.4. We leave the details to the reader. Q.E.D.

The reader can show that the conclusion need not hold if either $L = 0$ or $L = \infty$. However, there are some partial results that can be established in these cases, as will be seen in the exercises.

Exercises for Section 3.6

1. Show that if (x_n) is an unbounded sequence, then there exists a properly divergent subsequence.
2. Give examples of properly divergent sequences (x_n) and (y_n) with $y_n \neq 0$ for all $n \in \mathbf{N}$ such that:
   (a) (x_n/y_n) is convergent;
   (b) (x_n/y_n) is properly divergent.
3. Show that if $x_n > 0$ for all $n \in \mathbf{N}$, then $\lim (x_n) = 0$ if and only if $\lim (1/x_n) = \infty$.
4. Establish the proper divergence of the following sequences.
   (a) $(\sqrt{n})$, (b) $(\sqrt{n+1})$,
   (c) $(\sqrt{n-1})$, (d) $(n/\sqrt{n+1})$.
5. Is the sequence $(n \sin n)$ properly divergent?
6. Let (x_n) be properly divergent and let (y_n) be such that $\lim (x_n y_n)$ belongs to $\mathbf{R}$. Show that (y_n) converges to 0.
7. Let (x_n) and (y_n) be sequences of strictly positive numbers and suppose that $\lim (x_n/y_n) = 0$.
   (a) Show that if $\lim (x_n) = \infty$, then $\lim (y_n) = \infty$.
   (b) Show that if (y_n) is bounded, then $\lim (x_n) = 0$.
8. Investigate the convergence or the divergence of the following sequences:
   (a) $(\sqrt{n^2+2})$, (b) $(\sqrt{n}/(n^2+1))$,
   (c) $(\sqrt{n^2+1}/\sqrt{n})$, (d) $(\sin \sqrt{n})$,
9. Let (x_n) and (y_n) be sequences of strictly positive numbers and suppose that $\lim (x_n/y_n) = \infty$.
   (a) Show that if $\lim (y_n) = \infty$, then $\lim (x_n) = \infty$.
   (b) Show that if (x_n) is bounded, then $\lim (y_n) = 0$.

CHAPTER FOUR

LIMITS AND CONTINUITY

"Mathematical analysis" is generally understood to be that body of mathematics in which systematic use is made of various limiting concepts. We have already met one of these basic limiting concepts: the convergence of sequences of real numbers. In this chapter we shall encounter the notion of the limit of a function. We introduce this notion in Section 4.1 and study it further in Section 4.2. It will be seen that not only is the notion of the limit of a function closely parallel to that of the limit of a sequence, but also that questions concerning the existence of limits of functions often can be treated by considering certain related sequences. In Section 4.3 we introduce some extensions of the notion of a limit that are often useful.

In Section 4.4 we shall define what it means to say that a function is continuous at a point, and we initiate the study of continuous functions, which is continued into Section 4.5. In Section 4.6 we see that functions that are continuous on intervals possess a number of special properties that will be seen to be of considerable importance. In Section 4.7 we shall introduce the important notion of uniform continuity and we shall apply this notion to the problem of approximating continuous functions by more elementary functions. Monotone functions are an important class of functions and have strong continuity properties; they are discussed in Section 4.8. In particular it will be seen that continuous monotone functions have continuous monotone inverse functions. In Section 4.9 we give a brief introduction to the notion of compactness.

SECTION 4.1 Limits of Functions

In this section we shall define the important notion of the limit of a real valued function at a point. The reader will notice the very close parallel with the definition of the limit of a sequence. The intuitive idea behind the statement: "the function f approaches L at c" is that the values $f(x)$ will lie in an arbitrary, but pre-assigned neighborhood of L provided that we take x sufficiently close to c. More precisely, if V is a pre-assigned neighborhood of L, there will be a neighborhood U (depending on V) of c such that if $x \neq c$ belongs to U and if $f(x)$ is defined, then $f(x)$ belongs to V. In this case we do not want to be influenced by the

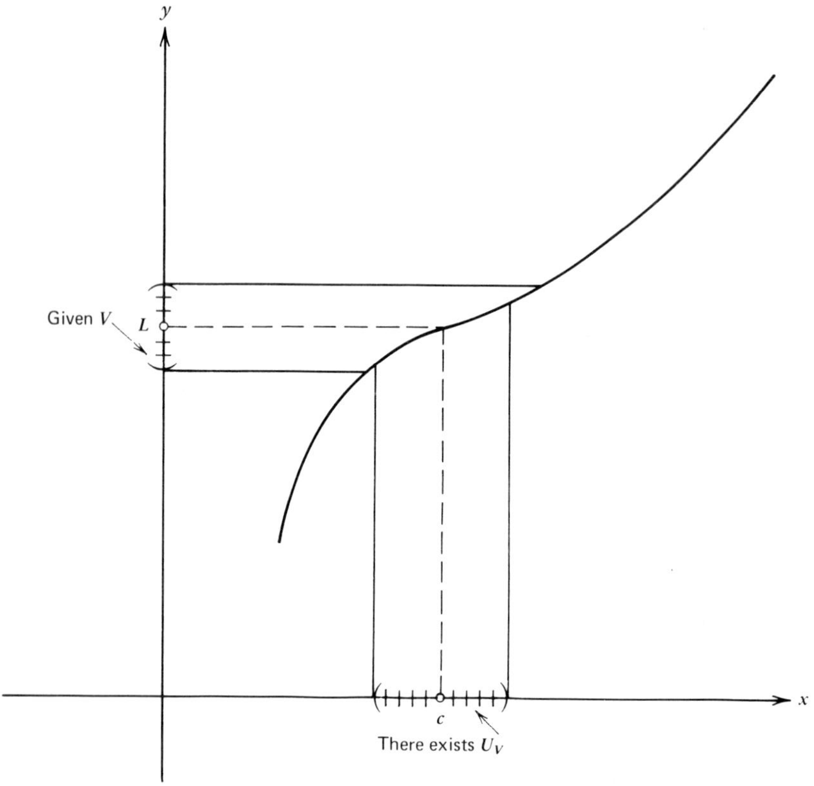

FIGURE 4.1.1 The limit of *f* at c is *L*.

value (if any) $f(c)$ of f at c—we wish to consider only the "trend" indicated by the values of f at points x near to (but different from) the point c.

We now state the precise definition of the limit of a function f at a point c. To assure the existence of points in the domain of f that are close to c, we shall always assume that c is a cluster point of the domain of f. [Recall, from Definition 2.5.2, that a point c is a cluster point of a set A if every neighborhood of c contains points of A different from c. The point c may or may not belong to A.]

4.1.1 Definition. Let $A \subseteq \mathbf{R}$, let $f : A \to \mathbf{R}$, and let $c \in \mathbf{R}$ be a cluster point of A. We say that a real number L is a **limit of** f **at** c if, given any neighborhood V of L, there exists a neighborhood U_V of c such that if $x \neq c$ is any point in $A \cap U_V$, then $f(x)$ belongs to V. (See Figure 4.1.1.)

If L is a limit of f at c, we sometimes say that f **converges to** L **at** c; we often write

$$L = \lim_{x \to c} f = \lim_{x \to c} f \qquad \text{or} \qquad L = \lim_{x \to c} f(x) = \lim_{x \to c} f(x) .$$

We also say that "$f(x)$ approaches L as x tends to c", or that "$f(x)$ approaches L as x approaches c". The symbolism

$$f(x) \to L \qquad \text{as} \qquad x \to c$$

is also used to express the statement that f has limit L at c.

If f does not have a limit at c, we sometimes say that f **diverges at** c.

Remark. It may happen that $c \in A$, but this is *not* required. In fact, if c does belong to A, we specifically ignore the value $f(c)$ in the determination of the limit. [There is another notion, called the **nondeleted limit,** in which the value of $f(c)$ is considered. This notion is discussed in *The Elements of Real Analysis*, Section 25.]

Our first result assures that the value of the limit is uniquely determined, when it exists.

4.1.2 Theorem. *If $f: A \to \mathbf{R}$ and if c is a cluster point of A, then f can have only one limit at c.*

Proof. Suppose, on the contrary, that there exist real numbers $L' \neq L''$ that satisfy Definition 4.1.1. We take neighborhoods V' and V'' of L' and L'', respectively, such that $V' \cap V'' = \emptyset$. By Definition 4.1.1 there exist neighborhoods U' and U'' of c such that: (i) if $x \neq c$ and $x \in A \cap U'$, then $f(x) \in V'$, and (ii) if $y \neq c$ and $y \in A \cap U''$, then $f(y) \in V''$. Now take $U := U' \cap U''$. The hypothesis that c is a cluster point guarantees that there exists a point $z \neq c$ with $z \in A \cap U = (A \cap U') \cap (A \cap U'')$. Consequently, $f(z) \in V'$ and $f(z) \in V''$; therefore $f(z) \in V' \cap V''$. But this contradicts the fact that $V' \cap V'' = \emptyset$. Thus, the supposition that $L' \neq L''$ leads to a contradiction. Q.E.D.

The ε-δ Criterion for Limits

We now give some equivalent formulations for Definition 4.1.1. We first express the neighborhood condition in terms of inequalities. Later we shall establish a sequential formulation for the limit of a function.

4.1.3 Theorem. *Let $A \subseteq \mathbf{R}$, let $f: A \to \mathbf{R}$, and let $c \in \mathbf{R}$ be a cluster point of A. Then:*

(i) $\lim_{x \to c} f = L$

if and only if

(ii) *for any given $\varepsilon > 0$ there exists a $\delta(\varepsilon) > 0$ such that if $0 < |x - c| < \delta(\varepsilon)$ and $x \in A$, then $|f(x) - L| < \varepsilon$.*

Proof. (i) $\Rightarrow$ (ii). Suppose f has limit L at c. Given $\varepsilon > 0$, we apply Definition 4.1.1 to the ε-neighborhood $V_\varepsilon := (L - \varepsilon, L + \varepsilon)$ of L. We thus obtain a neighborhood U_ε ($= U_{V_\varepsilon}$) of c such that if $x \neq c$ belongs to $A \cap U_\varepsilon$, then $f(x) \in V_\varepsilon$, whence it follows that $|f(x) - L| < \varepsilon$. Now since U_ε is a neighborhood of c, there exists a real number $\delta(\varepsilon) > 0$ such that

$$(c - \delta(\varepsilon), c + \delta(\varepsilon)) \subseteq U_\varepsilon .$$

We infer from this that if $0 < |x - c| < \delta(\varepsilon)$, then $x \neq c$ and $x \in U_\varepsilon$. Hence, if $0 < |x - c| < \delta(\varepsilon)$ and $x \in A$, then it follows that $|f(x) - L| < \varepsilon$. This establishes condition (ii).

(ii) $\Rightarrow$ (i). Suppose the condition stated in (ii) holds. If V is a given neighborhood of L, then V contains an ε-neighborhood $V_\varepsilon := (L - \varepsilon, L + \varepsilon)$ of L for some $\varepsilon > 0$. Applying (ii), we obtain a $\delta(\varepsilon) > 0$ such that $0 < |x - c| < \delta(\varepsilon)$ and $x \in A$ imply that $|f(x) - L| < \varepsilon$; that is, that $f(x) \in V_\varepsilon$. If we now let $U_V := (c - \delta(\varepsilon), c + \delta(\varepsilon))$, then $x \in A \cap U_V$ implies that $f(x) \in V_\varepsilon \subseteq V$. Therefore, f has limit L at c. Q.E.D.

Remark. The preceding formulation can be thought of as the $\delta(\varepsilon)$ Game, similar in spirit to the $K(\varepsilon)$ Game for limits of sequences. (See pp. 74–75.) Player A presents an $\varepsilon > 0$ to Player B. If Player B can always respond by producing a $\delta(\varepsilon) > 0$ that satisfies the condition of Theorem 4.1.3, then the limit of f at c is established to be L.

4.1.4 Examples. (a) $\lim\limits_{x \to c} b = b$.

To be more explicit, let $f(x) := b$ for all $x \in \boldsymbol{R}$; we claim that $\lim\limits_{x \to c} f = b$. Indeed, given $\varepsilon > 0$ let $\delta := 1$. Then if $0 < |x - c| < 1$, we have $|f(x) - b| = |b - b| = 0 < \varepsilon$. Since $\varepsilon > 0$ is arbitrary, we deduce from 4.1.3 that we have $\lim\limits_{x \to c} f = b$.

(b) $\lim\limits_{x \to c} x = c$.

Let $g(x) := x$ for all $x \in \boldsymbol{R}$. If $\varepsilon > 0$ let $\delta(\varepsilon) := \varepsilon$. Then if $0 < |x - c| < \delta(\varepsilon)$, we trivially have $|g(x) - c| = |x - c| < \varepsilon$. Since $\varepsilon > 0$ is arbitrary, we deduce that $\lim\limits_{x \to c} g = c$.

(c) $\lim\limits_{x \to c} x^2 = c^2$.

Let $h(x) := x^2$ for all $x \in \boldsymbol{R}$. We want to make the difference

$$|h(x) - c^2| = |x^2 - c^2|$$

less than a pre-assigned $\varepsilon > 0$ by taking x sufficiently close to c. To do so, we note that $x^2 - c^2 = (x + c)(x - c)$. Moreover, if $|x - c| < 1$, then

$$|x| \leq |c| + 1 \qquad \text{so that} \qquad |x + c| \leq |x| + |c| \leq 2|c| + 1 .$$

Therefore, if $|x - c| < 1$, we have

$$(*) \qquad |x^2 - c^2| = |x + c|\,|x - c| \leq (2|c| + 1)\,|x - c| .$$

Moreover this last term will be less than ε provided we take $|x - c| < \varepsilon/(2|c| + 1)$.

Consequently, if we choose

$$\delta(\varepsilon) := \inf\left\{1, \frac{\varepsilon}{2|c|+1}\right\},$$

then if $0 < |x - c| < \delta(\varepsilon)$, it will follow first that $|x - c| < 1$ so that $(*)$ is valid, and therefore, since $|x - c| < \varepsilon/(2|c| + 1)$ that

$$|x^2 - c^2| \leqslant (2|c| + 1)\,|x - c| < \varepsilon\,.$$

Since we have a way of choosing $\delta(\varepsilon) > 0$ for an arbitrary choice of $\varepsilon > 0$, we infer that $\lim\limits_{x\to c} h(x) = \lim\limits_{x\to c} x^2 = c^2$.

(d) $\lim\limits_{x\to c} \dfrac{1}{x} = \dfrac{1}{c}$ if $c > 0$.

Let $\varphi(x) := 1/x$ for $x > 0$ and let $c > 0$. To show that $\lim\limits_{x\to c} \varphi = 1/c$ we wish to make the difference

$$\left|\varphi(x) - \frac{1}{c}\right| = \left|\frac{1}{x} - \frac{1}{c}\right|$$

less than a pre-assigned $\varepsilon > 0$ by taking x sufficiently close to $c > 0$. We first note that

$$\frac{1}{x} - \frac{1}{c} = \frac{1}{cx}(c - x),$$

whence it follows that

$$\left|\varphi(x) - \frac{1}{c}\right| = \frac{1}{cx}\left|x - c\right|$$

for $x > 0$. It is useful to get an upper bound for the term $1/(cx)$ that holds in some neighborhood of c. In particular, if $|x - c| < \frac{1}{2}c$, then $\frac{1}{2}c < x < \frac{3}{2}c$ (why?), so that

$$0 < \frac{1}{cx} < \frac{2}{c^2} \qquad \text{for} \quad |x - c| < \tfrac{1}{2}c\,.$$

Therefore, for these values of x we have

$$(\dagger) \qquad \left|\varphi(x) - \frac{1}{c}\right| \leqslant \frac{2}{c^2}\left|x - c\right|.$$

In order to make this last term less than ε it suffices to take $|x - c| < \frac{1}{2}c^2\varepsilon$. Consequently, if we choose

$$\delta(\varepsilon) := \inf\{\tfrac{1}{2}c, \tfrac{1}{2}c^2\varepsilon\},$$

then if $0 < |x - c| < \delta(\varepsilon)$, it will follow first that $|x - c| < \frac{1}{2}c$ so that (†) is valid, and therefore, since $|x - c| < (\frac{1}{2}c^2)\varepsilon$, that

$$\left|\varphi(x) - \frac{1}{c}\right| = \left|\frac{1}{x} - \frac{1}{c}\right| < \varepsilon.$$

Since we have a way of choosing $\delta(\varepsilon) > 0$ for an arbitrary choice of $\varepsilon > 0$, we infer that $\lim_{x \to c} \varphi = 1/c$.

(e) $\displaystyle\lim_{x \to 2} \frac{x^3 - 4}{x^2 + 1} = \frac{4}{5}$.

Let $\psi(x) := (x^3 - 4)/(x^2 + 1)$ for $x \in \boldsymbol{R}$. Then a little algebraic manipulation gives us

$$\begin{aligned}\left|\psi(x) - \frac{4}{5}\right| &= \frac{|5x^3 - 4x^2 - 24|}{5(x^2 + 1)} \\ &= \frac{|5x^2 + 6x + 12|}{5(x^2 + 1)} \cdot |x - 2|.\end{aligned}$$

To get a bound on the coefficient of $|x - 2|$, we restrict x by the condition $1 < x < 3$. For x in this interval, we have $5x^2 + 6x + 12 \leqslant 5 \cdot 3^2 + 6 \cdot 3 + 12 = 75$ and $5(x^2 + 1) \geqslant 5(1 + 1) = 10$, so that

$$\left|\psi(x) - \frac{4}{5}\right| \leqslant \frac{75}{10}|x - 2| = \frac{15}{2}|x - 2|.$$

Now for given $\varepsilon > 0$, we choose

$$\delta(\varepsilon) := \inf\left\{1, \frac{2}{15}\varepsilon\right\}.$$

Then if $0 < |x - 2| < \delta(\varepsilon)$, we have $|\psi(x) - (4/5)| \leqslant (15/2)\,|x - 2| \leqslant \varepsilon$. Since $\varepsilon > 0$ is arbitrary, the assertion is proved.

Sequential Criterion for Limits

The following important formulation of limit of a function is in terms of limits of sequences. This characterization permits the theory of Chapter 3 to be applied to the study of limits of functions. Thus, as we shall see in the next section, many of the basic limit properties of functions can be established on the basis of corresponding properties for convergent sequences.

4.1.5 Theorem *(Sequential Criterion). Let $A \subseteq \mathbf{R}$, let $f: A \to \mathbf{R}$, and let $c \in \mathbf{R}$ be a cluster point of A. Then:*

(i) $L = \lim_{x \to c} f$

if and only if

(ii) *for every sequence (x_n) in A that converges to c such that $x_n \neq c$ for all $n \in \mathbf{N}$, the sequence $(f(x_n))$ converges to L.*

Proof. (i) $\Rightarrow$ (ii). Assuming that f has limit L at c, we must show that if $\lim (x_n) = c$ where $x_n \in A$ and $x_n \neq c$ for all $n \in \mathbf{N}$, then $\lim (f(x_n)) = L$. If V is an arbitrary neighborhood of L, we invoke the hypothesis (i) to obtain a neighborhood U_V of c such that if $x \in A \cap U_V$ and $x \neq c$, then $f(x) \in V$. Then from the definition of a convergent sequence, we have a natural number $K(U_V)$ such that if $n \geqslant K(U_V)$, then $x_n \in U_V$ and hence $f(x_n) \in V$. Therefore the sequence $(f(x_n))$ converges to L.

(ii) $\Rightarrow$ (i). The proof is by contradiction. If (i) is not true, then there exists a neighborhood V_0 of L such that no matter what neighborhood U of c we pick, there will exist at least one number x_U such that $x_U \neq c$ and $x_U \in A \cap U$, but such that $f(x_U)$ does not belong to V_0. For every $n \in \mathbf{N}$, let U_n be the $(1/n)$-neighborhood of c; that is,

$$U_n := (c - 1/n,\ c + 1/n).$$

For $n \in \mathbf{N}$, let x_n $(= x_{U_n})$ be a number such that $x_n \neq c$ and $x_n \in A \cap U_n$, but $f(x_n) \notin V_0$. We conclude that the sequence (x_n) converges to c (since $x_n \in U_n$ for all $n \in \mathbf{N}$) but the sequence $(f(x_n))$ does not converge to L. Thus the supposition that (i) does not hold implies that (ii) does not hold. Therefore (ii) implies (i).

Q.E.D.

The following useful divergence criteria is a consequence of the preceding theorem (and of its proof). We leave the details of its proof as an exercise.

4.1.6 Divergence Criteria. *Let $A \subseteq \mathbf{R}$, let $f: A \to \mathbf{R}$ and let $c \in \mathbf{R}$ be a cluster point of A.*

(a) *If $L \in \mathbf{R}$, then f does not have limit L at c if and only if there exists a sequence (x_n) in A with $x_n \neq c$ for all $n \in \mathbf{N}$ such that the sequence (x_n) converges to c but the sequence $(f(x_n))$ does* **not** *converge to L.*

(b) *The function f does not have a limit at c if and only if there exists a sequence (x_n) in A with $x_n \neq c$ for all $n \in \mathbf{N}$ such that the sequence (x_n) converges to c but the sequence $(f(x_n))$ does* **not** *converge in $\mathbf{R}$.*

We now give some applications of this result to show how it can be used.

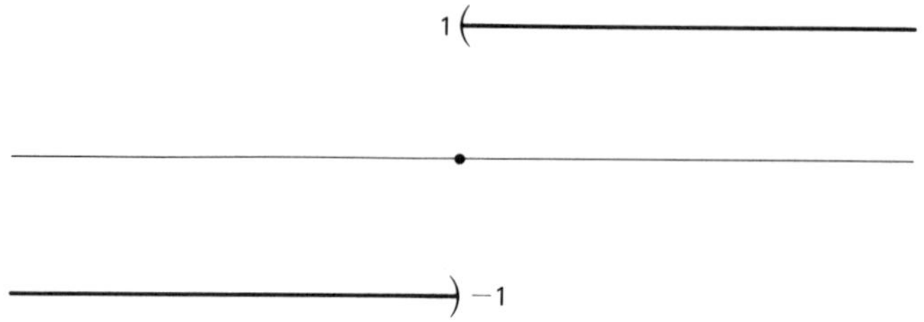

FIGURE 4.1.2 The signum function.

4.1.7 Examples. (a) $\lim_{x \to 0} (1/x)$ does not exist in $\boldsymbol{R}$.

As in Example 4.1.4(d), let $\varphi(x) := 1/x$ for $x > 0$. However, here we consider $c = 0$. The argument given in Example 4.1.4(d) breaks down if $c = 0$ since we cannot obtain a bound such as that in (†) of that example. Indeed, if we take the sequence (x_n) with $x_n := 1/n$ for $n \in \boldsymbol{N}$, then $\lim (x_n) = 0$, but $\varphi(x_n) = 1/(1/n) = n$. As we know, the sequence $(\varphi(x_n)) = (n)$ is not convergent in $\boldsymbol{R}$, since it is not bounded. Hence, by Theorem 4.1.6(b), $\lim_{x \to 0} (1/x)$ does not exist in $\boldsymbol{R}$. [However, see Example 4.3.9(a).]

(b) $\lim_{x \to 0} \operatorname{sgn}(x)$ does not exist.

Let the **signum function** sgn be defined by

$$\begin{aligned} \operatorname{sgn}(x) &:= +1 \quad &&\text{for} \quad x > 0, \\ &:= 0 \quad &&\text{for} \quad x = 0, \\ &:= -1 \quad &&\text{for} \quad x < 0. \end{aligned}$$

Note that $\operatorname{sgn}(x) = x/|x|$ for $x \neq 0$. (See Figure 4.1.2.) We shall show that sgn does not have a limit at $x = 0$. We shall do this by showing that there is a sequence (x_n) such that $\lim (x_n) = 0$ but such that $(\operatorname{sgn}(x_n))$ does not converge.

Indeed, let $x_n := (-1)^n/n$ for $n \in \boldsymbol{N}$ so that $\lim (x_n) = 0$. However, since

$$\operatorname{sgn}(x_n) = (-1)^n \qquad \text{for} \quad n \in \boldsymbol{N},$$

it follows from Example 3.4.5(a) that $(\operatorname{sgn}(x_n))$ does not converge. Therefore $\lim_{x \to 0} \operatorname{sgn}(x)$ does not exist.

†(c) $\lim_{x \to 0} \sin(1/x)$ does not exist in $\boldsymbol{R}$.

Let $g(x) := \sin(1/x)$ for $x \neq 0$. (See Figure 4.1.3.) We shall show that g does not have a limit at $c = 0$, by exhibiting two sequences (x_n) and (y_n) with $x_n \neq 0$ and $y_n \neq 0$ for all $n \in \boldsymbol{N}$ and such that $\lim (x_n) = 0$ and $\lim (y_n) = 0$, but such that $\lim (g(x_n)) \neq \lim (g(y_n))$. In view of Theorem 4.1.5 this implies that $\lim_{x \to 0} g$ cannot exist. (Explain why.)

† In order to have some interesting applications in this and later examples, we shall make use of well-known properties of trigonometric and exponential functions that will be established in Chapter 7.

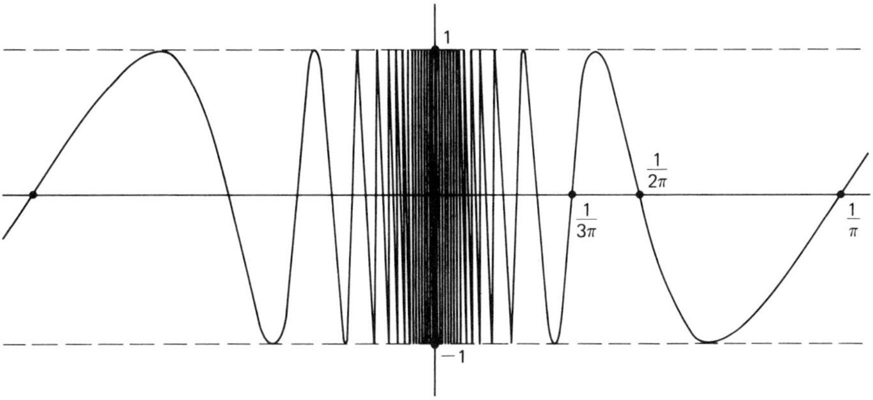

FIGURE 4.1.3 The function $g(x) = \sin\frac{1}{x}$ $(x \neq 0)$.

Indeed, we recall from calculus that $\sin t = 0$ if $t = n\pi$ for $n \in \mathbf{Z}$, and that $\sin t = +1$ if $t = \frac{1}{2}\pi + 2\pi n$ for $n \in \mathbf{Z}$. Now let $x_n := 1/n\pi$ for $n \in \mathbf{N}$; then $\lim(x_n) = 0$ and $g(x_n) = \sin n\pi = 0$ for all $n \in \mathbf{N}$, so that $\lim(g(x_n)) = 0$. On the other hand, let $y_n := (\frac{1}{2}\pi + 2\pi n)^{-1}$ for $n \in \mathbf{N}$; then $\lim(y_n) = 0$ and $g(y_n) = \sin(\frac{1}{2}\pi + 2\pi n) = 1$ for all $n \in \mathbf{N}$, so that $\lim(g(y_n)) = 1$. We conclude that $\lim_{x\to 0} \sin(1/x)$ does not exist.

Exercises for Section 4.1

1. Determine a condition on $|x - 1|$ that will assure that:
   (a) $|x^2 - 1| < 1/2$;
   (b) $|x^2 - 1| < 1/10^3$;
   (c) $|x^2 - 1| < 1/n$, for a given $n \in \mathbf{N}$;
   (d) $|x^3 - 1| < 1/n$, for a given $n \in \mathbf{N}$.
2. Let c be a cluster point of $A \subseteq \mathbf{R}$ and let $f : A \to \mathbf{R}$. Prove that $\lim_{x\to c} f(x) = L$ if and only if $\lim_{x\to c} |f(x) - L| = 0$.
3. Let $f : \mathbf{R} \to \mathbf{R}$ and let $c \in \mathbf{R}$. Show that $\lim_{x\to c} f(x) = L$ if and only if $\lim_{x\to 0} f(x+c) = L$.
4. Let $f : \mathbf{R} \to \mathbf{R}$; let $I \subseteq \mathbf{R}$ be an open interval, and let $c \in I$. If f_1 is the restriction of f to I, show that f_1 has a limit at c if and only if f has a limit at c, and that $\lim_{x\to c} f = \lim_{x\to c} f_1$.
5. Let $f : \mathbf{R} \to \mathbf{R}$, let $J \subseteq \mathbf{R}$ be a closed interval, and let $c \in J$. If f_2 is the restriction of f to J, show that if f has a limit at c then f_2 has a limit at c. Show that it does *not* follow that if f_2 has a limit at c, then f has a limit at c.
6. Let $I := (0,a)$, $a > 0$, and let $g(x) := x^2$ for $x \in I$. For any x, c in I, show that $|g(x) - c^2| \leq 2a\,|x - c|$. Use this inequality to prove that $\lim_{x\to c} x^2 = c^2$ for any $c \in I$.
7. Let $I \subseteq \mathbf{R}$ be an interval, let $f : I \to \mathbf{R}$, and let $c \in I$. Suppose there exist numbers K and L such that $|f(x) - L| \leq K\,|x - c|$ for $x \in I$. Show that $\lim_{x\to c} f = L$.

8. Show that $\lim_{x\to c} x^3 = c^3$ for any $c \in \mathbf{R}$.
9. Show that $\lim_{x\to c} \sqrt{x} = \sqrt{c}$ for any $c \geq 0$.
10. Use *both* the ε-δ and the sequential formulations of the notion of a limit to establish the following:

    (a) $\lim_{x\to 2} \dfrac{1}{1-x} = -1 \quad (x > 1)$; (b) $\lim_{x\to 1} \dfrac{x}{1+x} = \dfrac{1}{2} \quad (x > 0)$;

    (c) $\lim_{x\to 0} \dfrac{x^2}{|x|} = 0 \quad (x \neq 0)$; (d) $\lim_{x\to 1} \dfrac{x^2 - x + 1}{x+1} = \dfrac{1}{2} \quad (x > 0)$.

11. Show that the following limits do *not* exist in $\mathbf{R}$:

    (a) $\lim_{x\to 0} \dfrac{1}{x^2} \quad (x > 0)$; (b) $\lim_{x\to 0} \dfrac{1}{\sqrt{x}} \quad (x > 0)$;

    (c) $\lim_{x\to 0} (x + \operatorname{sgn}(x))$; (d) $\lim_{x\to 0} \sin(1/x^2) \quad (x \neq 0)$.

12. Suppose the function $f: \mathbf{R} \to \mathbf{R}$ has limit L at 0, and let $a > 0$. If $g: \mathbf{R} \to \mathbf{R}$ is defined by $g(x) := f(ax)$ for $x \in \mathbf{R}$, show that $\lim_{x\to 0} g = L$.
13. Let c be a cluster point of $A \subseteq \mathbf{R}$ and let $f: A \to \mathbf{R}$ be such that $\lim_{x\to c} (f(x))^2 = L$. Show that if $L = 0$, then $\lim_{x\to c} f(x) = 0$. Show by example that if $L \neq 0$, then f may not have a limit at c.
14. Let $f: \mathbf{R} \to \mathbf{R}$ be defined by setting $f(x) := x$ if x is rational, and $f(x) := 0$ if x is irrational. Show that f has a limit at $x = 0$. Use a sequential argument to show that if $c \neq 0$, then f does not have a limit at c.

SECTION 4.2 Limit Theorems

We shall now obtain results that are useful in calculating limits of functions. These results are parallel to the limit theorems established in Section 3.2 for sequences. In fact, in most cases these results can be proved by using Theorem 4.1.5 and results from Section 3.2. Alternatively, the results in this section can be proved by using arguments that are very similar to the ones employed in Section 3.2.

4.2.1 Definition. Let $A \subseteq \mathbf{R}$, let $f: A \to \mathbf{R}$, and let $c \in \mathbf{R}$ be a cluster point of A. We say that f is **bounded on a neighborhood of** c in case there exists a neighborhood U of c and a constant $M > 0$ such that we have $|f(x)| \leq M$ for all $x \in A \cap U$.

4.2.2 Theorem. *If $f: A \to \mathbf{R}$ has a limit at $c \in \mathbf{R}$, then f is bounded on some neighborhood of c.*

Proof. If $L = \lim_{x\to c} f$, then by Theorem 4.1.3, there exists $\delta > 0$ such that if $0 < |x - c| < \delta$ and $x \in A$, then $|f(x) - L| < 1$ so that

$$-1 < f(x) - L < 1.$$

Therefore, if $U := (c - \delta, c + \delta)$, and if $x \in A \cap U$, $x \neq c$, then

$$|f(x)| \leq |L| + 1.$$

If $c \notin A$ we take $M := |L| + 1$, while if $c \in A$ we set $M := \sup\{|f(c)|, |L| + 1\}$. It follows that if $x \in A \cap U$, then $|f(x)| \leq M$. This shows that f is bounded on a neighborhood of c. Q.E.D.

The next definition is similar to the Definition 3.1.3 for sums, differences, products, and quotients of sequences. The theorem following is exactly parallel to Theorem 3.2.3.

4.2.3 Definition. Let $A \subseteq \mathbf{R}$ and let f and g be functions defined on A to $\mathbf{R}$. We define the **sum** $f + g$, the **difference** $f - g$, and the **product** fg on A to $\mathbf{R}$ to be the functions given by

$$(f + g)(x) := f(x) + g(x), \quad (f - g)(x) := f(x) - g(x), \quad (fg)(x) := f(x)g(x),$$

for all $x \in A$. Further, if $b \in \mathbf{R}$, we define the **multiple** bf to be the function given by

$$(bf)(x) := bf(x) \qquad \text{for} \quad x \in A.$$

Finally, if $h(x) \neq 0$ for $x \in A$, we define the **quotient** f/h to be the function given by

$$\left(\frac{f}{h}\right)(x) := \frac{f(x)}{h(x)}$$

for all $x \in A$.

4.2.4 Theorem. *Let $A \subseteq \mathbf{R}$, let f and g be functions on A to $\mathbf{R}$, and let $c \in \mathbf{R}$ be a cluster point of A. Further, let $b \in \mathbf{R}$.*

(a) *If $L = \lim\limits_{x \to c} f$ and $M = \lim\limits_{x \to c} g$, then:*

$$L + M = \lim_{x \to c} (f + g), \qquad L - M = \lim_{x \to c} (f - g),$$

$$LM = \lim_{x \to c} (fg), \qquad bL = \lim_{x \to c} (bf).$$

(b) *If $h : A \to \mathbf{R}$, if $h(x) \neq 0$ for all $x \in A$, and if $H = \lim\limits_{x \to c} h \neq 0$, then*

$$\frac{L}{H} = \lim_{x \to c} \left(\frac{f}{h}\right).$$

Proof. One proof of this theorem is exactly similar to that of Theorem 3.2.3. Alternatively, it can be proved by making use of Theorem 3.2.3 and Theorem 4.1.5. For example, let (x_n) be any sequence in A such that $x_n \neq c$ for $n \in \boldsymbol{N}$, and $c = \lim (x_n)$. It follows from Theorem 4.1.5 that

$$L = \lim \big(f(x_n)\big), \quad M = \lim \big(g(x_n)\big).$$

On the other hand, Definition 4.2.3 implies that

$$(fg)(x_n) = f(x_n)g(x_n) \qquad \text{for} \quad n \in \boldsymbol{N}.$$

Therefore an application of Theorem 3.2.3 yields

$$\begin{aligned} \lim \big((fg)\,(x_n)\big) &= \lim \big(f(x_n)g(x_n)\big) \\ &= \big(\lim\,(f(x_n))\big)\big(\lim\,(g(x_n))\big) \\ &= LM. \end{aligned}$$

Consequently, it follows from Theorem 4.1.5 that

$$\lim_{x \to c} (fg) = \lim \big((fg)(x_n)\big) = LM.$$

The other parts of this theorem are proved in a similar manner. We leave the details to the reader. Q.E.D.

Remarks. (1) We note that, in part (b), the additional assumption that $H = \lim\limits_{x \to c} h \neq 0$ is made. If this assumption is not satisfied, then the limit

$$\lim_{x \to c} \frac{f(x)}{h(x)}$$

may not exist. But even if this limit does exist, we *cannot* use Theorem 4.2.4(b) to evaluate it.

(2) Let $A \subseteq \boldsymbol{R}$, and let $f_1, f_2, \ldots, f_n$ be functions on A to $\boldsymbol{R}$, and let c be a cluster point of A. If

$$L_k := \lim_{x \to c} f_k \qquad \text{for} \quad k = 1, \ldots, n,$$

then it follows from Theorem 4.2.4 by an induction argument that

$$L_1 + L_2 + \cdots + L_n = \lim_{x \to c} (f_1 + f_2 + \cdots + f_n),$$

and

$$L_1 \cdot L_2 \cdots L_n = \lim_{x \to c} (f_1 \cdot f_2 \cdots f_n).$$

(3) In particular, we deduce from (2) that if $L = \lim_{x \to c} f$ and $n \in \mathbf{N}$, then

$$L^n = \lim_{x \to c} \left(f(x)\right)^n.$$

4.2.5 Examples. (a) Some of the limits that were established in Section 4.1 can be proved by using Theorem 4.2.4. For example, it follows from this result that since $\lim_{x \to c} x = c$, then

$$\lim_{x \to c} x^2 = c^2,$$

and that if $c > 0$, then

$$\lim_{x \to c} \frac{1}{x} = \frac{1}{\lim_{x \to c} x} = \frac{1}{c}.$$

(Explain why.)

(b) $\lim_{x \to 2} (x^2 + 1)(x^3 - 4) = 20.$

It follows from Theorem 4.2.4 that

$$\begin{aligned} \lim_{x \to 2} (x^2 + 1)(x^3 - 4) &= \left(\lim_{x \to 2} (x^2 + 1)\right)\left(\lim_{x \to 2} (x^3 - 4)\right) \\ &= 5 \cdot 4 = 20. \end{aligned}$$

(c) $\lim_{x \to 2} \left(\dfrac{x^3 - 4}{x^2 + 1}\right) = \dfrac{4}{5}.$

If we apply Theorem 4.2.4(b), we have

$$\lim_{x \to 2} \frac{x^3 - 4}{x^2 + 1} = \frac{\lim_{x \to 2} (x^3 - 4)}{\lim_{x \to 2} (x^2 + 1)} = \frac{4}{5}.$$

Note that since the limit in the denominator [i.e., $\lim_{x \to 2} (x^2 + 1) = 5$] is not equal to 0, then Theorem 4.2.4(b) is applicable.

(d) $\lim_{x \to 2} \dfrac{x^2 - 4}{3x - 6} = \dfrac{4}{3}.$

If we let $f(x) := x^2 - 4$ and $h(x) := 3x - 6$ for $x \in R$, then we *cannot* use Theorem 4.2.4(b) to evaluate $\lim_{x \to 2} \left(f(x)/h(x)\right)$ because

$$\begin{aligned} H = \lim_{x \to 2} h(x) &= \lim_{x \to 2} (3x - 6) \\ &= 3 \lim_{x \to 2} x - 6 = 3 \cdot 2 - 6 = 0. \end{aligned}$$

However, if $x \neq 2$, then it follows that

$$\frac{x^2 - 4}{3x - 6} = \frac{(x + 2)(x - 2)}{3(x - 2)} = \frac{1}{3}(x + 2).$$

Therefore we have

$$\lim_{x\to 2} \frac{x^2 - x}{3x - 6} = \lim_{x\to 2} \frac{1}{3}(x + 2) = \frac{1}{3}(\lim_{x\to 2} x + 2)$$

$$= \frac{1}{3}(2 + 2) = \frac{4}{3}.$$

Note that the function $g(x) = (x^2 - 4)/(3x - 6)$ has a limit at $x = 2$ *even though it is not defined there.*

(e) $\lim_{x\to 0} \frac{1}{x}$ does not exist in $\boldsymbol{R}$.

Of course $\lim_{x\to 0} 1 = 1$ and $H := \lim_{x\to 0} x = 0$. However, since $H = 0$, we *cannot* use Theorem 4.2.4(b) to evaluate $\lim_{x\to 0} (1/x)$. In fact, as was seen in Example 4.1.7(a), the function $\varphi(x) := 1/x$ does not have a limit at $x = 0$. This conclusion also follows from Theorem 4.2.2 since the function $\varphi(x) = 1/x$ is not bounded on a neighborhood of $x = 0$. (Why?)

(f) If p is a polynomial function, then $\lim_{x\to c} p(x) = p(c)$.

Let p be a polynomial function on $\boldsymbol{R}$ so that $p(x) = a_n x^n + a_{n-1}x^{n-1} + \ldots + a_1 x + a_0$ for all $x \in \boldsymbol{R}$. It follows from Theorem 4.2.4 and the fact that $\lim_{x\to c} x^k = c^k$, that

$$\begin{aligned}\lim_{x\to c} p(x) &= \lim_{x\to c} [a_n x^n + a_{n-1}x^{n-1} + \cdots + a_1 x + a_0]\\ &= \lim_{x\to c} (a_n x^n) + \lim_{x\to c} (a_{n-1}x^{n-1}) + \cdots + \lim_{x\to c} (a_1 x) + \lim_{x\to c} a_0\\ &= a_n c^n + a_{n-1}c^{n-1} + \cdots + a_1 c + a_0\\ &= p(c).\end{aligned}$$

Hence $\lim_{x\to c} p(x) = p(c)$ for any polynomial function p.

(g) If p and q are polynomial functions on $\boldsymbol{R}$ and if $q(c) \neq 0$, then

$$\lim_{x\to c} \frac{p(x)}{q(x)} = \frac{p(c)}{q(c)}.$$

Since $q(x)$ is a polynomial function, it follows from a theorem in algebra that there are at most a finite number of real numbers $\alpha_1, \ldots, \alpha_m$ [the real zeroes of $q(x)$] such that $q(\alpha_j) = 0$ and such that if $x \notin \{\alpha_1, \ldots, \alpha_m\}$, then $q(x) \neq 0$. Hence, if $x \notin \{\alpha_1, \ldots, \alpha_m\}$, we can define

$$r(x) := \frac{p(x)}{q(x)}.$$

If c is not a zero of $q(x)$, then $q(c) \neq 0$, and it follows from part (f) that $\lim_{x \to c} q(x) = q(c) \neq 0$. Therefore we can apply Theorem 4.2.4(b) to conclude that

$$\lim_{x \to c} \frac{p(x)}{q(x)} = \frac{\lim_{x \to c} p(x)}{\lim_{x \to c} q(x)} = \frac{p(c)}{q(c)}.$$

The next result is a direct analog of Theorem 3.2.6, from which it follows.

4.2.6 Theorem. *Let $A \subseteq \mathbf{R}$, let $f : A \to \mathbf{R}$ and let $c \in \mathbf{R}$ be a cluster point of A. If*

$$a \leqslant f(x) \leqslant b \qquad \textit{for all } x \in A,\ x \neq c,$$

and if $\lim_{x \to c} f$ exists, then $a \leqslant \lim_{x \to c} f \leqslant b$.

Proof. Indeed, if $L = \lim_{x \to c} f$, then it follows from Theorem 4.1.5 that if (x_n) is any sequence of real numbers such that $c \neq x_n \in A$ for all $n \in \mathbf{N}$ and if the sequence (x_n) converges to c, then the sequence $(f(x_n))$ converges to L. Since $a \leqslant f(x_n) \leqslant b$ for all $n \in \mathbf{N}$, it follows from Theorem 3.2.6 that $a \leqslant L \leqslant b$. Q.E.D.

We now state an analog of the Squeeze Theorem 3.2.7. We leave its proof to the reader.

4.2.7 Squeeze Theorem. *Let $A \subseteq \mathbf{R}$, let $f, g, h : A \to \mathbf{R}$, and let $c \in \mathbf{R}$ be a cluster point of A. If*

$$f(x) \leqslant g(x) \leqslant h(x) \qquad \textit{for all } x \in A,\ x \neq c,$$

and if $\lim_{x \to c} f = L = \lim_{x \to c} h$, then we have $\lim_{x \to c} g = L$.

4.2.8 Examples. (a) $\lim_{x \to 0} x^{3/2} = 0\ (x > 0)$.

Let $f(x) := x^{3/2}$ for $x > 0$. Since the inequality $x < x^{1/2} \leqslant 1$ holds for $0 < x \leqslant 1$, it follows that $x^2 \leqslant f(x) = x^{3/2} \leqslant x$ for $0 < x \leqslant 1$. Since

$$\lim_{x \to 0} x^2 = 0 \qquad \text{and} \qquad \lim_{x \to 0} x = 0,$$

it follows from the Squeeze Theorem 4.2.7 that $\lim_{x \to 0} x^{3/2} = 0$.

(b) $\lim_{x \to 0} \sin x = 0.$

It will be proved later (see Theorem 7.4.8), that

$$-x \leq \sin x \leq x \qquad \text{for all} \quad x \geq 0.$$

Since $\lim_{x \to 0} (\pm x) = 0$, it follows from the Squeeze Theorem that $\lim_{x \to 0} \sin x = 0$.

(c) $\lim_{x \to 0} \cos x = 1.$

It will be proved later (see Theorem 7.4.8) that

$$(*) \qquad 1 - \tfrac{1}{2}x^2 \leq \cos x \leq 1 \qquad \text{for all} \quad x \in \boldsymbol{R}.$$

Since $\lim_{x \to 0} (1 - \frac{1}{2}x^2) = 1$, it follows from the Squeeze Theorem that $\lim_{x \to 0} \cos x = 1$.

(d) $\lim_{x \to 0} \left(\dfrac{\cos x - 1}{x} \right) = 0.$

We cannot use Theorem 4.2.4(b) (directly, at least) to evaluate this limit. (Why?) However, it follows from the inequality $(*)$ in part (c) that

$$-\tfrac{1}{2}x \leq (\cos x - 1)/x \leq 0 \qquad \text{for} \quad x > 0$$

and that

$$0 \leq (\cos x - 1)/x \leq \tfrac{1}{2}x \qquad \text{for} \quad x < 0.$$

Now let $f(x) := -x/2$ for $x \geq 0$ and $f(x) := 0$ for $x < 0$, and let $h(x) := 0$ for $x \geq 0$ and $h(x) := -x/2$ for $x < 0$. Then we have

$$f(x) \leq (\cos x - 1)/x \leq h(x) \qquad \text{for} \quad x \neq 0.$$

Since it is readily seen (how?) that $\lim_{x \to 0} f = 0 = \lim_{x \to 0} h$, it follows from the Squeeze Theorem that $\lim_{x \to 0} (\cos x - 1)/x = 0$.

(e) $\lim_{x \to 0} \left(\dfrac{\sin x}{x} \right) = 1.$

Again we cannot use Theorem 4.2.4(b) to evaluate this limit. However, it will be proved later (see Theorem 7.4.8) that

$$x - \tfrac{1}{6}x^3 \leq \sin x \leq x \qquad \text{for} \quad x \geq 0$$

and

$$x \leq \sin x \leq x - \tfrac{1}{6}x^3 \qquad \text{for} \quad x \leq 0.$$

Therefore it follows (why?) that

$$1 - \tfrac{1}{6}x^2 \leq (\sin x)/x \leq 1 \qquad \text{for all} \quad x \neq 0.$$

But since $\lim_{x\to 0}(1 - \frac{1}{6}x^2) = 1 - \frac{1}{6}\lim_{x\to 0} x^2 = 1$, we infer from the Squeeze Theorem that $\lim_{x\to 0}(\sin x)/x = 1$.

(f) $\lim_{x\to 0}(x \sin(1/x)) = 0$.

Let $f(x) := x\sin(1/x)$ for $x \neq 0$. Since $-1 \leq \sin z \leq 1$ for all $z \in \mathbf{R}$, we have the inequality

$$-|x| \leq f(x) = x \sin(1/x) \leq |x|$$

for all $x \in \mathbf{R}$, $x \neq 0$. Since $\lim_{x\to 0}|x| = 0$, it follows from the Squeeze Theorem that $\lim_{x\to 0} f = 0$.

The next result is parallel to Theorem 3.2.9.

4.2.9 Theorem. *Let $A \subseteq \mathbf{R}$, let $f : A \to \mathbf{R}$ and let $c \in \mathbf{R}$ be a cluster point of A. If $\lim_{x\to c} f$ exists, and if $|f|$ denotes the function defined for $x \in A$ by $|f|(x) := |f(x)|$, then we have $\lim_{x\to c}|f| = |\lim_{x\to c} f|$.*

Proof. Let $L = \lim_{x\to c} f$. It follows from the Triangle Inequality that $||f(x)| - |L|| \leq |f(x) - L|$ for all $x \in A$. The assertion of the theorem follows directly from this inequality. (Why?) Q.E.D.

The next result is parallel to Theorem 3.2.10; we leave its proof to the reader.

4.2.10 Theorem. *Let $A \subseteq \mathbf{R}$, let $f : A \to \mathbf{R}$, and let $c \in \mathbf{R}$ be a cluster point of A. In addition suppose that $f(x) \geq 0$ for all $x \in A$, and let $\sqrt{f}$ be the function defined for $x \in A$ by $(\sqrt{f})(x) := \sqrt{f(x)}$. If $\lim_{x\to c} f$ exists, then we have $\lim_{x\to c}\sqrt{f} = \sqrt{\lim_{x\to c} f}$.*

The final result is, in some sense, a partial converse to Theorem 4.2.6.

4.2.11 Theorem. *Let $A \subseteq \mathbf{R}$, let $f : A \to \mathbf{R}$ and let $c \in \mathbf{R}$ be a cluster point of A. If*

$$\lim_{x\to c} f > 0 \qquad [\textit{respectively, } \lim_{x\to c} f < 0],$$

then there exists a neighborhood U of c such that $f(x) > 0$ [respectively, $f(x) < 0$] for all $x \in A \cap U$, $x \neq c$.

Proof. Let $L := \lim_{x\to c} f$ and suppose that $L > 0$. Then $V := (\frac{1}{2}L, 2L)$ is a neighborhood of L; hence by Definition 4.1.1 there exists a neighborhood U of c such that if $x \in A \cap U$ and $x \neq c$, then $f(x) \in V$, in which case $f(x) > 0$.

If $L < 0$, a similar argument applies. Q.E.D.

Exercises for Section 4.2

1. Apply Theorem 4.2.4 to determine the following limits:

   (a) $\lim_{x \to 1} (x + 1)(2x + 3) \quad (x \in \mathbf{R})$; (b) $\lim_{x \to 1} \dfrac{x^2 + 2}{x^2 - 2} \quad (x > 0)$;

   (c) $\lim_{x \to 2} \left(\dfrac{1}{x + 1} - \dfrac{1}{2x}\right) \quad (x > 0)$; (d) $\lim_{x \to 0} \dfrac{|x + 1|}{x^2 + 2} \quad (x \in \mathbf{R})$.

2. Determine the following limits and state which theorems are used in each case.

   (a) $\lim_{x \to 2} \sqrt{\dfrac{2x + 1}{x + 3}} \quad (x > 0)$; (b) $\lim_{x \to 2} \dfrac{x^2 - 4}{x - 2} \quad (x > 0)$;

   (c) $\lim_{x \to 0} \dfrac{(x + 1)^2 - 1}{x} \quad (x > 0)$; (d) $\lim_{x \to 1} \dfrac{\sqrt{x} - 1}{x - 1} \quad (x > 0)$.

3. Find $\lim_{x \to 0} \dfrac{\sqrt{1 + 2x} - \sqrt{1 + 3x}}{x + 2x^2}$ where $x > 0$.
4. Prove that $\lim_{x \to 0} \cos(1/x)$ does not exist but that $\lim_{x \to 0} x \cos(1/x) = 0$.
5. Let f, g be defined on $A \subseteq \mathbf{R}$ to $\mathbf{R}$, and let c be a cluster point of A. Suppose that f is bounded on a neighborhood of c and that $\lim_{x \to c} g = 0$. Prove that $\lim_{x \to c} fg = 0$.
6. Complete the proof of Theorem 4.2.4 by using the sequential criterion for limits.
7. Use the ε-δ formulation of limit to prove Theorem 4.2.4.
8. Let $n \in \mathbf{N}$ be such that $n \geq 3$. Derive the inequality $-x^2 \leq x^n \leq x^2$ for $-1 < x < 1$. Then use the fact that $\lim_{x \to 0} x^2 = 0$ to show that $\lim_{x \to 0} x^n = 0$.
9. Let f, g be defined on A to $\mathbf{R}$ and let c be a cluster point of A.
   (a) Show that if both $\lim_{x \to c} f$ and $\lim_{x \to c} (f + g)$ exist, then $\lim_{x \to c} g$ exists.
   (b) If $\lim_{x \to c} f$ and $\lim_{x \to c} fg$ exist, does it follow that $\lim_{x \to c} g$ exists?
10. Give examples of functions f and g such that f and g do not have limits at a point c, but such that both $f + g$ and fg have limits at c.
11. Determine whether the following limits exist in $\mathbf{R}$.

    (a) $\lim_{x \to 0} \sin(1/x^2) \quad (x \neq 0)$; (b) $\lim_{x \to 0} x \sin(1/x^2) \quad (x \neq 0)$;

    (c) $\lim_{x \to 0} \operatorname{sgn} \sin(1/x) \quad (x \neq 0)$; (d) $\lim_{x \to 0} \sqrt{x} \sin(1/x^2) \quad (x > 0)$.

12. Let $f : \mathbf{R} \to \mathbf{R}$ be such that $f(x + y) = f(x) + f(y)$ for all x, y in $\mathbf{R}$. Assume that $\lim_{x \to 0} f = L$ exists. Prove that $L = 0$, and then prove that f has a limit at every point $c \in R$. [*Hint*: First note that $f(2x) = f(x) + f(x) = 2f(x)$ for $x \in \mathbf{R}$. Also note that $f(x) = f(x - c) + f(c)$ for x, c in $\mathbf{R}$.]

SECTION 4.3 Some Extensions of the Limit Concept†

In this section, we shall present three types of extensions of the notion of a limit of a function that often occur.

One-sided Limits

There are times when a function f may not possess a limit at a point c, yet a limit does exist when the function is restricted to an interval on one side of the cluster point c.

For example, the signum function considered in Example 4.1.7(b), and illustrated in Figure 4.1.2, has no limit at $c = 0$. However, if we restrict the signum function to the interval $(0,\infty)$, the resulting function has a limit of 1 at $c = 0$. Similarly, if we restrict the signum function to the interval $(-\infty,0)$, the resulting function has a limit of -1 at $c = 0$. These are elementary examples of right-hand and left-hand limits at $c = 0$.

The definitions of the right-hand and the left-hand limits are direct modifications of Definition 4.1.1. In fact, replacing the set A in Definition 4.1.1 by the set $A \cap (c,\infty)$ leads to the definition of the right-hand limit at a point c that is a cluster point of $A \cap (c,\infty)$. Similarly, replacing A by $A \cap (-\infty,c)$ leads to the definition of the left-hand limit at a point c that is a cluster point of $A \cap (-\infty,c)$. However, rather than formulating these definitions in terms of neighborhoods, we shall state the ε-δ forms analogous to Theorem 4.1.3.

4.3.1 Definition. Let $A \subseteq \mathbf{R}$ and let $f : A \to \mathbf{R}$.

(i) If $c \in \mathbf{R}$ is a cluster point of the set $A \cap (c,\infty) = \{x \in A : x > c\}$, then we say that $L \in \mathbf{R}$ is a **right-hand limit of f at c** and we write

$$\lim_{x \to c+} f = L$$

if given any $\varepsilon > 0$ there exists a $\delta(\varepsilon) > 0$ such that if $0 < x - c < \delta(\varepsilon)$ and $x \in A$, then $|f(x) - L| < \varepsilon$.

(ii) If $c \in \mathbf{R}$ is a cluster point of the set $A \cap (-\infty,c) = \{x \in A : x < c\}$, then we say that $L \in \mathbf{R}$ is a **left-hand limit of f at c** and we write

$$\lim_{x \to c-} f = L$$

if given any $\varepsilon > 0$ there exists a $\delta(\varepsilon) > 0$ such that if $0 < c - x < \delta(\varepsilon)$ and $x \in A$, then $|f(x) - L| < \varepsilon$.

† This section can be largely omitted on a first reading of this chapter. Indeed only the notion of one-sided limits is used, and not until Section 4.8.

Note. (1) If L is a right-hand limit of f at c, we sometimes say that L is a **limit of f from the right at c.** We sometimes write

$$\lim_{x \to c+} f(x) = L.$$

Similar terminology and notation are used for left-hand limits.

(2) The limits

$$\lim_{x \to c+} f \quad \text{and} \quad \lim_{x \to c-} f$$

are called **one-sided limits of f at c.** It is possible that neither one-sided limit may exist. Also, one of them may exist without the other existing. Similarly, as is the case for $f(x) := \operatorname{sgn}(x)$ at $c = 0$, they may both exist and be different.

(3) If A is an interval with left end point c, then it is readily seen that $f : A \to \mathbf{R}$ has a limit at c if and only if it has a right-hand limit at c. Moreover, in this case the limit $\lim_{x \to c} f$ and the right-hand limit $\lim_{x \to c+} f$ are equal.

A similar situation occurs for the left-hand limit when A is an interval with right end point c.

We leave it to the reader to show that f can have only one right-hand (respectively, left-hand) limit at a point. There are results analogous to those established in Sections 4.1 and 4.2 for two-sided limits. In particular, the existence of one-sided limits can be reduced to sequential considerations.

4.3.2 Theorem. *Let $A \subseteq \mathbf{R}$, let $f : A \to \mathbf{R}$, and let $c \in \mathbf{R}$ be a cluster point of $A \cap (c,\infty)$. Then:*

(i) $\lim_{x \to c+} f = L \in \mathbf{R}$

if and only if

(ii) *for every sequence (x_n) that converges to c such that $x_n \in A$ and $x_n > c$ for all $n \in \mathbf{N}$, it follows that the sequence $(f(x_n))$ converges to $L \in \mathbf{R}$.*

We leave the proof of this result (and the formulation and proof of the analogous result for left-hand limits) to the reader. We will not take the space to write out the formulations of the one-sided version of the other results in Sections 4.1 and 4.2.

The following result relates the notion of the limit of a function to one-sided limits.

4.3.3 Theorem. *Let $A \subseteq \mathbf{R}$, let $f : A \to \mathbf{R}$, and let $c \in \mathbf{R}$ be a cluster point of both of the sets*

$$A \cap (c,\infty) \quad \textit{and} \quad A \cap (-\infty,c).$$

Then $\lim_{x \to c} f = L \in \mathbf{R}$ if and only if $\lim_{x \to c+} f = L = \lim_{x \to c-} f$.

Proof. ($\Rightarrow$) If $\lim_{x \to c} f = L$, it is an easy exercise to show that $\lim_{x \to c+} f = L$ and $\lim_{x \to c-} f = L$.

($\Leftarrow$) If $L = \lim_{x \to c+} f$, given $\varepsilon > 0$ there exists $\delta_1(\varepsilon) > 0$ such that if $x \in A$ and $0 < x - c < \delta_1(c)$, then $|f(x) - L| < \varepsilon$. Similarly, if $L = \lim_{x \to c-} f$, there exists $\delta_2(\varepsilon) > 0$ such that if $x \in A$ and $0 < c - x < \delta_2(\varepsilon)$, then $|f(x) - L| < \varepsilon$. Now let $\delta(\varepsilon) := \inf\{\delta_1(\varepsilon), \delta_2(\varepsilon)\}$; then if $x \in A$ and $0 < |x - c| < \delta(\varepsilon)$, it follows that $|f(x) - L| < \varepsilon$. Therefore f has limit L at c. Q.E.D.

4.3.4 Examples. (a) Let $f(x) := \operatorname{sgn}(x)$.

We have seen in Example 4.1.7(b) that sgn does not have a limit at 0. It is clear that $\lim_{x \to 0+} \operatorname{sgn}(x) = +1$ and that $\lim_{x \to 0-} \operatorname{sgn}(x) = -1$. Since these one-sided limits are different, it also follows from Theorem 4.3.3 that $\operatorname{sgn}(x)$ does not have a limit at 0.

(b) Let $g(x) := e^{1/x}$ for $x \neq 0$. (See Figure 4.3.1.)

We first show that g does not have a finite right-hand limit at $c = 0$ since it is not bounded on any right-hand neighborhood $(0,\delta)$ of 0. We shall make use of the inequality

$$(*) \qquad 0 < t < e^t \quad \text{for} \quad t > 0,$$

which will be proved later (see Corollary 7.3.3). It follows from $(*)$ that if $x > 0$ then $0 < 1/x < e^{1/x}$. Hence, if we take $x_n = 1/n$, then $g(x_n) > n$ for all $n \in \mathbf{N}$. Therefore $\lim_{x \to 0+} e^{1/x}$ does not exist in $\boldsymbol{R}$.

However, $\lim_{x \to 0-} e^{1/x} = 0$. Indeed, if $x < 0$ and we take $t = -1/x$ in $(*)$ we

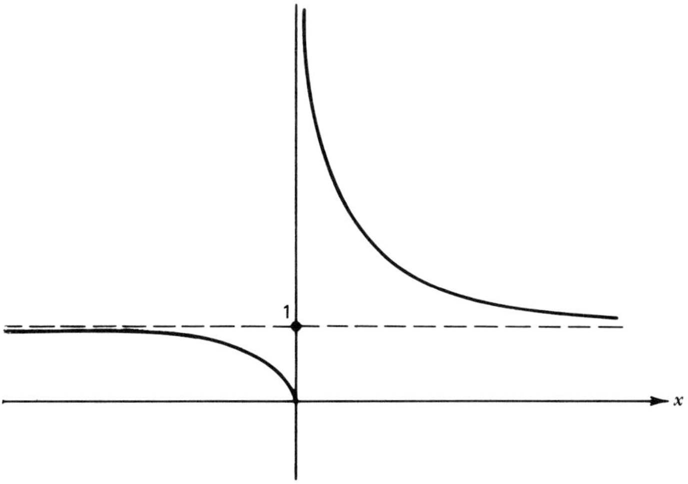

FIGURE 4.3.1 Graph of $g(x) = e^{1/x}$ ($x \neq 0$).

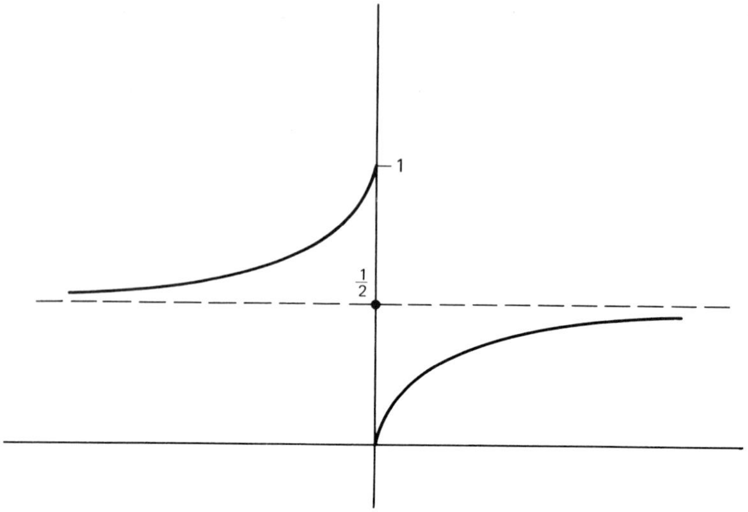

FIGURE 4.3.2 Graph of $h(x) = \dfrac{1}{e^{1/x} + 1}$ $(x \neq 0)$.

obtain $0 < -1/x < e^{-1/x}$. Since $x < 0$, this implies that $0 < e^{1/x} < -x$ for all $x < 0$. It follows from this inequality that $\lim_{x\to 0-} e^{1/x} = 0$.

(c) Let $h(x) := 1/(e^{1/x} + 1)$ for $x \neq 0$. (See Figure 4.3.2.)

We have seen in part (b) that $0 < 1/x < e^{1/x}$ for $x > 0$, whence it follows that

$$0 < \frac{1}{e^{1/x} + 1} < \frac{1}{e^{1/x}} < x,$$

which implies that $\lim_{x\to 0+} h = 0$.

Since we have seen in part (b) that $\lim_{x\to 0-} e^{1/x} = 0$, it follows from the analog of Theorem 4.2.4(b) for left-hand limits that

$$\lim_{x\to 0-}\left(\frac{1}{e^{1/x} + 1}\right) = \frac{1}{\lim_{x\to 0-} e^{1/x} + 1} = \frac{1}{0 + 1} = 1.$$

Note that for this function, both one-sided limits exist in $\boldsymbol{R}$, but they are unequal.

Infinite Limits

The function $f(x) := 1/x^2$ for $x \neq 0$ (see Figure 4.3.3) is not bounded on a neighborhood of 0 so it cannot have a limit in the sense of Definition 4.1.1. While the symbols ∞ $(= +\infty)$ and $-\infty$ do not represent real numbers, it is sometimes useful to be able to say that "$f(x) = 1/x^2$ tends to ∞ as $x \to 0$". This use of $\pm\infty$ will not cause any difficulties, provided we exercise caution and *never* interpret ∞ or $-\infty$ as being real numbers.

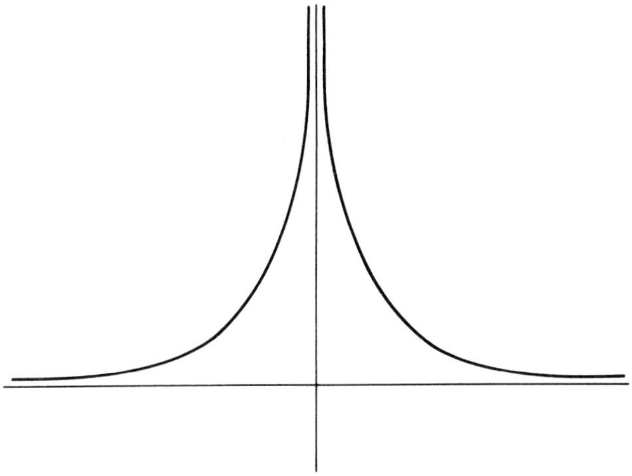

FIGURE 4.3.3 **Graph of** $f(x) = \frac{1}{x^2}$ $(x \neq 0)$.

4.3.5 Definition. Let $A \subseteq \boldsymbol{R}$, let $f\colon A \to \boldsymbol{R}$, and let $c \in \boldsymbol{R}$ be a cluster point of A.

(i) We say that f **tends to** ∞ **as** $x \to c$, and write

$$\lim_{x \to c} f = \infty,$$

if for every $\alpha \in \boldsymbol{R}$ there exists $\delta(\alpha) > 0$ such that if $0 < |x - c| < \delta(\alpha)$ and $x \in A$, then $f(x) > \alpha$.

(ii) We say that f **tends to** $-\infty$ **as** $x \to c$, and write

$$\lim_{x \to c} f = -\infty,$$

if for every $\beta \in \boldsymbol{R}$ there exists $\delta(\beta) > 0$ such that if $0 < |x - c| < \delta(\beta)$ and $x \in A$, then $f(x) < \beta$.

4.3.6 Examples. (a) $\lim_{x \to 0} (1/x^2) = \infty$.

For, if $\alpha > 0$ is given, let $\delta(\alpha) := 1/\sqrt{\alpha}$. It follows that if $0 < |x| < \delta(\alpha)$ then $x^2 < 1/\alpha$ so that $1/x^2 > \alpha$.

(b) Let $g(x) := 1/x$ for $x \neq 0$. (See Figure 4.3.4 on the next page.)

The function g does *not* tend to either ∞ or $-\infty$ as $x \to 0$. For, if $\alpha > 0$ then $g(x) < \alpha$ for all $x < 0$, so that g does not tend to ∞ as $x \to 0$. Similarly, if $\beta < 0$ then $g(x) > \beta$ for all $x > 0$, so that g does not tend to $-\infty$ as $x \to 0$.

While many of the results in Sections 4.1 and 4.2 have extensions to this limiting notion, not all of them do since $\pm\infty$ are not real numbers. The following result is an analog of the Squeeze Theorem 4.2.7. (See also Theorem 3.6.4.)

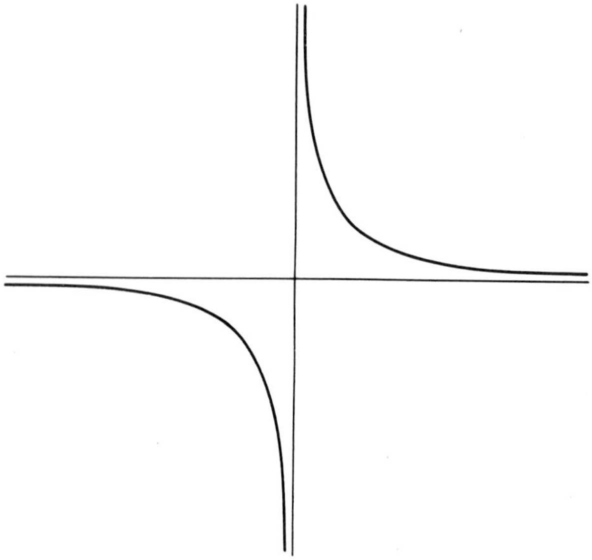

FIGURE 4.3.4 Graph of $g(x) = \dfrac{1}{x}$ $(x \neq 0)$.

4.3.7 Theorem. *Let $A \subseteq \boldsymbol{R}$, let $f, g : A \to \boldsymbol{R}$, and let $c \in \boldsymbol{R}$ be a cluster point of A. Suppose that $f(x) \leqslant g(x)$ for all $x \in A$, $x \neq c$.*

(a) *If $\lim\limits_{x \to c} f = \infty$, then $\lim\limits_{x \to c} g = \infty$.*

(b) *If $\lim\limits_{x \to c} g = -\infty$, then $\lim\limits_{x \to c} f = -\infty$.*

Proof. (a) If $\lim\limits_{x \to c} f = \infty$ and $\alpha \in \boldsymbol{R}$ is given, then there exists $\delta(\alpha) > 0$ such that if $0 < |x - c| < \delta(\alpha)$ and $x \in A$, then $f(x) > \alpha$. But since $f(x) \leqslant g(x)$ for all $x \in A$, $x \neq c$, it follows that if $0 < |x - c| < \delta(\alpha)$ and $x \in A$, then $g(x) > \alpha$. Therefore $\lim\limits_{x \to c} g = \infty$.

The proof of (b) is similar. Q.E.D.

The function $g(x) = 1/x$ considered in Example 4.3.6(b) suggests that it might be useful to consider one-sided infinite limits.

4.3.8 Definition. Let $A \subseteq \boldsymbol{R}$ and let $f : A \to \boldsymbol{R}$.

(i) If $c \in \boldsymbol{R}$ is a cluster point of the set $A \cap (c, \infty) = \{x \in A : x > c\}$, then we say that f **tends to** ∞ [respectively, $-\infty$] **as** $x \to c+$, and we write

$$\lim_{x \to c+} f = \infty \qquad \left[\text{respectively, } \lim_{x \to c+} f = -\infty\right],$$

if for every $\alpha \in \boldsymbol{R}$ there is $\delta(\alpha) > 0$ such that if $0 < x - c < \delta(\alpha)$ and $x \in A$, then $f(x) > \alpha$ [respectively, $f(x) < \alpha$].

(ii) If $c \in \boldsymbol{R}$ is a cluster point of the set $A \cap (-\infty, c) = \{x \in A : x < c\}$, then we say that f **tends to** ∞ [respectively, $-\infty$] **as** $x \to c-$, and we write

$$\lim_{x \to c-} f = \infty \qquad [\text{respectively, } \lim_{x \to c-} f = -\infty],$$

if for every $\alpha \in \boldsymbol{R}$ there exists $\delta(\alpha) > 0$ such that if $0 < c - x < \delta(\alpha)$ and $x \in A$, then $f(x) > \alpha$ [respectively, $f(x) < \alpha$].

4.3.9 Examples. (a) Let $g(x) := 1/x$ for $x \neq 0$. We have noted in Example 4.3.6(b) that $\lim_{x \to 0} g$ does not exist. However, it is an easy exercise to show that

$$\lim_{x \to 0+} (1/x) = \infty \qquad \text{and} \qquad \lim_{x \to 0-} (1/x) = -\infty.$$

(b) It was seen in Example 4.3.4(b) that the function $g(x) := e^{1/x}$ for $x \neq 0$ is not bounded on any interval $(0, \delta)$, $\delta > 0$. Hence the right-hand limit of $e^{1/x}$ as $x \to 0+$ does not exist in the sense of Definition 4.3.1(i). However, since

$$\frac{1}{x} < e^{1/x} \qquad \text{for} \quad x > 0,$$

it is readily seen that $\lim_{x \to 0+} e^{1/x} = \infty$ in the sense of Definition 4.3.8.

Limits at Infinity

It is also desirable to define the notion of the limit of a function as $x \to \infty$ (or as $x \to -\infty$).

4.3.10 Definition. Let $A \subseteq \boldsymbol{R}$ and let $f : A \to \boldsymbol{R}$.

(i) Suppose that $(a, \infty) \subseteq A$ for some $A \in \boldsymbol{R}$. We say that $L \in \boldsymbol{R}$ is a **limit of** f **as** $x \to \infty$, and write

$$\lim_{x \to \infty} f = L,$$

if given any $\varepsilon > 0$ there exists $K(\varepsilon) > a$ such that if $x > K(\varepsilon)$ then $|f(x) - L| < \varepsilon$.

(ii) Suppose that $(-\infty, b) \subseteq A$ for some $b \in \boldsymbol{R}$. We say that $L \in \boldsymbol{R}$ **is a limit of** f **as** $x \to -\infty$, and write

$$\lim_{x \to -\infty} f = L,$$

if given any $\varepsilon > 0$ there exists $K(\varepsilon) < b$ such that if $x < K(\varepsilon)$ then $|f(x) - L| < \varepsilon$.

The reader should note the close resemblance between 4.3.10(i) and the definition of a limit of a sequence.

We leave it to the reader to show that the limits of f as $x \to \pm\infty$ are unique whenever they exist. We also have sequential criteria for these limits; we shall only state the criterion as $x \to \infty$. This uses the notion of the limit of a properly divergent sequence (see Definition 3.6.1).

4.3.11 Theorem. *Let $A \subseteq \mathbf{R}$, let $f : A \to \mathbf{R}$, and suppose that $(a,\infty) \subseteq A$ for some $a \in \mathbf{R}$. Then:*

(i) $L = \lim_{x\to\infty} f$

if and only if

(ii) *for every sequence (x_n) in $A \cap (a,\infty)$ such that* $\lim (x_n) = \infty$, *then the sequence $(f(x_n))$ converges to L.*

We leave it to the reader to prove this theorem and to formulate and prove the companion result concerning the limit as $x \to -\infty$.

4.3.12 Examples. (a) Let $g(x) := 1/x$ for $x \neq 0$.

It is an elementary exercise to show that $\lim_{x\to\infty} (1/x) = 0 = \lim_{x\to-\infty} (1/x)$. (See Figure 4.3.4 on p. 134.)

(b) Let $f(x) := 1/x^2$ for $x \neq 0$.

The reader may show that $\lim_{x\to\infty} (1/x^2) = 0 = \lim_{x\to-\infty} (1/x^2)$. (See Figure 4.3.3.) One way to do this is to show that if $x \geqslant 1$ then $0 \leqslant 1/x^2 \leqslant 1/x$. In view of part (a), this implies that $\lim_{x\to\infty} (1/x^2) = 0$.

Just as it is convenient to be able to say that $f(x) \to \pm\infty$ as $x \to c$ for $c \in \mathbf{R}$, it is convenient to have the corresponding notion as $x \to \pm\infty$.

4.3.13 Definition. Let $A \subseteq \mathbf{R}$ and let $f : A \to \mathbf{R}$.

(i) Suppose that $(a,\infty) \subseteq A$ for some $a \in A$. We say that f **tends to** ∞ [respectively, $-\infty$] **as** $x \to \infty$, and write

$$\lim_{x\to\infty} f = \infty \qquad \text{[respectively, } \lim_{x\to\infty} f = -\infty\text{]},$$

if given any $\alpha \in \mathbf{R}$ there exists $K(\alpha) > a$ such that if $x > K(\alpha)$, then $f(x) > \alpha$ [respectively, $f(x) < \alpha$]. (See Figure 4.3.5.)

(ii) Suppose that $(-\infty,b) \subseteq A$ for some $b \in \mathbf{R}$. We say that f **tends to** ∞ [respectively, $-\infty$] **as** $x \to -\infty$, and write

$$\lim_{x\to-\infty} f = \infty \qquad \text{[respectively, } \lim_{x\to-\infty} f = -\infty\text{]},$$

if given any $\alpha \in \mathbf{R}$ there exists $K(\alpha) < b$ such that if $x < K(\alpha)$, then $f(x) > \alpha$ [respectively, $f(x) < \alpha$].

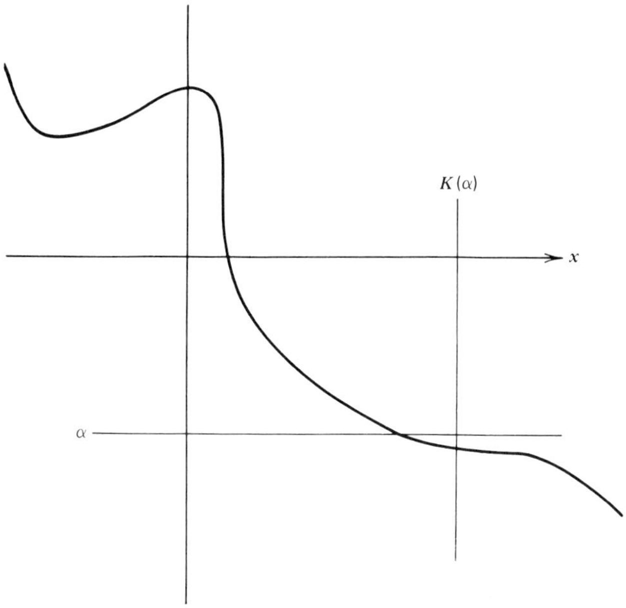

FIGURE 4.3.5 $\lim_{x \to \infty} f = -\infty$.

As before there is a sequential criterion for this limit. We will formulate the one as $x \to \infty$.

4.3.14 Theorem. *Let $A \subseteq \mathbf{R}$, let $f : A \to \mathbf{R}$, and suppose that $(a, \infty) \subseteq A$ for some $a \in \mathbf{R}$. Then:*

(i) $\lim_{x \to \infty} f = \infty$ *[respectively,* $\lim_{x \to \infty} f = -\infty$*]*

if and only if

(ii) *for every sequence (x_n) in (a, ∞) such that* $\lim (x_n) = \infty$, *we have* $\lim \big(f(x_n)\big) = \infty$ *[respectively,* $\lim \big(f(x_n)\big) = -\infty$*].*

The next result is an analogue of Theorem 3.6.5.

4.3.15 Theorem. *Let $A \subseteq \mathbf{R}$, let $f, g : A \to \mathbf{R}$, and suppose that $(a, \infty) \subseteq A$ for some $a \in \mathbf{R}$. Suppose further that $g(x) > 0$ for $x > a$ and that*

$$\lim_{x \to \infty} \frac{f(x)}{g(x)} = L$$

for some $L \in \mathbf{R}$, $L \neq 0$.

(i) *If $L > 0$, then $\lim_{x \to \infty} f = \infty$ if and only if $\lim_{x \to \infty} g = \infty$.*

(ii) *If $L < 0$, then $\lim_{x \to \infty} f = -\infty$ if and only if $\lim_{x \to \infty} g = \infty$.*

Proof. (i) Since $L > 0$, the hypothesis implies that there exists $a_1 > a$ such that

$$0 < \tfrac{1}{2}L \leq \frac{f(x)}{g(x)} < 2L \qquad \text{for} \quad x > a_1.$$

Therefore we have $(\frac{1}{2}L)g(x) < f(x) < (2L)g(x)$ for all $x > a_1$, from which the conclusion follows readily.

The proof of (ii) is similar. Q.E.D.

We leave it to the reader to formulate the analogous result as $x \to -\infty$.

4.3.16 Examples. (a) $\lim_{x\to\infty} x^n = \infty$ for $n \in \boldsymbol{N}$.

Let $g(x) := x^n$ for $x \in (0,\infty)$. Given $\alpha \in \boldsymbol{R}$, let $K(\alpha) := \sup\{1,\alpha\}$. Then if $x > K(\alpha)$ we have $g(x) = x^n \geq x > \alpha$. Since $\alpha \in \boldsymbol{R}$ is arbitrary, it follows that $\lim_{x\to\infty} g = \infty$.

(b) $\lim_{x\to-\infty} x^n = \infty$ for $n \in \boldsymbol{N}$, n even, and $\lim_{x\to-\infty} x^n = -\infty$ for $n \in \boldsymbol{N}$, n odd.

We shall treat the case n odd, say $n = 2k + 1$ with $k = 0, 1, \ldots$. Given $\alpha \in \boldsymbol{R}$, let $K(\alpha) := \inf\{\alpha, -1\}$. If $x < K(\alpha)$, then since $(x^2)^k \geq 1$, we have $x^n = (x^2)^k x \leq x < \alpha$. Since $\alpha \in \boldsymbol{R}$ is arbitrary, it follows that $\lim_{x\to-\infty} x^n = -\infty$.

(c) Let $p : \boldsymbol{R} \to \boldsymbol{R}$ be the polynomial function $p(x) := a_n x^n + a_{n-1}x^{n-1} + \cdots + a_1 x + a_0$. Then $\lim_{x\to\infty} p = \infty$ if $a_n > 0$ and $\lim_{x\to\infty} p = -\infty$ if $a_n < 0$.

Indeed, let $g(x) := x^n$ and apply Theorem 4.3.15. Since

$$\frac{p(x)}{g(x)} = a_n + a_{n-1}\left(\frac{1}{x}\right) + \cdots + a_1\left(\frac{1}{x^{n-1}}\right) + a_0\left(\frac{1}{x^n}\right),$$

it follows that $\lim_{x\to\infty} (p(x)/g(x)) = a_n$. Since $\lim_{x\to\infty} g = \infty$, the assertion follows from Theorem 4.3.15.

(d) Let p be the polynomial function in part (c). Then $\lim_{x\to-\infty} p = \infty$ [respectively, $-\infty$] if n is even [respectively, odd] and $a_n > 0$.

We leave the details to the reader.

Exercises for Section 4.3

1. Prove Theorem 4.3.2.
2. Give an example of a function that has a right-hand limit but not a left-hand limit at a point.
3. Let $f(x) := |x|^{-1/2}$ for $x \neq 0$. Show that $\lim_{x\to0+} f(x) = \lim_{x\to0-} f(x) = +\infty$.
4. Let $c \in \boldsymbol{R}$ and let f be defined for $x \in (c,\infty)$ and $f(x) > 0$ for all $x \in (c,\infty)$. Show that $\lim_{x\to c} f = \infty$ if and only if $\lim_{x\to c} 1/f = 0$.

5. Evaluate the following limits, or show that they do not exist.

   (a) $\lim_{x\to 1+} \dfrac{x}{x-1}$ $(x \neq 1)$; (b) $\lim_{x\to 1} \dfrac{x}{x-1}$ $(x \neq 1)$;

   (c) $\lim_{x\to 0+} (x+2)/\sqrt{x}$ $(x > 0)$; (d) $\lim_{x\to\infty} (x+2)/\sqrt{x}$ $(x > 0)$;

   (e) $\lim_{x\to 0} (\sqrt{x+1})/x$ $(x > -1)$; (f) $\lim_{x\to\infty} (\sqrt{x+1})/x$ $(x > 0)$;

   (g) $\lim_{x\to\infty} \dfrac{\sqrt{x}-5}{\sqrt{x}+3}$ $(x > 0)$; (h) $\lim_{x\to\infty} \dfrac{\sqrt{x}-x}{\sqrt{x}+x}$ $(x > 0)$.

6. Prove Theorem 4.3.11.
7. Suppose that f and g have limits in $\boldsymbol{R}$ as $x \to \infty$ and that $f(x) \leqslant g(x)$ for $x \in (a,\infty)$. Prove that $\lim_{x\to\infty} f \leqslant \lim_{x\to\infty} g$.
8. Let f be defined on $(0,\infty)$ to $\boldsymbol{R}$. Prove that $\lim_{x\to\infty} f(x) = L$ if and only if $\lim_{x\to 0+} f(1/x) = L$.
9. Show that if $f : (a,\infty) \to \boldsymbol{R}$ is such that $\lim_{x\to\infty} xf(x) = L$ where $L \in \boldsymbol{R}$, then $\lim_{x\to\infty} f(x) = 0$.
10. Prove Theorem 4.3.14.
11. Complete the proof of Theorem 4.3.15.
12. Suppose that $\lim_{x\to c} f(x) = L$ where $L > 0$, and that $\lim_{x\to c} g(x) = \infty$. Show that $\lim_{x\to c} f(x)g(x) = \infty$. If $L = 0$, show by example that this conclusion may fail.
13. Find functions f and g defined on $(0,\infty)$ with $\lim_{x\to\infty} f = \infty$, and $\lim_{x\to\infty} g = \infty$, and $\lim_{x\to\infty} (f - g) = 0$. Can you find such functions, with $g(x) > 0$ for $x \in (0,\infty)$, such that $\lim_{x\to\infty} f/g = 0$?
14. Let f and g be defined on (a,∞) and suppose $\lim_{x\to\infty} f = L$ and $\lim_{x\to\infty} g = \infty$. Prove that $\lim_{x\to\infty} f\circ g = L$.

SECTION 4.4 Continuous Functions

In this section, which is very similar to Section 4.1, we shall define what it means to say that a function is continuous at a point, or on a set. This notion of continuity is one of the central notions of mathematical analysis and will be used in almost all of the following material in this book. Consequently, it is essential that the reader master it.

4.4.1 Definition. Let $A \subseteq \boldsymbol{R}$, let $f : A \to \boldsymbol{R}$, and let $c \in A$. We say that f is **continuous at** c if, given any neighborhood V of $f(c)$ there exists a neighborhood U_V of c such that if $x \in A \cap U_V$, then $f(x)$ belongs to V. (See Figure 4.4.1 on the next page.)

Remarks. (1) If $c \in A$ is a cluster point of A, then a comparison of Definitions 4.1.1 and 4.4.1 show that f is continuous at c if and only if

(1) $$f(c) = \lim_{x\to c} f.$$

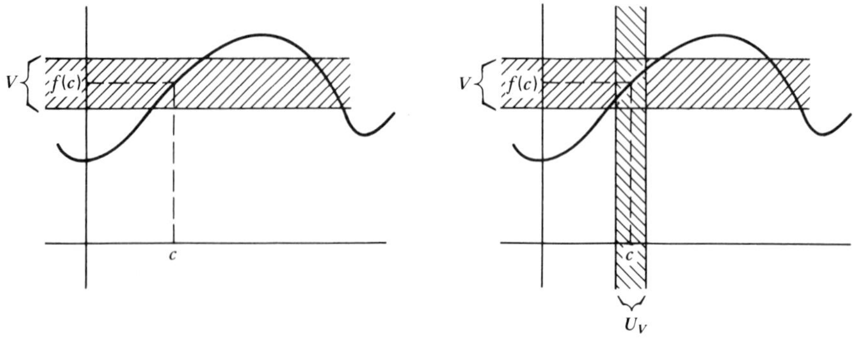

FIGURE 4.4.1 Given *V*, the neighborhood U_V is determined so that $f(U_V) \subseteq V$.

Thus, if c is a cluster point of A, then in order for (1) to hold, three conditions must be met: (i) f must be defined at c (so that $f(c)$ makes sense), (ii) the limit of f at c must exist in $\boldsymbol{R}$ (so that $\lim_{x \to c} f$ makes sense), and (iii) these values $f(c)$ and $\lim_{x \to c} f$ must be equal.

(2) If $c \in A$ is not a cluster point of A, then there exists a neighborhood U of c such that $A \cap U = \{c\}$. Thus we conclude that a function f is automatically continuous at a point $c \in A$ that is not a cluster point of A. Such points are often called "isolated points" of A; they are of little practical interest to us, since they are "far from the action". Since continuity is automatic for such points, we shall generally test for continuity only at cluster points. Thus we can regard the condition (1) as being characteristic for continuity at c.

In the next definition we define continuity of f on a set.

4.4.2 Definition. Let $A \subseteq \boldsymbol{R}$ and let $f : A \to \boldsymbol{R}$. If $B \subseteq A$, we say that f **is continuous on** B if f is continuous at every point of B.

We now give some equivalent formulations for Definition 4.4.1.

4.4.3 Theorem. *Let $A \subseteq \boldsymbol{R}$, let $f : A \to \boldsymbol{R}$, and let $c \in A$. Then the following conditions are equivalent.*

(i) *f is continuous at c; that is, given any neighborhood V of $f(c)$ there exists a neighborhood U_V of c such that if x is any point of $A \cap U_V$, then $f(x)$ belongs to V.*

(ii) *Given any $\varepsilon > 0$ there exists a $\delta(\varepsilon) > 0$ such that if $|x - c| < \delta(\varepsilon)$ and $x \in A$, then $|f(x) - f(c)| < \varepsilon$.*

(iii) *If (x_n) is any sequence of real numbers such that $x_n \in A$ for all $n \in \mathbf{N}$ and (x_n) converges to c, then the sequence $(f(x_n))$ converges to $f(c)$.*

The proof of this theorem requires only slight modifications in the proofs of Theorems 4.1.3 and 4.1.5. We leave the details as an important exercise for the reader.

The following Discontinuity Criterion is a consequence of the equivalence of (i) and (iii) of the last theorem; it should be compared with the Divergence Criterion 4.1.6(a) with $L = f(c)$. Its proof should be written out in detail by the reader.

4.4.4 Discontinuity Criterion. *Let $A \subseteq \boldsymbol{R}$, let $f : A \to \boldsymbol{R}$, and let $c \in A$. Then f is discontinuous at c if and only if there exists a sequence (x_n) in A such that (x_n) converges to c, but the sequence $(f(x_n))$ does* **not** *converge to $f(c)$.*

4.4.5 Examples. (a) $f(x) := b$ is continuous on $\boldsymbol{R}$. It was seen in Example 4.1.4(a) that if $c \in \boldsymbol{R}$, then $\lim_{x\to c} f = b$. Since $f(c) = b$, this shows that f is continuous at every point $c \in \boldsymbol{R}$. Thus f is continuous on $\boldsymbol{R}$.

(b) $g(x) := x$ is continuous on $\boldsymbol{R}$.

It was seen in Example 4.1.4(b) that if $c \in \boldsymbol{R}$, then $\lim_{x\to c} g = c$. Since $g(c) = c$, this shows that g is continuous at every point $c \in \boldsymbol{R}$. Thus h is continuous on $\boldsymbol{R}$.

(c) $h(x) := x^2$ is continuous on $\boldsymbol{R}$.

It was seen in Example 4.1.4(c) that if $c \in \boldsymbol{R}$, then $\lim_{x\to c} h = c^2$. Since $h(c) = c^2$, this shows that h is continuous at every point $c \in \boldsymbol{R}$. Thus h is continuous on $\boldsymbol{R}$.

(d) $\varphi(x) := 1/x$ is continuous on $A := \{x \in \boldsymbol{R} : x > 0\}$.

It was seen in Example 4.1.4(d) that if $c \in A$, then $\lim_{x\to c} \varphi = 1/c$. Since $\varphi(c) = 1/c$, this shows that f if continuous at every point $c \in A$. Thus f is continuous on A.

(e) $\varphi(x) := 1/x$ is not continuous at $x = 0$.

Indeed, if $\varphi(x) = 1/x$ for $x > 0$, then φ is not defined for $x = 0$, so it cannot be continuous there. Alternatively, it was seen in Example 4.1.7(a) that $\lim_{x\to 0} \varphi$ does not exist in $\boldsymbol{R}$, so φ cannot be continuous at $x = 0$.

(f) The signum function sgn is not continuous at 0.

The signum function was defined in Example 4.1.7(b), where it was also shown that $\lim_{x\to 0} \operatorname{sgn}(x)$ does not exist in $\boldsymbol{R}$. Therefore sgn is not continuous at $x = 0$ (even though sgn 0 is defined).

It is an exercise to show that sgn is continuous at every point $c \neq 0$.

(g) Let $A := \boldsymbol{R}$ and let f be Dirichlet's "discontinuous function" defined by

$$
\begin{aligned}
f(x) &:= 1 \quad \text{if } x \text{ is rational,} \\
&:= 0 \quad \text{if } x \text{ is irrational.}
\end{aligned}
$$

We claim that f is *not continuous at any point of* $\mathbf{R}$.

Indeed, if c is a rational number, let (x_n) be a sequence of irrational numbers that converges to c. (Corollary 2.4.11 to the Density Theorem 2.4.10 assures us that such a sequence does exist.) Since $f(x_n) = 0$ for all $n \in \mathbf{N}$, we have $\lim (f(x_n)) = 0$, while $f(c) = 1$. Therefore f is not continuous at the rational number c.

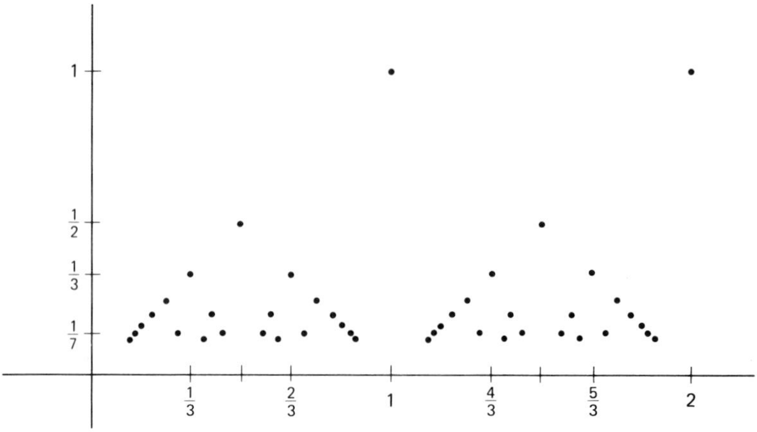

FIGURE 4.4.2 The function in Example 4.4.5(h).

On the other hand, if b is an irrational number, let (y_n) be a sequence of rational numbers that converge to b. (The Density Theorem 2.4.10 assures us that such a sequence does exist.) Since $f(y_n) = 1$ for all $n \in \boldsymbol{N}$, we have $\lim (f(y_n)) = 1$, while $f(b) = 0$. Therefore f is not continuous at the irrational number b.

Since every real number is either rational or irrational, we deduce that f is not continuous at any point in $\boldsymbol{R}$.

(h) Let $A := \{x \in \boldsymbol{R} : x > 0\}$. For any irrational number $x > 0$ we define $h(x) := 0$. For a rational number in A of the form m/n, with natural numbers m, n having no common factors except 1, we define $h(m/n) := 1/n$. (See Figure 4.4.2.) We claim that *h is continuous at every irrational number in A, and is discontinuous at every rational number in A.*

Indeed, if $a > 0$ is rational, let (x_n) be a sequence of irrational numbers in A that converges to a. Then $\lim (h(x_n)) = 0$, while $h(a) > 0$. Hence h is discontinuous at a.

On the other hand, if b is an irrational number and $\varepsilon > 0$, then (by the Archimedean Property) there is a natural number n_0 such that $1/n_0 < \varepsilon$. There are only a finite number of rationals with denominator less than n_0 in the interval $(b-1, b+1)$. Hence $\delta > 0$ can be chosen so small that the neighborhood $(b - \delta, b + \delta)$ contains no rational numbers with denominator less than n_0. It then follows that for $|x - b| < \delta$, $x \in A$, we have $|h(x) - h(b)| = |h(x)| \leq 1/n_0 < \varepsilon$. Thus h is continuous at the irrational number b.

Consequently, we deduce that h is continuous precisely at the irrational points in A.

4.4.6 Remarks. (a) Sometimes a function $f : A \to \boldsymbol{R}$ is not continuous at a point c because it is not defined at this point. However, if the function f has a limit L at the point c and if we define F on $A \cup \{c\} \to \boldsymbol{R}$ by

$$\begin{aligned} F(x) &:= L && \text{for} \quad x = c, \\ &:= f(x) && \text{for} \quad x \in A, \end{aligned}$$

then F is continuous at c. To see this, one needs to check that $\lim_{x \to c} F = L$. (We leave this to the reader.)

(b) If a function $g : A \to \boldsymbol{R}$ does not have a limit at c, then there is no way that we can obtain a function $G : A \cup \{c\} \to \boldsymbol{R}$ that is continuous at c by defining

$$\begin{aligned} G(x) &:= C && \text{for} \quad x = c, \\ &:= g(x) && \text{for} \quad x \in A. \end{aligned}$$

To see this, observe that if $\lim_{x \to c} G$ exists and equals C, then $\lim_{x \to c} g$ must also exist and equal C.

4.4.7 Examples. (a) The function $g(x) := \sin (1/x)$ for $x \neq 0$ (see Figure 4.1.3) does not have a limit at $x = 0$ (see Example 4.1.7(c)). Thus there is no value that we can assign at $x = 0$ to obtain a continuous extension of g at $x = 0$.

(b) Let $f(x) := x \sin (1/x)$ for $x \neq 0$. (See Figure 4.4.3.) Since f is not defined

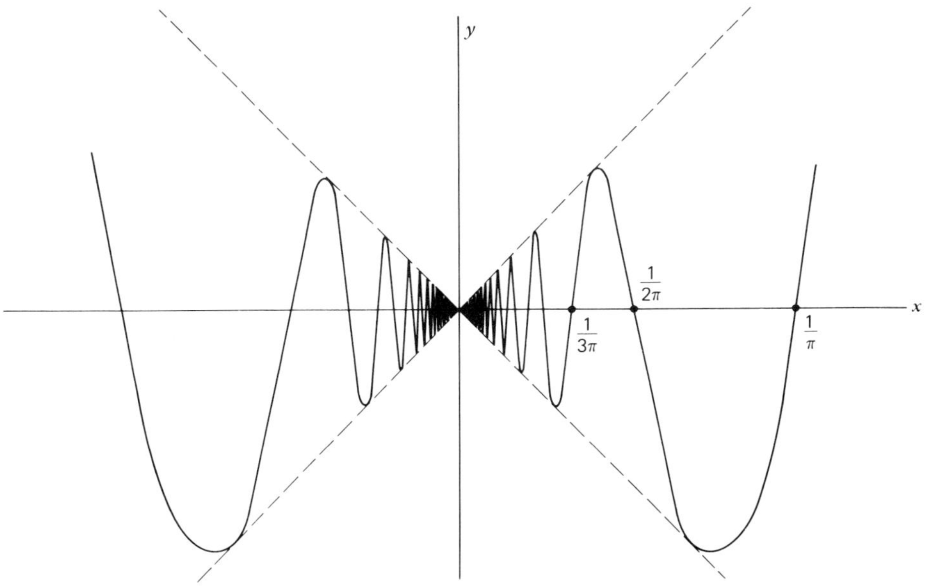

FIGURE 4.4.3 Graph of $f(x) = x \sin \frac{1}{x}$ $(x \neq 0)$.

at $x = 0$, the function f cannot be continuous at this point. However, it was seen in Example 4.2.8(f) that $\lim_{x \to 0} (x \sin (1/x)) = 0$. Therefore it follows from Remark 4.4.6(a) that if we define $F : \boldsymbol{R} \to \boldsymbol{R}$ by

$$
\begin{aligned}
F(x) &:= 0 && \text{for } x = 0, \\
&:= x \sin (1/x) && \text{for } x \neq 0,
\end{aligned}
$$

then F is continuous at $x = 0$.

The Types of Isolated Discontinuities†

We have seen in Example 4.4.5(g) that a function may be discontinuous at every point, and in Example 4.4.5(h) that it may be discontinuous at every rational number in its domain. However, most functions that we encounter are continuous at "most" points. It is instructive to consider the types of "isolated discontinuities" that can occur. We shall first consider the case where the discontinuity takes place at the left end point of an interval, and then the case where the discontinuity takes place at an interior point of the interval.

(I) Discontinuity at an end point. Let c be the left end point of an interval $I := [c, b)$ and suppose that f is continuous on (c, b) but discontinuous at c. We distinguish three cases: (A) $\lim_{x \to c} f$ exists, (B) $\lim_{x \to c} f$ does not exist, but f is bounded on some neighborhood of c, and (C) f is not bounded on any neighborhood of c.

(A) If $\lim_{x \to c} f$ exists, then f will be discontinuous at c if either (i) f is not defined at c, or (ii) f is defined at c but $f(c) \neq \lim_{x \to c} f$. In either case, we say that f has a **removable discontinuity**, since if we define $F : I \to \boldsymbol{R}$ by

$$
\begin{aligned}
F(x) &:= f(x) && \text{for } c < x < b, \\
&:= \lim_{x \to c} f && \text{for } x = c,
\end{aligned}
$$

then F is continuous at c. Hence by either (i) assigning a value at the point c, or (ii) changing the value at the point c, we "remove" the discontinuity of f at c. (See Figure 4.4.4.)

(B) If $\lim_{x \to c} f$ does not exist, but there exists a neighborhood U of c such that f is bounded on $U \cap I$, then it can be shown that there are two sequences (x_n), (y_n) in (c, b) such that $\lim (x_n) = c = \lim (y_n)$ and such that $\alpha := \lim (f(x_n)) <$

† The remainder of this section can be omitted on a first reading.

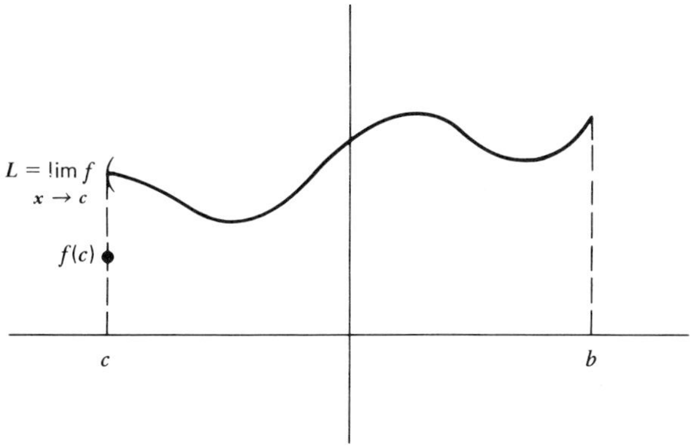

FIGURE 4.4.4 A removable discontinuity.

$\lim (g(x_n)) =: \beta$. An example of such a function is the function $f(x) := \sin(1/x)$ $(x > 0)$, whose graph is given in Figure 4.4.5, where we can find sequences such that $\alpha = -1$ and $\beta = +1$. There is no way to define (or re-define) f at 0 to obtain a function that is continuous at 0.

In this case, we say that f has an **oscillatory discontinuity at** c.

(C) If f is not bounded on any neighborhood of c, we say that f has an **infinite discontinuity** at c. A variety of different things can happen in this case. Perhaps the simplest is that we might have $\lim_{x \to c} f = \infty$ or $\lim_{x \to c} f = -\infty$ (in the sense of Definition 4.3.5). Such is the case for $f_1(x) := 1/x$ $(x > 0)$ or $f_2(x) := \log x$ $(x > 0)$. (See Figures 4.4.6 and 4.4.7 on the next page.)

Considerably more complicated is the case of a function such as $f_3(x) := (1/x) \sin(1/x)$ for $x > 0$. (See Figure 4.4.8.) For this function there exist se-

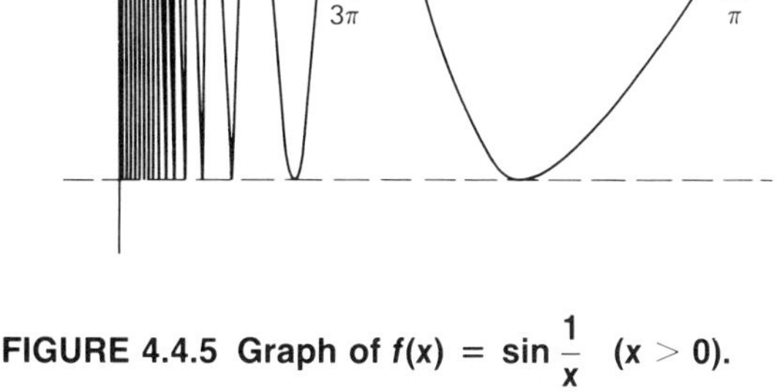

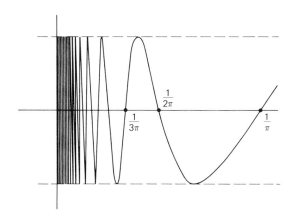

FIGURE 4.4.5 Graph of $f(x) = \sin \frac{1}{x}$ $(x > 0)$.

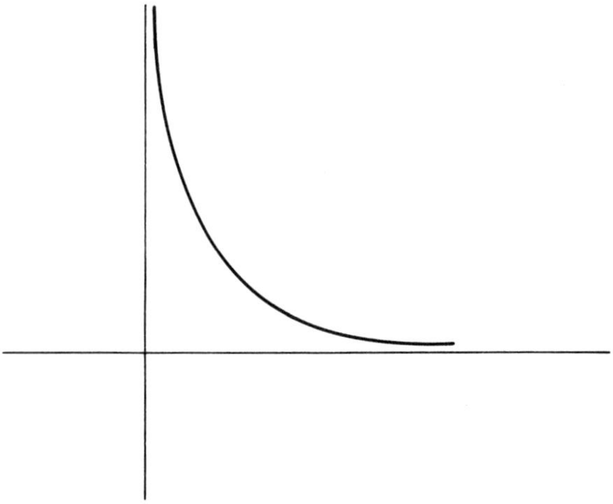

FIGURE 4.4.6 Graph of $f_1(x) = \frac{1}{x}$ $(x > 0)$.

quences (x_n) and (y_n) such that $\lim (x_n) = 0 = \lim (y_n)$ and such that $\lim \big(f_3(x_n)\big) = -\infty$ and $\lim \big(f_3(y_n)\big) = \infty$. In fact, for *any* $L \in \boldsymbol{R}$ there exists a sequence (z_n) such that $\lim (z_n) = 0$ and $\lim \big(f_3(z_n)\big) = L$.

Somewhat less complicated, but still exhibiting a form of "infinite oscillations" is $f_4(x) := (1/x)\,|\sin(1/x)|$ for $x > 0$. (See Figure 4.4.9.) This function is bounded below, but not above. Here, if $L \in \boldsymbol{R}$ is any number such that $L \geqslant 0$, then there exists a sequence (z_n) such that $\lim (z_n) = 0$ and $\lim \big(f_4(z_n)\big) = L$.

(II) Discontinuity at an interior point. Let c be a point of $I := (a,b)$ and suppose that f is continuous on (a,c) and (c,b), but discontinuous at the point c.

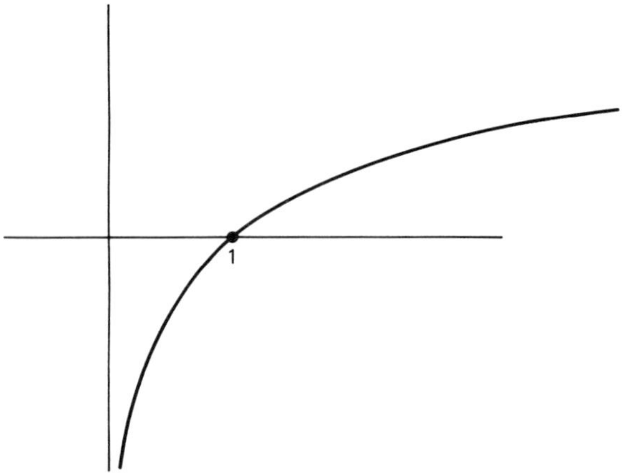

FIGURE 4.4.7 Graph of $f_2(x) = \log x$ $(x > 0)$.

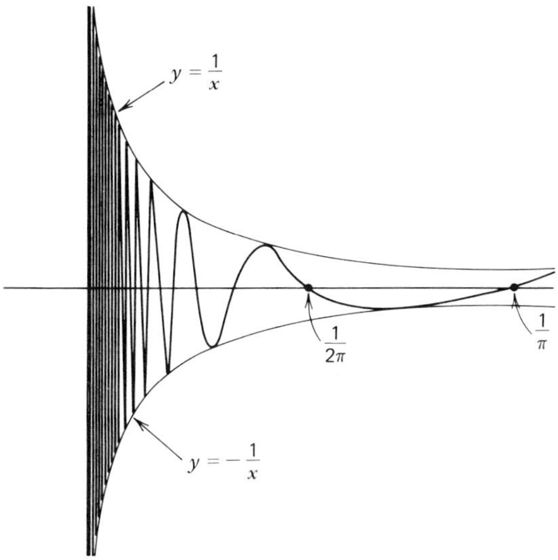

FIGURE 4.4.8 Graph of $f_3(x) = \frac{1}{x} \sin \frac{1}{x}$ $(x > 0)$.

If we examine f on both of the intervals $(a,c]$ and $[c,b)$, then any combination of the cases (A), (B), and (C) can occur on these two subintervals. Whenever case (B) or case (C) occurs on one or both of the intervals $(a,c]$ or $[c,b)$, the function f is irrevocably discontinuous at c. However, if case (A) occurs on both of the intervals $(a,c]$ and $[c,b)$, by which we mean that the one-sided limits $\lim_{x \to c-} f$ and

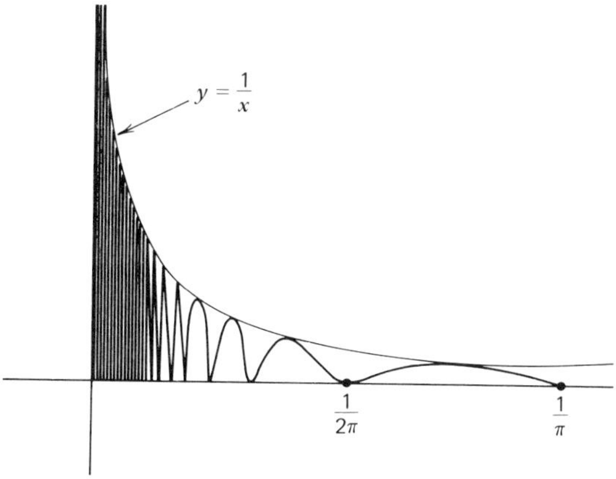

FIGURE 4.4.9 Graph of $f_4(x) = \frac{1}{x} \left|\sin \frac{1}{x}\right|$ $(x > 0)$.

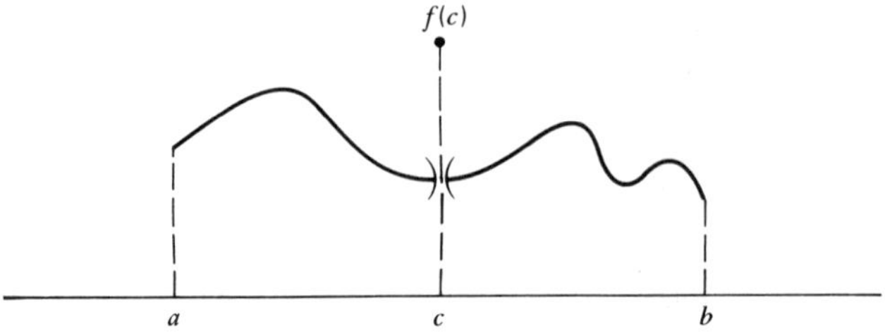

FIGURE 4.4.10 A removable discontinuity.

$\lim_{x\to c+} f$ exist in the sense of Definition 4.3.1, then two distinct situations can occur.

(A1) If $\lim_{x\to c-} f = \lim_{x\to c+} f$, then if we define $F : (a,b) \to \boldsymbol{R}$ by

$$F(x) := f(x) \qquad \text{for} \quad x \in (a,b),\ x \neq c,$$
$$:= \lim_{x\to c-} f \qquad \text{for} \quad x = c,$$

then it is readily seen that $\lim_{x\to c} F = \lim_{x\to c-} f = F(c)$, so that F is continuous at c. In this case we say that f has a **removable discontinuity** at c. (See Figure 4.4.10.)

(A2) If $\lim_{x\to c-} f \neq \lim_{x\to c+} f$, then the function cannot be defined at c to have a limit (see Theorem 4.3.3), and hence is irrevocably discontinuous. In this case, we say that f has a **jump discontinuity** at c. (See Figure 4.4.11.)

For example, the signum function sgn, introduced in Example 4.1.7(b), has a jump discontinuity at 0 but is continuous elsewhere. The greatest integer function (see Exercise 4.4.4) has a jump discontinuity at every integer, but is continuous at every other point.

If the one-sided limits $\lim_{x\to c-} f$ and $\lim_{x\to c+} f$ exist, we define the **jump of f at c** to be

$$j_f(c) := \lim_{x\to c+} f - \lim_{x\to c-} f.$$

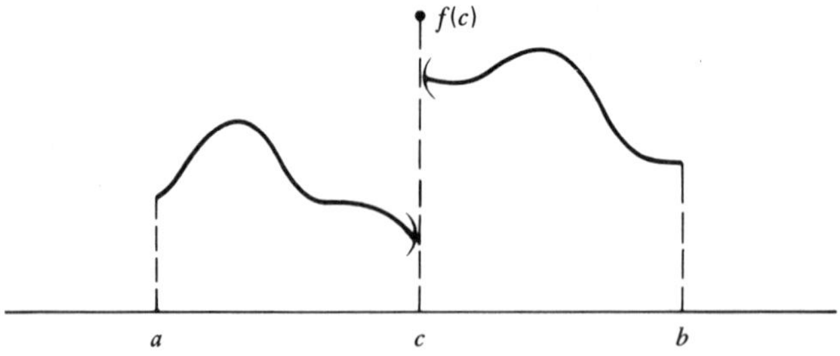

FIGURE 4.4.11 A jump discontinuity.

Clearly, f has a removable discontinuity (or perhaps no discontinuity at all) if $j_f(c) = 0$, and f has a jump discontinuity at c if $j_f(c) \neq 0$.

Exercises for Section 4.4

1. Prove Theorem 4.4.3.
2. Establish the Discontinuity Criterion 4.4.4.
3. Let $a < b < c$. Suppose that f is continuous on $[a,b]$, that g is continuous on $[b,c]$, and that $f(b) = g(b)$. Define h on $[a,c]$ by $h(x) := f(x)$ for $x \in [a,b]$ and $h(x) := g(x)$ for $x \in (b,c]$. Prove that h is continuous on $[a,c]$.
4. If $x \in \mathbf{R}$, we define $[x]$ to be the greatest integer $n \in \mathbf{Z}$ such that $n \leqslant x$. (Thus, for example, $[8.3] = 8$, $[\pi] = 3$, $[-\pi] = -4$.) The function $x \mapsto [x]$ is called the **greatest integer function**. Determine the points of continuity of the following functions:
   (a) $f(x) := [x] \quad (x \in \mathbf{R})$; (b) $g(x) := x[x] \quad (x \in \mathbf{R})$;
   (c) $h(x) := [\sin x] \quad (x \in \mathbf{R})$; (d) $k(x) := [1/x] \quad (x \neq 0)$.
5. Let f be defined for all $x \in \mathbf{R}$, $x \neq 2$, by $f(x) := (x^2 + x - 6)/(x-2)$. Can f be defined at $x = 2$ in such a way that f is continuous at this point?
6. Let $A \subseteq \mathbf{R}$ and let $f : A \to \mathbf{R}$ be continuous at a point $c \in A$. Show that for any $\varepsilon > 0$, there exists a neighborhood U of c such that if $x, y \in A \cap U$, then $|f(x) - f(y)| < \varepsilon$.
7. Let $f : \mathbf{R} \to \mathbf{R}$ be continuous at c and let $f(c) > 0$. Show that there exists a neighborhood U of c such that if $x \in U$ then $f(x) > 0$.
8. Let $f : \mathbf{R} \to \mathbf{R}$ be continuous on $\mathbf{R}$ and let $S := \{x \in \mathbf{R} : f(x) = 0\}$ be the "zero set" of f. Prove that S is a closed set in $\mathbf{R}$. [*Hint*: Theorem 3.4.7 can be used.]
9. Let $A \subseteq B \subseteq \mathbf{R}$, let $f : B \to \mathbf{R}$ and let g be the restriction of f to A (that is, $g(x) := f(x)$ for $x \in A$).
   (a) If f is continuous at $c \in A$, show that g is continuous at c.
   (b) Show by example that if g is continuous at c, it need not follow that f is continuous at c.
10. Show that the absolute value function $f(x) := |x|$ is continuous at every point $c \in \mathbf{R}$.
11. Let $K > 0$ and let $f : \mathbf{R} \to \mathbf{R}$ satisfy the condition $|f(x) - f(y)| \leqslant K\,|x - y|$ for all $x, y \in \mathbf{R}$. Show that f is continuous at every point $c \in \mathbf{R}$.
12. Suppose that $f : \mathbf{R} \to \mathbf{R}$ is continuous on $\mathbf{R}$ and that $f(r) = 0$ for every rational number r. Prove that $f(x) = 0$ for all $x \in \mathbf{R}$.
13. Define $g : \mathbf{R} \to \mathbf{R}$ by $g(x) := 2x$ for x rational, and $g(x) := x + 3$ for x irrational. Find all points for which g is continuous.
14. Let $A := (0,\infty)$ and let $k : A \to \mathbf{R}$ be defined as follows. For $x \in A$, x irrational, we define $k(x) := 0$; for $x \in A$ rational and of the form $x = m/n$ with natural numbers m,n having no common factors except 1, we define $k(x) := n$. Prove that k is unbounded on every open interval in $\mathbf{R}$. Conclude that k is not continuous at any point of A.
15. Let $f : (0,1) \to \mathbf{R}$ be bounded but such that $\lim_{x\to 0} f$ does not exist. Show that there are two sequences (x_n) and (y_n) in $(0,1)$ with $\lim (x_n) = 0 = \lim (y_n)$, but such that $\lim \big(f(x_n)\big)$ and $\lim \big(f(y_n)\big)$ exist but are not equal.

SECTION 4.5 Combinations of Continuous Functions

Let $A \subseteq \boldsymbol{R}$ and let f and g be functions that are defined on A to $\boldsymbol{R}$ and let $b \in \boldsymbol{R}$. In Definition 4.2.3 we defined the sum, difference, product, and multiple functions denoted by $f + g$, $f - g$, fg, bf. In addition, if $h : A \to \boldsymbol{R}$ is such that $h(x) \neq 0$ for all $x \in A$, then we defined the quotient function denoted by f/h.

The next result is similar to Theorem 4.2.4, from which it follows.

4.5.1 Theorem. *Let $A \subseteq \boldsymbol{R}$, let f and g be functions on A to $\boldsymbol{R}$, and let $b \in \boldsymbol{R}$. Suppose that $c \in A$ and that f and g are continuous at c.*

(a) *Then $f + g$, $f - g$, fg, and bf are continuous at c.*

(b) *If $h : A \to \boldsymbol{R}$ is continuous at $c \in A$ and if $h(x) \neq 0$ for all $x \in A$, then the quotient f/h is continuous at c.*

Proof. If $c \in A$ is not a cluster point of A, then the conclusion is automatic. Hence we assume that c is a cluster point of A.

(a) Since f and g are continuous at c, then

$$f(c) = \lim_{x \to c} f \qquad \text{and} \qquad g(c) = \lim_{x \to c} g.$$

Hence it follows from Theorem 4.2.4(a) that

$$(f + g)(c) = f(c) + g(c) = \lim_{x \to c} (f + g).$$

Therefore $f + g$ is continuous at c. The remaining assertions in part (a) are proved in a similar fashion.

(b) Since $c \in A$, then $h(c) \neq 0$. But since $h(c) = \lim_{x \to c} h$, it follows from Theorem 4.2.4(b) that

$$\frac{f}{h}(c) = \frac{f(c)}{h(c)} = \frac{\lim\limits_{x \to c} f}{\lim\limits_{x \to c} h} = \lim_{x \to c} \left(\frac{f}{h}\right).$$

Therefore f/h is continuous at c. Q.E.D.

The next result is an immediate consequence of Theorem 4.5.1, applied to every point of A. However, since it is an extremely important result, we shall state it formally.

4.5.2 Theorem. *Let $A \subseteq \boldsymbol{R}$, let f and g be continuous on A to $\boldsymbol{R}$, and let $b \in \boldsymbol{R}$.*

(a) *The functions $f + g$, $f - g$, fg, and bf are continuous on A.*

(b) *If $h : A \to \boldsymbol{R}$ is continuous on A and $h(x) \neq 0$ for $x \in A$, then the quotient f/h is continuous on A.*

4.5.3 Remark. To define quotients, it is sometimes more convenient to proceed as follows. If $\varphi : A \to \boldsymbol{R}$, let

$$A_1 := \{x \in A : \varphi(x) \neq 0\}.$$

We can define the quotient f/φ on the set A_1 by

$$(*) \qquad \left(\frac{f}{\varphi}\right)(x) := \frac{f(x)}{\varphi(x)} \quad \text{for} \quad x \in A_1.$$

If φ is continuous at a point $c \in A_1$, it is clear that the restriction φ_1 of φ to A_1 is also continuous at c. Therefore it follows from Theorem 4.5.1(b) applied to φ_1 that f/φ_1 is continuous at $c \in A_1$. Since $(f/\varphi)(x) = (f/\varphi_1)(x)$ for $x \in A_1$ it follows that f/φ is continuous at $c \in A_1$. Similarly, *if f and φ are continuous on A, then the function f/φ, defined on A_1 by $(*)$, is continuous on A_1.*

4.5.4 Examples. (a) Polynomial functions.

If p is a polynomial function, so that $p(x) = a_n x^n + a_{n-1} x^{n-1} + \cdots + a_1 x + a_0$ for all $x \in \boldsymbol{R}$, then it follows from Example 4.2.5(f) that $p(c) = \lim_{x \to c} p$ for any $c \in \boldsymbol{R}$. Thus *a polynomial function is continuous on $\boldsymbol{R}$.*

(b) Rational functions.

If p and q are polynomial functions on $\boldsymbol{R}$, then there are at most a finite number $\alpha_1, \ldots, \alpha_m$ of real roots of q. If $x \notin \{\alpha_1, \ldots, \alpha_m\}$ then $q(x) \neq 0$ so that we can define the rational function r by

$$r(x) := \frac{p(x)}{q(x)} \quad \text{for} \quad x \notin \{\alpha_1, \ldots, \alpha_m\}.$$

It was seen in Example 4.2.5(g) that if $q(c) \neq 0$, then

$$r(c) = \frac{p(c)}{q(c)} = \lim_{x \to c} \frac{p(x)}{q(x)} = \lim_{x \to c} r(x).$$

In other words, r is continuous at c. Since c is any real number that is not a root of q, we infer that *a rational function is continuous at every real number for which it is defined.*

(c) We shall show that the sine function sin is continuous on $\boldsymbol{R}$.

To do so we make use of the following properties of the sine and cosine function that will be proved later. For all $x, y, z \in \boldsymbol{R}$ we have:

$$|\sin z| \leq |z|, \qquad |\cos z| \leq 1,$$

$$\sin x - \sin y = 2 \sin[\tfrac{1}{2}(x-y)] \cos[\tfrac{1}{2}(x+y)].$$

Hence if $c \in \boldsymbol{R}$, then we have

$$|\sin x - \sin c| \leq 2 \cdot \tfrac{1}{2} |x - c| \cdot 1 = |x - c|.$$

Therefore sin is continuous at c. Since $c \in \boldsymbol{R}$ is arbitrary, it follows that sin is continuous on $\boldsymbol{R}$.

(d) The cosine function is continuous on $\boldsymbol{R}$.

We make use of the following properties of the sine and cosine functions that will be proved later. For all $x, y, z \in \boldsymbol{R}$ we have:

$$|\sin z| \leqslant |z|, \qquad |\sin z| \leqslant 1,$$

$$\cos x - \cos y = 2 \sin [\tfrac{1}{2}(x + y)] \sin [\tfrac{1}{2}(y - x)].$$

Hence if $c \in \boldsymbol{R}$, then we have

$$|\cos x - \cos c| \leqslant 2 \cdot 1 \cdot \tfrac{1}{2} |c - x| = |x - c|.$$

Therefore cos is continuous at c. Since $c \in \boldsymbol{R}$ is arbitrary, it follows that cos is continuous on $\boldsymbol{R}$.

(e) The functions tan, cot, sec, csc are continuous where they are defined. For example, the cotangent function is defined by

$$\cot x := \frac{\cos x}{\sin x}$$

provided $\sin x \neq 0$ (that is, provided $x \neq n\pi$, $n \in \boldsymbol{Z}$). Since sin and cos are continuous on $\boldsymbol{R}$, it follows from Remark 4.5.3 that the function cot is continuous on its domain. The other trigonometric functions are treated similarly.

4.5.5 Theorem. *Let $A \subseteq \boldsymbol{R}$, let $f : A \to \boldsymbol{R}$, and let $|f|$ be defined for $x \in A$ by $|f|(x) := |f(x)|$.*

(a) *If f is continuous at a point $c \in A$, then $|f|$ is continuous at c.*

(b) *If f is continuous on A, then $|f|$ is continuous on A.*

Proof. This is an immediate consequence of Theorem 4.2.9. Q.E.D.

4.5.6 Theorem. *Let $A \subseteq \boldsymbol{R}$, let $f : A \to \boldsymbol{R}$, and let $f(x) \geqslant 0$ for all $x \in A$. We let $\sqrt{f}$ be defined for $x \in A$ by $(\sqrt{f})(x) := \sqrt{f(x)}$.*

(a) *If f is continuous at a point $c \in A$, then $\sqrt{f}$ is continuous at c.*

(b) *If f is continuous on A, then $\sqrt{f}$ is continuous on A.*

Proof. This is an immediate consequence of Theorem 4.2.10. Q.E.D.

4.5.7 Definition. Let $A \subseteq \boldsymbol{R}$ and let $f, g : A \to \boldsymbol{R}$. We define the functions $\sup(f,g) : A \to \boldsymbol{R}$ and $\inf(f,g) : A \to \boldsymbol{R}$ by

$$\sup(f,g)(x) := \sup\{f(x), g(x)\}, \qquad \inf(f,g)(x) := \inf\{f(x), g(x)\},$$

for each $x \in A$. (Sometimes $\sup(f,g)$ is denoted by $\max(f,g)$, and $\inf(f,g)$ is denoted by $\min(f,g)$.)

4.5.8 Theorem. *Let $A \subseteq \mathbf{R}$ and let $f,g : A \to \mathbf{R}$.*

(a) *If f and g are continuous at a point $c \in A$, then* $\sup(f,g)$ *and* $\inf(f,g)$ *are continuous at c.*

(b) *If f and g are continuous on A, then* $\sup(f,g)$ *and* $\inf(f,g)$ *are continuous on A.*

Proof. If $x \in A$, then the reader should show that

$$\begin{aligned} \sup(f,g)(x) &= \sup\{f(x), g(x)\} \\ &= \tfrac{1}{2}(f(x) + g(x)) + \tfrac{1}{2}|f(x) - g(x)|. \end{aligned}$$

Hence we have

$$\sup(f,g) = \tfrac{1}{2}(f + g) + \tfrac{1}{2}|f - g|.$$

It follows from Theorem 4.5.5 that $\sup(f,g)$ is continuous at any point where both f and g are continuous.

Similarly, the reader can show that

$$\begin{aligned} \inf(f,g)(x) &= \inf\{f(x), g(x)\} \\ &= \tfrac{1}{2}(f(x) + g(x)) - \tfrac{1}{2}|f(x) - g(x)| \end{aligned}$$

for all $x \in A$, so that $\inf(f,g)$ is continuous at any point where both f and g are continuous. Q.E.D.

Composition of Continuous Functions

We shall now show that if the function $f : A \to \mathbf{R}$ is continuous at a point c and if $g : B \to \mathbf{R}$ is continuous at $b = f(c)$, then the composition $g \circ f$ is continuous at c. In order to assure that $g \circ f$ is defined on all of A, we also need to assume that $f(A) \subseteq B$.

4.5.9 Theorem. *Let $A, B \subseteq \mathbf{R}$ and let $f : A \to \mathbf{R}$ and $g : B \to \mathbf{R}$ be functions such that $f(A) \subseteq B$. If f is continuous at a point $c \in A$ and g is continuous at $b = f(c) \in B$, then the composition $g \circ f : A \to \mathbf{R}$ is continuous at c.*

Proof. Let W be a neighborhood of $g(b)$. Since g is continuous at b, there is a neighborhood V of $b = f(c)$ such that if $y \in B \cap V$ then $g(y) \in W$. Since f is continuous at c, there is a neighborhood U of c such that if $x \in A \cap U$, then $f(x) \in V$. (See Figure 4.5.1.) Since $f(A) \subseteq B$, it follows that if $x \in A \cap U$, then $f(x) \in B \cap V$ so that $g \circ f(x) = g(f(x)) \in W$. But since W is an arbitrary neighborhood of $g(b)$, this implies that $g \circ f$ is continuous at c. Q.E.D.

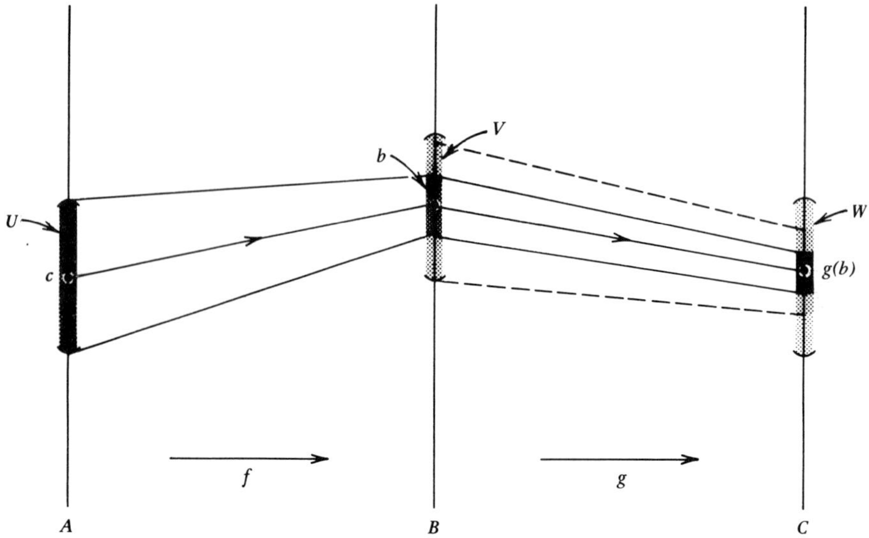

FIGURE 4.5.1 The composition of *f* and *g*.

4.5.10 Theorem. *Let $A, B \subseteq \boldsymbol{R}$, let $f : A \to \boldsymbol{R}$ be continuous on A, and let $g : B \to \boldsymbol{R}$ be continuous on B. If $f(A) \subseteq B$, then the composite function $g \circ f : A \to \boldsymbol{R}$ is continuous on A.*

Proof. The theorem follows immediately from the preceding result, if f and g are continuous at every point of A and B, respectively. Q.E.D.

Theorems 4.5.9 and 4.5.10 are very useful in establishing that certain functions are continuous. They can be used in many situations where it would be difficult to apply the definition directly.

4.5.11 Examples. (a) Let $g_1(x) := |x|$ for $x \in \boldsymbol{R}$. It follows from the Triangle Inequality (see Corollary 2.3.4) that

$$|g_1(x) - g_1(c)| \leqslant |x - c|$$

for all $x, c \in \boldsymbol{R}$. Hence g_1 is continuous at $c \in \boldsymbol{R}$. If $f : A \to \boldsymbol{R}$ is any function that is continuous on A, then Theorem 4.5.10 implies that $g_1 \circ f = |f|$ is continuous on A. This gives another proof of Theorem 4.5.5.

(b) Let $g_2(x) := \sqrt{x}$ for $x \geqslant 0$. It follows from Theorems 3.2.10 and 4.4.3(iv) that g_2 is continuous at any number $c \geqslant 0$. If $f : A \to \boldsymbol{R}$ is continuous on A and if $f(x) \geqslant 0$ for all $x \in A$, then it follows from Theorem 4.5.10 that $g_2 \circ f = \sqrt{f}$ is continuous on A. This gives another proof of Theorem 4.5.6.

(c) Let $g_3(x) := \sin x$ for $x \in \boldsymbol{R}$. We have seen in Example 4.5.4(c) that g_3

is continuous on $\boldsymbol{R}$. If $f: A \to \boldsymbol{R}$ is continuous on A, then it follows from Theorem 4.5.10 that $g_3 \circ f$ is continuous on A.

In particular, if $f(x) := 1/x$ for $x \neq 0$, then the function $g(x) := \sin(1/x)$ is continuous at every point $c \neq 0$. [We have seen, in Example 4.4.7(a), that g is *not* continuous at 0.]

Exercises for Section 4.5

1. Determine the points of continuity of the following functions and state which theorems are used in each case.

   (a) $f(x) := \dfrac{x^2 + 2x + 1}{x^2 + 1} \quad (x \in \boldsymbol{R})$;

   (b) $g(x) := \sqrt{x + \sqrt{x}} \quad (x \geq 0)$;

   (c) $h(x) := \dfrac{\sqrt{1 + |\sin x|}}{x} \quad (x \neq 0)$;

   (d) $k(x) := \cos\sqrt{1 + x^2} \quad (x \in \boldsymbol{R})$.

2. Show that if $f: A \to \boldsymbol{R}$ is continuous on $A \subseteq \boldsymbol{R}$ and if $n \in \boldsymbol{N}$, then the function f^n defined by $f^n(x) := (f(x))^n$ for $x \in A$, is continuous on A.
3. Give an example of functions f and g that are both discontinuous at a point c in $\boldsymbol{R}$ such that: (a) the sum $f + g$ is continuous at c, (b) the product fg is continuous at c.
4. Let $x \mapsto [x]$ denote the greatest integer function (see Exercise 4.4.4). Determine the points of continuity of the function $f(x) := x - [x]$, $x \in \boldsymbol{R}$.
5. Let g be defined on $\boldsymbol{R}$ by $g(1) := 0$, and $g(x) := 2$ if $x \neq 1$, and let $f(x) := x + 1$ for all $x \in \boldsymbol{R}$. Show that $\lim_{x \to 0} g \circ f \neq g \circ f(0)$. Why doesn't this contradict Theorem 4.5.9?
6. Let f, g be defined on $\boldsymbol{R}$ and let $c \in \boldsymbol{R}$. Suppose that $\lim_{x \to c} f = b$ and that g is continuous at b. Show that $\lim_{x \to c} g \circ f = g(b)$. (Compare this result with Theorem 4.5.9 and the preceding exercise.)
7. Give an example of a function $f: [0,1] \to \boldsymbol{R}$ that is discontinuous at every point of $[0,1]$ but such that $|f|$ is continuous on $[0,1]$.
8. Let f, g be continuous from $\boldsymbol{R}$ to $\boldsymbol{R}$, and suppose that $f(r) = g(r)$ for all rational numbers r. Is it true that $f(x) = g(x)$ for all $x \in \boldsymbol{R}$?
9. Let $h: \boldsymbol{R} \to \boldsymbol{R}$ be continuous on $\boldsymbol{R}$ satisfying $h(m/2^n) = 0$ for all $m \in \boldsymbol{Z}$, $n \in \boldsymbol{N}$. Show that $h(x) = 0$ for all $x \in \boldsymbol{R}$.
10. If $f: \boldsymbol{R} \to \boldsymbol{R}$ is continuous on $\boldsymbol{R}$, show that the set $P := \{x \in \boldsymbol{R}: f(x) > 0\}$ is an open set in $\boldsymbol{R}$.
11. If f and g are continuous on $\boldsymbol{R}$, show that the set $S := \{x \in \boldsymbol{R}: f(x) > g(x)\}$ is an open set in $\boldsymbol{R}$.
12. A function $f: \boldsymbol{R} \to \boldsymbol{R}$ is said to be **additive** if $f(x + y) = f(x) + f(y)$ for all x, y in $\boldsymbol{R}$. Prove that if f is continuous at some point x_0, then it is continuous at every point of $\boldsymbol{R}$. (See Exercise 4.2.12.)
13. Suppose that f is a continuous additive function on $\boldsymbol{R}$. If $c := f(1)$, show that we have $f(x) = cx$ for all $x \in \boldsymbol{R}$. [*Hint*: First show that if r is a rational number, then $f(r) = cr$.]

14. Let $g : \mathbf{R} \to \mathbf{R}$ satisfy the relation $g(x + y) = g(x)\,g(y)$ for all x, y in $\mathbf{R}$. Show that if g is continuous at $x = 0$, then g is continuous at every point of $\mathbf{R}$. Also if we have $g(a) = 0$ for some $a \in \mathbf{R}$, then $g(x) = 0$ for all $x \in \mathbf{R}$.

SECTION 4.6 Continuous Functions on Intervals

Functions that are continuous on bounded intervals have a number of important properties that are not possessed by general continuous functions. In this section we shall establish some rather deep results that are of considerable importance and that will be applied later.

4.6.1 Definition. A function $f : A \to \mathbf{R}$ is said to be **bounded on** A if there exists a constant $M > 0$ such that $|f(x)| \leq M$ for all $x \in A$.

In other words, a function is bounded if its range is a bounded set in $\mathbf{R}$. We note that a continuous function is not necessarily bounded. For example, the function $f(x) := 1/x$ is continuous on the set $A := \{x \in \mathbf{R} : x > 0\}$. However, f is not bounded on A. In fact, $f(x) = 1/x$ is not even bounded when restricted to the set $B := \{x \in \mathbf{R} : 0 < x < 1\}$. However, $f(x) = 1/x$ *is* bounded when restricted to the set $C := \{x \in \mathbf{R} : 1 \leq x\}$, even though the set C is *not* bounded.

4.6.2 Boundedness Theorem.† *Let $I := [a,b]$ be a closed bounded interval and let $f : I \to \mathbf{R}$ be continuous on I. Then f is bounded on I.*

Proof. Suppose that f is not bounded on I. Then, for any $n \in \mathbf{N}$ there is a number $x_n \in I$ such that $|f(x_n)| > n$. Since I is bounded, the sequence $X := (x_n)$ is bounded. Therefore, it follows from the Bolzano-Weierstrass Theorem 3.4.6 that there is a subsequence $X' = (x_{n_r})$ of X that converges to a number x. Since I is closed and the elements of X' belong to I, it follows from Theorem 3.4.7 that $x \in I$. Hence f is continuous at x, so that $\big(f(x_{n_r})\big)$ converges to $f(x)$. We then conclude from Theorem 3.2.2 that the convergent sequence $\big(f(x_{n_r})\big)$ must be bounded. But this is a contradiction since

$$|f(x_{n_r})| > n_r \geq r \qquad \text{for} \quad r \in \mathbf{N}.$$

Therefore the supposition that the continuous function f is not bounded on the closed bounded interval I leads to a contradiction. Q.E.D.

4.6.3 Definition. Let $A \subseteq \mathbf{R}$ and let $f : A \to \mathbf{R}$. We say that f **has an absolute maximum on** A if there is a point $x^* \in A$ such that

$$f(x^*) \geq f(x) \qquad \text{for all} \quad x \in A.$$

† This theorem, as well as 4.6.4, is true for an arbitrary closed bounded set. For further developments, see Section 4.9.

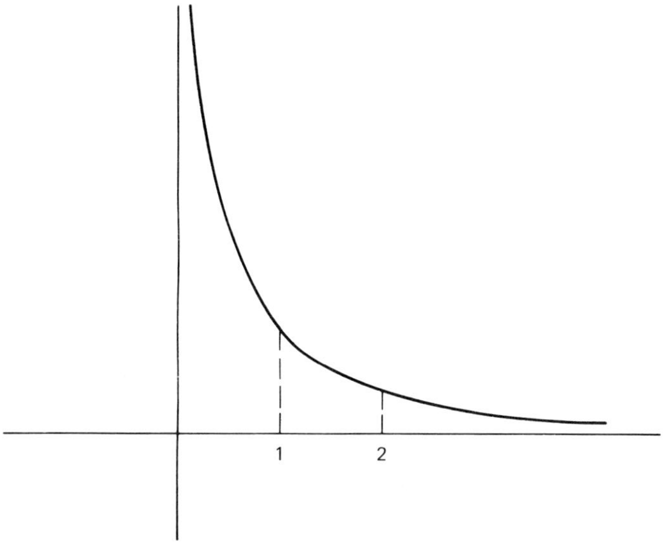

FIGURE 4.6.1 The function $f(x) = \frac{1}{x}$ $(x > 0)$.

We say that f **has an absolute minimum on** A if there is a point $x_* \in A$ such that

$$f(x_*) \leqslant f(x) \qquad \text{for all} \quad x \in A.$$

We say that x^* is an **absolute maximum point** for f on A, and that x_* is an **absolute minimum point** for f on A, if they exist.

We note that a continuous function on a set A does not necessarily have an absolute maximum or an absolute minimum on the set. For example, $f(x) := 1/x$ has neither an absolute maximum nor an absolute minimum on the set $A := \{x \in \boldsymbol{R} : x > 0\}$. (See Figure 4.6.1). There can be no absolute maximum for f on A since f is not bounded above on A, and there is no point at which f attains the value $0 = \inf\{f(x) : x \in A\}$. The same function has neither an absolute maximum nor an absolute minimum when it is restricted to the set $\{x \in \boldsymbol{R} : 0 < x < 1\}$, while it has *both* an absolute maximum and an absolute minimum when it is restricted to the set $\{x \in \boldsymbol{R} : 1 \leqslant x \leqslant 2\}$. In addition $f(x) = 1/x$ has an absolute maximum but no absolute minimum when restricted to the set $\{x \in \boldsymbol{R} : x \geqslant 1\}$, but no absolute maximum and no absolute minimum when restricted to the set $\{x \in \boldsymbol{R} : x > 1\}$.

It is readily seen that if a function has an absolute maximum point, then this point is not necessarily uniquely determined. For example, the function $g(x) := x^2$ defined for $x \in A := [-1, +1]$ has the two points $x = \pm 1$ giving the absolute maximum on A, and the single point $x = 0$ yields its absolute minimum on A. (See Figure 4.6.2.) To pick an extreme example, the constant function $h(x) := 1$ for $x \in \boldsymbol{R}$ is such that *every point* of $\boldsymbol{R}$ is both an absolute maximum and an absolute minimum point for f.

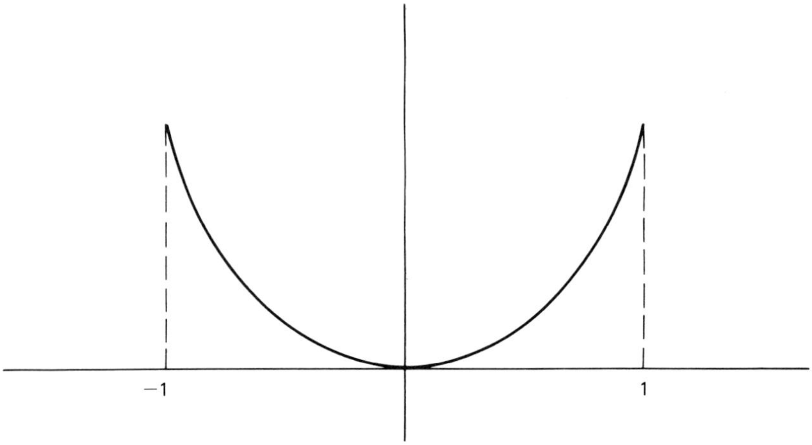

FIGURE 4.6.2 The function $g(x) = x^2$ $(|x| \le 1)$.

4.6.4 Maximum-Minimum Theorem. *Let $I := [a,b]$ be a closed bounded interval and let $f : I \to \mathbf{R}$ be continuous on I. Then f has an absolute maximum and an absolute minimum on I.*

Proof. Consider the non-empty set $f(I) := \{f(x) : x \in I\}$ of values of f on I. In the previous Theorem 4.6.2 it was established that $f(I)$ is a bounded subset of $\mathbf{R}$. Let $s^* := \sup f(I)$ and $s_* := \inf f(I)$. We claim that there exist points x^* and x_* in I such that $s^* = f(x^*)$ and $s_* = f(x_*)$. We shall establish the existence of the point x^*, leaving the proof of the existence of x_* to the reader.

Since $s^* = \sup f(I)$, if $n \in \mathbf{N}$, then the number $s^* - 1/n$ is not an upper bound of the set $f(I)$. Consequently there exists a number $x_n \in I$ such that

$$(\dagger) \qquad s^* - \frac{1}{n} < f(x_n) \le s^* \qquad \text{for} \quad n \in \mathbf{N}.$$

Since I is bounded, the sequence $X := (x_n)$ is bounded. Therefore, by the Bolzano-Weierstrass Theorem 3.4.6, there is a subsequence $X' = (x_{n_r})$ of X that converges to some number x^*. Since I is closed and the elements of X' belong to I, it follows from Theorem 3.4.7 that $x^* \in I$. Therefore f is continuous at x^* so that $\lim \big(f(x_{n_r})\big) = f(x^*)$. Since it follows from $(\dagger)$ that

$$s^* - \frac{1}{n_r} < f(x_{n_r}) \le s^* \qquad \text{for} \quad r \in \mathbf{N},$$

we conclude from the Squeeze Theorem 3.2.7 that $\lim \big(f(x_{n_r})\big) = s^*$. Therefore we have

$$f(x^*) = \lim \big(f(x_{n_r})\big) = s^* = \sup f(I).$$

We conclude that x^* is an absolute maximum point of f on I. Q.E.D.

The next result gives a basis for the location of roots of continuous functions. The proof given also provides an algorithm for the calculation of the root and can be readily programmed for a computer. An alternative proof is indicated in Exercise 4.6.8.

4.6.5 Location of Roots Theorem. *Let I be an interval and let $f : I \to \mathbf{R}$ be continuous on I. If $\alpha < \beta$ are numbers in I such that $f(\alpha) < 0 < f(\beta)$ (or such that $f(\alpha) > 0 > f(\beta)$), then there exists a number $c \in (\alpha,\beta)$ such that $f(c) = 0$.*

Proof. Let $I_1 := [\alpha,\beta]$ and $\gamma := \frac{1}{2}(\alpha + \beta)$. If $f(\gamma) = 0$ we take $c := \gamma$ and the proof is complete. If $f(\gamma) > 0$ we set $\alpha_2 := \alpha$, $\beta_2 := \gamma$, while if $f(\gamma) < 0$ we set $\alpha_2 := \gamma$, $\beta_2 := \beta$. In either case we let $I_2 := [\alpha_2,\beta_2]$, whence $f(\alpha_2) < 0$ and $f(\beta_2) > 0$. We continue this bisection process.

Suppose that the intervals $I_1, I_2, \ldots, I_k := [\alpha_k,\beta_k]$ have been obtained by successive bisection and are such that $f(\alpha_k) < 0$ and $f(\beta_k) > 0$. Let $\gamma_k := \frac{1}{2}(\alpha_k + \beta_k)$. If $f(\gamma_k) = 0$, we take $c := \gamma_k$ and the proof is complete. If $f(\gamma_k) > 0$ we set $\alpha_{k+1} := \alpha_k$, $\beta_{k+1} := \gamma_k$, while if $f(\gamma_k) < 0$ we set $\alpha_{k+1} := \gamma_k$, $\beta_{k+1} := \beta_k$. In either case we let $I_{k+1} := [\alpha_{k+1},\beta_{k+1}]$ whence

$$f(\alpha_{k+1}) < 0 \qquad \text{and} \qquad f(\beta_{k+1}) > 0.$$

If this process terminates by locating a point γ_n such that $f(\gamma_n) = 0$, the proof is complete. If this process does not terminate, we obtain a nested sequence of closed bounded intervals $I_n = [\alpha_n,\beta_n]$, $n \in \mathbf{N}$. Since these intervals are found by repeated bisection, we have $\beta_n - \alpha_n = (\beta - \alpha)/2^{n-1}$. It follows from the Nested Intervals Property 2.5.1 that there exists a point c belonging to I_n for all $n \in \mathbf{N}$. Since $\alpha_n \leq c \leq \beta_n$ for all $n \in \mathbf{N}$, we have $0 \leq c - \alpha_n \leq \beta_n - \alpha_n = (\beta - \alpha)/2^{n-1}$, and $0 \leq \beta_n - c \leq \beta_n - \alpha_n = (\beta - \alpha)/2^{n-1}$. Hence it follows that $c = \lim(\alpha_n)$ and $c = \lim(\beta_n)$. Since f is continuous at c, we have

$$\lim\big(f(\alpha_n)\big) = f(c) = \lim\big(f(\beta_n)\big).$$

On the other hand, since $f(\beta_n) \geq 0$ for all $n \in \mathbf{N}$, it follows from Theorem 3.2.4 that $f(c) = \lim\big(f(\beta_n)\big) \geq 0$. Also, since $f(\alpha_n) \leq 0$ for all $n \in \mathbf{N}$, it follows from the same result that $f(c) = \lim\big(f(\alpha_n)\big) \leq 0$. Therefore we must have $f(c) = 0$. Consequently c is a root of f. Q.E.D.

The next result is a generalization of the previous one. It assures that a continuous function on an interval takes on (at least once) any number that lies between two of its values.

4.6.6 Bolzano's Intermediate Value Theorem. *Let I be an interval and let $f : I \to \mathbf{R}$ be continuous on I. If $a, b \in I$ and if $k \in \mathbf{R}$ satisfies $f(a) < k < f(b)$, then there exists a point $c \in I$ between a and b and such that $f(c) = k$.*

Proof. Suppose that $a < b$ and let $g(x) := f(x) - k$; then $g(a) < 0 < g(b)$. By the Location of Roots Theorem 4.6.5 there exists a point c with $a < c < b$ such that $0 = g(c) = f(c) - k$. Therefore $f(c) = k$.

If $b < a$, let $h(x) := k - f(x)$ so that $h(b) < 0 < h(a)$. Therefore there exists a point c with $b < c < a$ such that $0 = h(c) = k - f(c)$, whence $f(c) = k$.

Q.E.D.

4.6.7 Corollary. *Let $I = [a,b]$ be a closed, bounded interval and let $f : I \to \mathbf{R}$ be continuous on I. If $k \in \mathbf{R}$ is any number satisfying*

$$\inf f(I) \leqslant k \leqslant \sup f(I),$$

then there exists a number $c \in I$ such that $f(c) = k$.

Proof. It follows from the Maximum-Minimum Theorem 4.6.4 that there are points c_* and c^* in I such that

$$\inf f(I) = f(c_*) \leqslant k \leqslant f(c^*) = \sup f(I).$$

The conclusion now follows from Theorem 4.6.6. Q.E.D.

The next theorem summarizes the main results of this section. It states that the image of a closed bounded interval under a continuous function is also a closed bounded interval. The end points of the image interval are the absolute minimum and absolute maximum values of the function, and the statement that all values between the absolute minimum and the absolute maximum values belong to the image is a way of describing Bolzano's Intermediate Value Theorem.

4.6.8 Theorem. *Let I be a closed bounded interval and let $f : I \to \mathbf{R}$ be continuous on I. Then the set $f(I) := \{f(x) : x \in I\}$ is a closed bounded interval.*

Proof. If we let $m := \inf f(I)$ and $M := \sup f(I)$, then we know from the Maximum-Minimum Theorem 4.6.4 that m and M belong to $f(I)$. Moreover, we have $f(I) \subseteq [m,M]$. On the other hand, if k is any element of $[m,M]$, then it follows from the preceding corollary that there exists a point $c \in I$ such that $k = f(c)$. Hence, $k \in f(I)$ and we conclude that $[m,M] \subseteq f(I)$. Therefore, $f(I)$ is the interval $[m,M]$. Q.E.D.

Warning. If $I := [a,b]$ is an interval and $f : I \to \mathbf{R}$ is continuous on I, we have proved that $f(I)$ is the interval $[m,M]$. We have *not* proved that $f(I)$ is the interval $[f(a), f(b)]$. (See Figure 4.6.3.)

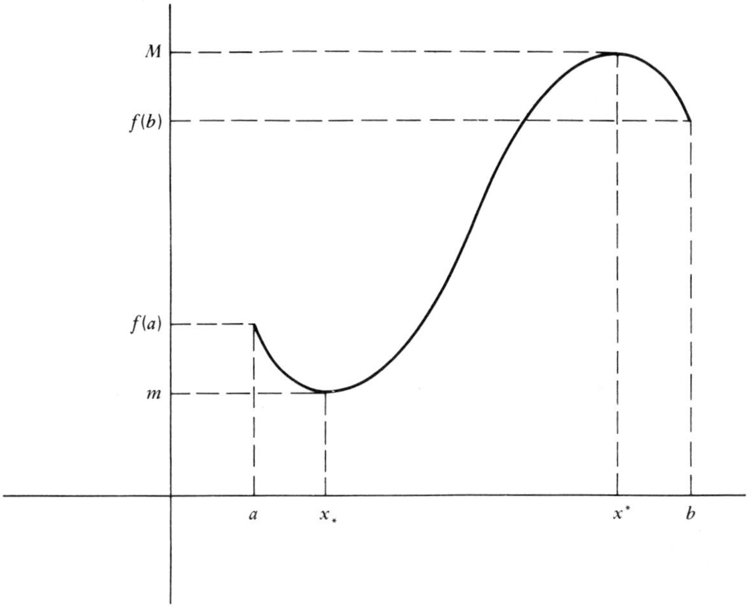

FIGURE 4.6.3 $f(I) = [m,M]$.

The preceding theorem is a "preservation" theorem in the sense that it states that the continuous image of a closed bounded interval is a set of the same type. The next theorem extends this result to general intervals. However, it should be noted that although the continuous image of an interval is shown to be an interval, it is *not* true that the image interval necessarily has the same form as the domain interval. For example, the continuous image of an open interval need not be an open interval, and the continuous image of an unbounded closed interval need not be a closed interval. Indeed, if $f(x) := 1/(x^2 + 1)$ for $x \in \boldsymbol{R}$, then f is continuous on $\boldsymbol{R}$ [see Example 4.5.4(b)]. It is easy to see that if $I_1 := (-1,1)$, then $f(I_1) = (\frac{1}{2},1]$, which is not an open interval. Also, if $I_2 := [0,\infty)$, then $f(I_2) = (0,1]$ which is not a closed interval. (See Figure 4.6.4 on the next page.)

In order to prove the Preservation of Intervals Theorem 4.6.10, we need the following lemma characterizing intervals.

4.6.9 Lemma. *Let $S \subseteq \boldsymbol{R}$ be a non-void set with the property*

$$(*) \qquad \text{if } x, y \in S \text{ and } x < y, \qquad \text{then } [x,y] \subseteq S.$$

Then S is an interval.

Proof. We shall assume that S has at least two points. There are four main cases to be considered: (i) S is bounded, (ii) S is bounded above but not bounded

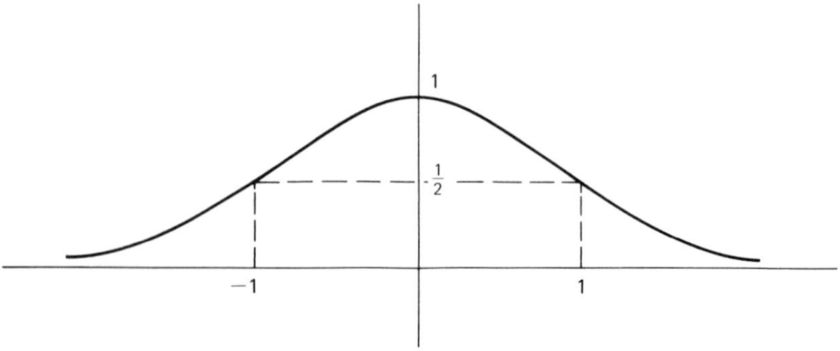

FIGURE 4.6.4 Graph of $f_1(x) = \dfrac{1}{x^2 + 1}$ $(x \in \boldsymbol{R})$.

below, (iii) S is bounded below but not bounded above, and (iv) S is neither bounded above nor bounded below.

(i) Let $a := \inf S$ and $b := \sup S$. If $s \in S$ then $a \leq s \leq b$ so that $s \in [a,b]$; since $s \in S$ is arbitrary, we conclude that $S \subseteq [a,b]$.

On the other hand we claim that $(a,b) \subseteq S$. For if $z \in (a,b)$, then z is not a lower bound of S so there exists $x \in S$ with $x < z$. Also z is not an upper bound of S so there exists $y \in S$ with $z < y$. Consequently $z \in [x,y]$ and property $(*)$ implies that $z \in [x,y] \subseteq S$. Since z is an arbitrary element of (a,b), we deduce that $(a,b) \subseteq S$.

If $a \notin S$ and $b \notin S$, then we have $S = (a,b)$; if $a \notin S$ and $b \in S$, we have $S = (a,b]$; if $a \in S$ and $b \notin S$, we have $S = [a,b)$; and if $a \in S$ and $b \in S$,we have $S = [a,b]$.

(ii) Let $b := \sup S$. If $s \in S$ then $s \leq b$ so we must have $S \subseteq (-\infty,b]$.

We claim that $(-\infty,b) \subseteq S$. For, if $z \in (-\infty,b)$, the argument given above implies that there exist x, $y \in S$ such that $z \in [x,y] \subseteq S$. Therefore $(-\infty,b) \subseteq S$.

If $b \notin S$, then we have $S = (-\infty,b)$; if $b \in S$, we have $S = (-\infty,b]$.

(iii) Let $a := \inf S$ and argue as in (ii). In this case we have $S = (a,\infty)$ if $a \notin S$, and $S = [a,\infty)$ if $a \in S$.

(iv) If $z \in \boldsymbol{R}$, the argument given above implies that there exist $x,y \in S$ such that $z \in [x,y] \subseteq S$. Therefore $\boldsymbol{R} \subseteq S$, so that $S = (-\infty,\infty)$.

Thus, in any case, S is an interval. Q.E.D.

4.6.10 Preservation of Intervals Theorem. *Let I be an interval and let $f : I \to \boldsymbol{R}$ be continuous on I. Then the set $f(I)$ is an interval.*

Proof. Let $\alpha,\beta \in f(I)$ with $\alpha < \beta$; then there exist points a, $b \in I$ such that $\alpha = f(a)$ and $\beta = f(b)$. Further, it follows from Bolzano's Intermediate Value Theorem 4.6.6 that if $k \in [\alpha,\beta]$ then there exists a number $c \in I$ with $k = f(c) \in f(I)$. Therefore $[\alpha,\beta] \subseteq f(I)$, showing that $f(I)$ possesses property $(*)$ of the preceding lemma. Therefore $f(I)$ is an interval. Q.E.D.

We have noted above that the continuous image of an open [respectively, closed] set is not necessarily open [respectively, closed]. Hence it is rather surprising that the "inverse images" of continuous functions do preserve open and closed sets.

Let $A \subseteq \mathbf{R}$ and let $f : A \to \mathbf{R}$. If $H \subseteq \mathbf{R}$ we recall from Definition 1.2.9 that the set

$$\{x \in A : f(x) \in H\}$$

is called the **inverse image** of H under f, and is denoted by $f^{-1}(H)$.

4.6.11 Global Continuity Theorem. *If $f : \mathbf{R} \to \mathbf{R}$, then the following statements are equivalent.*

(a) *f is continuous on $\mathbf{R}$.*

(b) *If F is a closed subset of $\mathbf{R}$, then $f^{-1}(F)$ is closed in $\mathbf{R}$.*

(c) *If G is an open subset of $\mathbf{R}$, then $f^{-1}(G)$ is open in $\mathbf{R}$.*

Proof. (a) $\Rightarrow$ (b) Let $F \subseteq \mathbf{R}$ be a closed set and let $F_1 := f^{-1}(F)$. We must show that F_1 is a closed set; we shall make use of Theorem 3.4.7. Let $X = (x_n)$ be a sequence of points in F_1 that converges to x_0. Since f is continuous at x_0, we have $f(x_0) = \lim (f(x_n))$. But since $x_n \in F_1$, we have $f(x_n) \in F$; it follows from Theorem 3.4.7 that $f(x_0) = \lim (f(x_n)) \in F$ so that $x_0 \in F_1$. Since (x_n) is an arbitrary sequence in F_1, we deduce from Theorem 3.4.7 that F_1 is closed.

(b) $\Rightarrow$(c) Let $G \subseteq \mathbf{R}$ be open; then $F := \mathscr{C}(G)$ is closed and it follows from (b) that $F_1 := f^{-1}(F)$ is closed. We claim that $G_1 := f^{-1}(G)$ is the complement of F_1. To prove this, note that $y \in G_1 = f^{-1}(G) \Leftrightarrow f(y) \in G = \mathscr{C}(F) \Leftrightarrow f(y) \notin F \Leftrightarrow y \notin f^{-1}(F) = F_1 \Leftrightarrow y \in \mathscr{C}(F_1)$. Thus $G_1 = \mathscr{C}(F_1)$ showing that G_1 is open.

(c) $\Rightarrow$ (a) Let $c \in \mathbf{R}$ be given and let V_ε be an ε-neighborhood of $f(c)$. Since V_ε is an open set it follows from (c) that $U_\varepsilon := f^{-1}(V_\varepsilon)$ is an open set and that $f(U_\varepsilon) \subseteq V_\varepsilon$. Since $c \in U_\varepsilon$ (because $f(c) \in V_\varepsilon$), we conclude that U_ε is a neighborhood of c. Therefore f is continuous at c. Q.E.D.

In courses in topology, continuous functions are often defined as those functions whose inverse images of open sets are open. The Global Continuity Theorem justifies that definition.

Exercises for Section 4.6

1. Let $I := [a,b]$ and let $f : I \to \mathbf{R}$ be a continuous function such that $f(x) > 0$ for each x in I. Prove that there exists a number $\alpha > 0$ such that $f(x) \geqslant \alpha$ for all $x \in I$.
2. Let $I := [a,b]$ and let $f : I \to \mathbf{R}$ and $g : I \to \mathbf{R}$ be continuous functions on I. Show that the set $\{x \in I : f(x) = g(x)\}$ is closed.
3. Let $I := [a,b]$ and let $f : I \to \mathbf{R}$ be a continuous function on I such that for each x in I there exists y in I such that $|f(y)| \leqslant \frac{1}{2}|f(x)|$. Prove there exists a point c in I such that $f(c) = 0$.

4. Show that every polynomial of odd degree with real coefficients has at least one real root.
5. Show that the polynomial $p(x) := x^4 + 7x^3 - 9$ has at least two real roots. Use a calculator to locate these roots to within two decimal places.
6. Let f be continuous on the interval $[0,1]$ to $\boldsymbol{R}$ and such that $f(0) = f(1)$. Prove that there exists a point c in $[0,\frac{1}{2}]$ such that $f(c) = f(c + \frac{1}{2})$. [*Hint*: Consider $g(x) := f(x) - f(x + \frac{1}{2})$.] Conclude that there are, at any time, antipodal points on the earth's equator that have the same temperature.
7. Show that the equation $x = \cos x$ has a solution in the interval $[0,\pi/2]$. Use the bisection procedure in the proof of the Location of Roots Theorem and a calculator to find an approximate solution of this equation, accurate to within two decimal places.
8. Let $I := [a,b]$, let $f : I \to \boldsymbol{R}$, and let $f(a) < 0$, $f(b) > 0$. Let $W := \{x \in I : f(x) < 0\}$, and let $w := \sup W$. Prove that $f(w) = 0$. (This provides an alternative proof of Theorem 4.6.5.)
9. Let $I := [0,\pi/2]$ and let $f : I \to \boldsymbol{R}$ be defined by $f(x) := \sup \{x^2, \cos x\}$ for $x \in I$. Show there exists an absolute minimum point $x_0 \in I$ for f on I. Show that x_0 is a solution to the equation $\cos x = x^2$.
10. Suppose that $f : \boldsymbol{R} \to \boldsymbol{R}$ is continuous on $\boldsymbol{R}$ and that $\lim_{x\to-\infty} f = 0$ and $\lim_{x\to\infty} f = 0$. Prove that f is bounded on $\boldsymbol{R}$ and attains either a maximum or minimum on $\boldsymbol{R}$. Give an example to show that both a maximum and a minimum need not be attained.
11. Let $f : \boldsymbol{R} \to \boldsymbol{R}$ be continuous on $\boldsymbol{R}$ and let $\beta \in \boldsymbol{R}$. Show that if $x_0 \in \boldsymbol{R}$ is such that $f(x_0) < \beta$, then there exists a neighborhood U of x_0 such that $f(x) < \beta$ for all $x \in U$.
12. Give an example of a function $f : \boldsymbol{R} \to \boldsymbol{R}$ such that the set $\{x \in \boldsymbol{R}: f(x) = 1\}$ is neither open nor closed in $\boldsymbol{R}$.
13. Examine which open [respectively, closed] intervals are mapped by $f(x) := x^2$ for $x \in \boldsymbol{R}$ onto open [respectively, closed] intervals. Note that the image of an open [respectively, closed] set under f is not necessarily an open [respectively, closed] set.
14. Examine the mapping of open [respectively, closed] intervals and sets under the functions $g(x) := 1/(x^2 + 1)$ and $h(x) := x^3$ for $x \in \boldsymbol{R}$.
15. If $f : [0,1] \to \boldsymbol{R}$ is continuous and has only rational [respectively, irrational] values, must f be constant?
16. Let $I := [a,b]$ and let $f : I \to \boldsymbol{R}$ be a (not necessarily continuous) function with the property that for every $x \in I$, the function f is bounded on a neighborhood U_x of x (in the sense of Definition 4.2.1). Prove that f is bounded on I.
17. Let $J := (a,b)$ and let $g : J \to \boldsymbol{R}$ be a continuous function with the property that for every $x \in J$, the function g is bounded on a neighborhood V_x of x. Show that g is not necessarily bounded on J.

SECTION 4.7 Uniform Continuity and Approximation

Let $A \subseteq \boldsymbol{R}$ and let $f : A \to \boldsymbol{R}$. It is readily seen from Theorem 4.4.3 that the following statements are equivalent:

(i) f is continuous at every point $u \in A$;

(ii) given $\varepsilon > 0$ and $u \in A$, there is a $\delta(\varepsilon,u) > 0$ such that if $x \in A$ and $|x - u| < \delta(\varepsilon,u)$, then $|f(x) - f(u)| < \varepsilon$.

The point we wish to emphasize here is that δ depends, in general, on *both* $\varepsilon > 0$ and $u \in A$. The fact that δ depends on u is a reflection of the fact that the function f may change its values rapidly near certain points and slowly near other points. [For example, consider $f(x) := \sin(1/x)$ for $x > 0$; see Figure 4.4.5 on p. 145.]

Now it often happens that the function f is such that the number δ can be chosen to be independent of the point $u \in A$ and to depend only on ε. For example, if $f(x) := 2x$ for all $x \in \boldsymbol{R}$, then

$$|f(x) - f(u)| = 2\,|x - u|,$$

and so we can choose $\delta(\varepsilon,u) := \varepsilon/2$ for all $\varepsilon > 0$, $u \in \boldsymbol{R}$.

On the other hand if we consider $g(x) := 1/x$ for $x \in A := \{x \in \boldsymbol{R} : x > 0\}$, then

(1) $$g(x) - g(u) = \frac{u - x}{ux}.$$

If $u \in A$ is given and if we take

(2) $$\delta(\varepsilon,u) := \inf\{\tfrac{1}{2}u, \tfrac{1}{2}u^2\varepsilon\},$$

then if $|x - u| < \delta(\varepsilon,u)$ we have $|x - u| < \frac{1}{2}u$ so that $\frac{1}{2}u < x < \frac{3}{2}u$ whence it follows that $1/x < 2/u$. Thus, if $|x - u| < \frac{1}{2}u$, the equality (1) yields the inequality

(3) $$|g(x) - g(u)| \leqslant (2/u^2)\,|x - u|.$$

Consequently, if $|x - u| < \delta(\varepsilon,u)$, the inequality (3) implies that

$$|g(x) - g(u)| < (2/u^2)\,(\tfrac{1}{2}u^2\varepsilon) = \varepsilon.$$

We have seen that the selection of $\delta(\varepsilon,u)$ by the formula (2) "works" in the sense that it enables us to give a value of δ that will insure that $|g(x) - g(u)| < \varepsilon$ when $|x - u| < \delta$ and $x, u \in A$. We note that the value of $\delta(\varepsilon,u)$ given in (2) certainly depends on the point $u \in A$. If we wish to consider all $u \in A$, formula (2) does not lead to one value $\delta(\varepsilon) > 0$ that will "work" simultaneously for all $u > 0$, since $\inf\{\delta(\varepsilon,u) : u > 0\} = 0$.

An alert reader will have observed that there are other selections that can be made for δ. (For example we could also take $\delta_1(\varepsilon,u) := \inf\{\frac{1}{3}u, \frac{2}{3}u^2\varepsilon\}$, as the reader can show; however, we still have $\inf\{\delta_1(\varepsilon,u) : u > 0\} = 0$.) In fact, there is no way of choosing one value of δ that will "work" for all $u > 0$ for the function $g(x) = 1/x$, as we shall see.

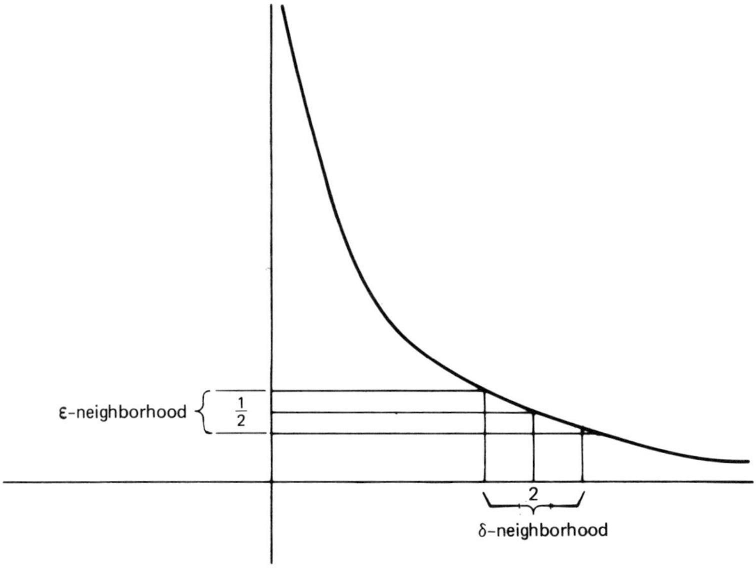

FIGURE 4.7.1 $g(x) = \frac{1}{x}$ $(x > 0)$.

The situation is exhibited graphically in Figures 4.7.1 and 4.7.2 where, for a given ε-neighborhood about each of $f(2) = \frac{1}{2}$ and $f(\frac{1}{2}) = 2$, the corresponding maximum values of δ are seen to be considerably different. As u tends to 0, the permissible values of δ tend to 0.

4.7.1 Definition. Let $A \subseteq \mathbf{R}$ and let $f: A \to \mathbf{R}$. We say that f is **uniformly continuous** on A if for each $\varepsilon > 0$ there is a $\delta(\varepsilon) > 0$ such that if $x, u \in A$ and if $|x - u| < \delta(\varepsilon)$, then $|f(x) - f(u)| < \varepsilon$.

It is clear that if f is uniformly continuous on A, then it is continuous at every point of A. In general, however, the converse does not hold, as is shown by the function $g(x) := 1/x$ on the set $A := \{x \in \mathbf{R}: x > 0\}$.

It is useful to formulate a condition equivalent to saying that f is *not* uniformly continuous on A. We give such criteria in the next result, leaving the proof to the reader as an exercise.

4.7.2 Non-Uniform Continuity Criteria. *Let $A \subseteq \mathbf{R}$ and let $f: A \to \mathbf{R}$. Then the following statements are equivalent.*

(i) *f is not uniformly continuous on A.*

(ii) *There exists an $\varepsilon_0 > 0$ such that for every $\delta > 0$ there are points x_δ, u_δ in A such that $|x_\delta - u_\delta| < \delta$ and $|f(x_\delta) - f(u_\delta)| \geq \varepsilon_0$.*

(iii) *There exists an $\varepsilon_0 > 0$ and two sequences (x_n) and (u_n) in A such that $\lim (x_n - u_n) = 0$ and $|f(x_n) - f(u_n)| \geq \varepsilon_0$ for all $n \in \mathbf{N}$.*

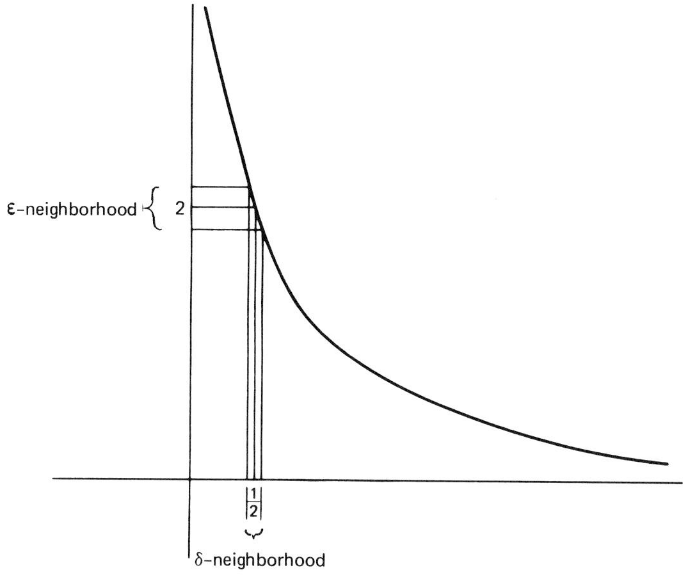

FIGURE 4.7.2 $g(x) = \frac{1}{x}$ $(x > 0)$.

We can apply this result to show that $g(x) := 1/x$ is not uniformly continuous on $A := \{x \in \boldsymbol{R} : x > 0\}$. For, if $x_n := 1/n$ and $u_n := 1/(n + 1)$, then we have $\lim (x_n - u_n) = 0$, but $|g(x_n) - g(u_n)| = 1$ for all $n \in \boldsymbol{N}$.

We now present an important result that assures that a continuous function on a closed bounded interval I is uniformly continuous on I. Another proof of this theorem is given in Section 4.9.

4.7.3 Uniform Continuity Theorem. *Let I be a closed bounded interval and let $f : I \to \boldsymbol{R}$ be continuous on I. Then f is uniformly continuous on I.*

Proof. If f is not uniformly continuous on I then, by the preceding result, there exists $\varepsilon_0 > 0$ and two sequences (x_n) and (u_n) in I such that $|x_n - u_n| < 1/n$ and $|f(x_n) - f(u_n)| \geq \varepsilon_0$ for all $n \in \boldsymbol{N}$. Since I is bounded, the sequence (x_n) is bounded; by the Bolzano-Weierstrass Theorem 3.4.6 there is a subsequence (x_{n_k}) of (x_n) that converges to an element z. Since I is closed, the limit z belongs to I, by Theorem 3.4.7. It is clear that the corresponding subsequence (u_{n_k}) also converges to z, since

$$|u_{n_k} - z| \leq |u_{n_k} - x_{n_k}| + |x_{n_k} - z|.$$

Now if f is continuous at the point z, then both of the sequences $\big(f(x_{n_k})\big)$ and $\big(f(u_{n_k})\big)$ must converge to $f(z)$. But this is not possible since

$$|f(x_n) - f(u_n)| \geq \varepsilon_0$$

for all $n \in \mathbf{N}$. Thus the hypothesis that f is not uniformly continuous on the closed bounded interval I implies that f is not continuous at some point of I. Consequently, if f is continuous at every point of I, then f is uniformly continuous on I. Q.E.D.

Lipschitz Functions

If a uniformly continuous function is given on a set that is not a closed bounded interval, then it is often difficult to establish its uniform continuity. However, there is a condition that frequently occurs that is sufficient to guarantee uniform continuity.

4.7.4 Definition. Let $A \subseteq \mathbf{R}$ and let $f : A \to \mathbf{R}$. If there exists a constant $K > 0$ such that

$$|f(x) - f(u)| \leq K\,|x - u| \tag{4}$$

for all $x, u \in A$, then f is said to be a **Lipschitz function** (or to satisfy a **Lipschitz condition**).

4.7.5 Theorem. *If $f : A \to \mathbf{R}$ is a Lipschitz function, then f is uniformly continuous on A.*

Proof. Indeed, if the inequality (4) holds, then it is clear that one can take $\delta(\varepsilon) := \varepsilon/K$ and that $|x - u| < \delta(\varepsilon)$ will imply that $|f(x) - f(u)| < K(\varepsilon/K) = \varepsilon$. Q.E.D.

4.7.6 Examples. (a) Not every uniformly continuous function is a Lipschitz function.

Let $g(x) := \sqrt{x}$ for x in the closed bounded interval $I := [0,2]$. Since g is continuous on I, it follows from the Uniform Continuity Theorem 4.7.3, that g is uniformly continuous on I. However, there is no number $K > 0$ such that $|g(x)| \leq K|x|$ for all $x \in I$. (Why not?) Thus g is not a Lipschitz function on I.

(b) The Uniform Continuity Theorem 4.7.3 and Theorem 4.7.5 can sometimes be combined to establish the uniform continuity of a function on a set. We consider $g(x) := \sqrt{x}$ on the set $A := [0,\infty)$. The uniform continuity of g on the interval $I := [0,2]$ follows from the Uniform Continuity Theorem 4.7.3 as noted in (a). Also, if both $x, u \geq 1$, then from

$$|g(x) - g(u)| = |\sqrt{x} - \sqrt{u}| = \frac{|x - u|}{\sqrt{x} + \sqrt{u}} \leq \tfrac{1}{2}\,|x - u|,$$

we see that g is a Lipschitz function on the set $J := [1,\infty)$, and hence (by Theorem 4.7.5) g is uniformly continuous on $[1,\infty)$. Since $A = I \cup J$, it is an easy matter to show [by taking $\delta(\varepsilon) := \inf\{1, \delta_I(\varepsilon), \delta_J(\varepsilon)\}$] that g is uniformly continuous on A. We leave the details to the reader.

Approximation†

In many applications it is important to be able to "approximate" continuous functions by functions of an elementary nature. Although there are a variety of definitions that can be used to make the word "approximately" more precise, one of the most natural (as well as one of the most important) is to require that, at every point of the given domain, the approximating function shall not differ from the given function by more than the pre-assigned error.

4.7.7 Definition. Let $I \subseteq \boldsymbol{R}$ be an interval and let $s : I \to \boldsymbol{R}$. Then s is called a **step function** if it has only a finite number of distinct values, each value being assumed on one or more intervals in I.

For example, the function $s : [-2,4] \to \boldsymbol{R}$ defined by

$$
\begin{aligned}
s(x) &:= 0, & -2 &\leqslant x < -1, \\
&:= 1, & -1 &\leqslant x \leqslant 0, \\
&:= \tfrac{1}{2}, & 0 &< x < \tfrac{1}{2}, \\
&:= 3, & \tfrac{1}{2} &\leqslant x < 1, \\
&:= -2, & 1 &\leqslant x \leqslant 3, \\
&:= 2, & 3 &< x \leqslant 4,
\end{aligned}
$$

is a step function. (See Figure 4.7.3 on the next page.)

We shall now show that a continuous function on a closed bounded interval I can be approximated arbitrarily closely by step functions.

4.7.8 Theorem. *Let I be a closed bounded interval and let $f : I \to \boldsymbol{R}$ be continuous on I. If $\varepsilon > 0$, then there exists a step function $s_\varepsilon : I \to \boldsymbol{R}$ such that $|f(x) - s_\varepsilon(x)| < \varepsilon$ for all $x \in I$.*

Proof. Since (by the Uniform Continuity Theorem 4.7.3) the function f is uniformly continuous, it follows that given $\varepsilon > 0$ there is a number $\delta(\varepsilon) > 0$ such that if $x, y \in I$ and $|x - y| < \delta(\varepsilon)$, then $|f(x) - f(y)| < \varepsilon$. Let $I := [a,b]$ and let $m \in \mathbf{N}$ be sufficiently large so that $h := (b - a)/m < \delta(\varepsilon)$. We now divide $I = [a,b]$ into m disjoint intervals of length h; namely, $I_1 = [a, a + h]$, and $I_k = (a + (k - 1)h, a + kh]$ for $k = 2, \ldots, m$. Since the length of each subinterval I_k is $h < \delta(\varepsilon)$, the difference between any two values of f in I_k is less than ε. We now define

$$
(5) \qquad s_\varepsilon(x) := f(a + kh) \qquad \text{for} \quad x \in I_k,\ k = 1, \ldots, m,
$$

† The rest of this section can be omitted on a first reading of this chapter.

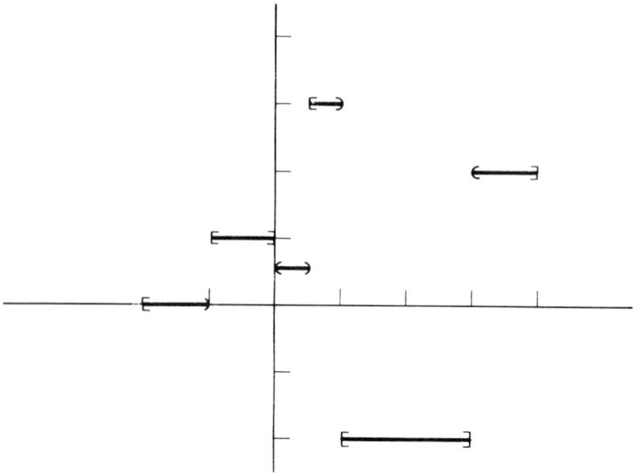

FIGURE 4.7.3 Graph of $y = s(x)$.

so that s_ε is constant on each interval I_k. (In fact the value of s_ε on I_k is the value of f at the right end point of I_k. See Figure 4.7.4.) Consequently if $x \in I_k$, then

$$|f(x) - s_\varepsilon(x)| = |f(x) - f(a + kh)| < \varepsilon.$$

Therefore we have $|f(x) - s_\varepsilon(x)| < \varepsilon$ for all $x \in I$. Q.E.D.

Note that the proof of the preceding theorem establishes somewhat more than was announced in the statement of the theorem. In fact, we have proved the following, somewhat more precise, assertion.

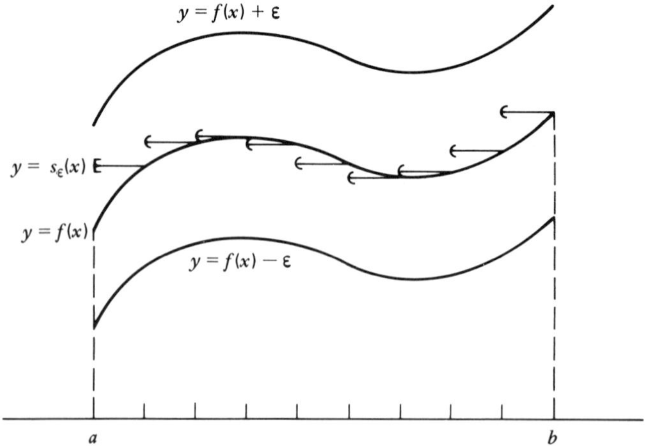

FIGURE 4.7.4 Approximation by step functions.

4.7.9 Corollary. *Let $I := [a,b]$ be a closed bounded interval and let $f : I \to \mathbf{R}$ be continuous on I. If $\varepsilon > 0$, there exists a natural number m such that if we divide I into m disjoint intervals I_k having length $h := (b - a)/m$, then the step function s_ε defined in equation (5) satisfies $|f(x) - s_\varepsilon(x)| < \varepsilon$ for all $x \in I$.*

Step functions are extremely elementary in character, but they are not continuous (except in trivial cases). Since it is often desirable to approximate continuous functions by simple continuous functions, we now shall show that we can approximate continuous functions by continuous piecewise linear functions.

4.7.10 Definition. Let $I := [a,b]$ be an interval. Then a function $g : I \to \mathbf{R}$ is said to be **piecewise linear** on I if I is the union of a finite number of disjoint intervals $I_1, \ldots, I_m$, such that the restriction of g to each interval I_k is a linear function.

Remark. It is evident that in order for a piecewise linear function g to be continuous on I, the line segments that form the graph of g meet at the end points of adjacent subintervals I_k, I_{k+1} $(k = 1, \ldots, m - 1)$.

4.7.11 Theorem. *Let I be a closed bounded interval and let $f : I \to \mathbf{R}$ be continuous on I. If $\varepsilon > 0$, then there exists a continuous piecewise linear function $g_\varepsilon : I \to \mathbf{R}$ such that $|f(x) - g_\varepsilon(x)| < \varepsilon$ for all $x \in I$.*

Proof. Since f is uniformly continuous on $I := [a,b]$, there is a number $\delta(\varepsilon) > 0$ such that if $x, y \in I$ and $|x - y| < \delta(\varepsilon)$, then $|f(x) - f(y)| < \varepsilon$. Let $m \in \mathbf{N}$ be sufficiently large so that $h := (b - a)/m < \delta(\varepsilon)$. Divide $I = [a,b]$ into m disjoint intervals of length h; namely $I_1 = [a, a + h]$, and $I_k = (a + (k - 1)h, a + kh]$ for $k = 2, \ldots, m$. On each interval I_k we define g_ε to be the linear function joining the points

$$(a + (k - 1)h, f(a + (k - 1)h)) \quad \text{and} \quad (a + kh, f(a + kh)).$$

Then g_ε is a continuous piecewise linear function on I. Since, for $x \in I_k$ the value $f(x)$ is within ε of $f(a + (k - 1)h)$ and $f(a + kh)$, it is an exercise to show that $|f(x) - g_\varepsilon(x)| < \varepsilon$ for all $x \in I_k$; therefore this inequality holds for all $x \in I$. (See Figure 4.7.5 on the next page.) Q.E.D.

We shall close this section by stating the important theorem of Weierstrass concerning the approximation of continuous functions by polynomial functions. As would be expected, in order to obtain an approximation within an arbitrarily pre-assigned $\varepsilon > 0$, we must be prepared to use polynomials of arbitrarily high degree.

4.7.12 Weierstrass Approximation Theorem. *Let $I = [a,b]$ and let $f : I \to \mathbf{R}$ be continuous. If $\varepsilon > 0$ is given, then there exists a polynomial function p_ε such that $|f(x) - p_\varepsilon(x)| < \varepsilon$ for all $x \in I$.*

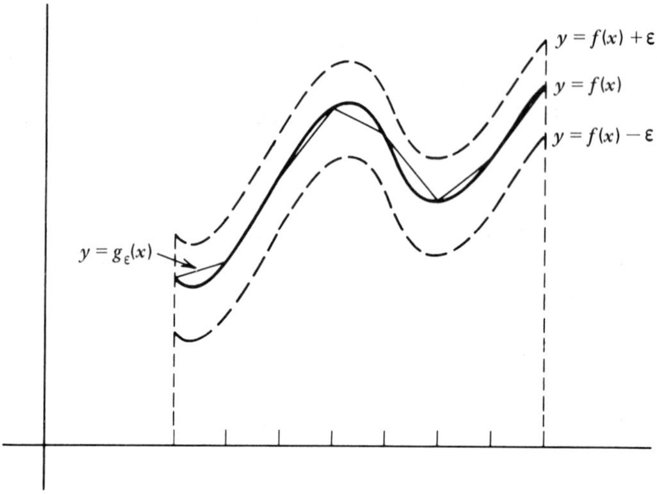

FIGURE 4.7.5 Approximation by piecewise linear functions.

There are a number of proofs of this result. Unfortunately, all of them are rather intricate, or employ results that are not yet at our disposal. One of the most elementary proofs is based on the following theorem due to Serge Bernšteĭn, for continuous functions on $[0,1]$. Given $f : [0,1] \to \boldsymbol{R}$, Bernšteĭn defined the sequence of polynomials:

$$B_n(x) = \sum_{k=0}^{n} f\left(\frac{k}{n}\right) \binom{n}{k} x^k(1 - x)^{n-k}. \tag{6}$$

The polynomial function B_n, as defined in (6), is called the **nth Bernšteĭn polynomial for** f; it is a polynomial of degree at most n and its coefficients depend on the values of the function f at the $n + 1$ equally spaced points

$$0, \frac{1}{n}, \frac{2}{n}, \ldots, \frac{k}{n}, \ldots, 1,$$

and on the binomial coefficients

$$\binom{n}{k} = \frac{n!}{k!\,(n-k)!} = \frac{n(n-1)\cdots(n-k+1)}{1\cdot 2\cdots k}$$

4.7.13 Bernšteĭn's Approximation Theorem. *Let $f : [0,1] \to \boldsymbol{R}$ be continuous and let $\varepsilon > 0$. There exists an $n_\varepsilon \in \boldsymbol{N}$ such that if $n \geq n_\varepsilon$, then we have $|f(x) - B_n(x)| < \varepsilon$ for all $x \in [0,1]$.*

The proof of Bernšteĭn's Approximation Theorem is given in *ERA*, pp. 169–172. It is shown there that if $\delta(\varepsilon) > 0$ is such that $|f(x) - f(y)| < \varepsilon$ for all $x,y \in [0,1]$ with

$|x - y| < \delta(\varepsilon)$, and if $M \geqslant |f(x)|$ for all $x \in [0,1]$, then we can take

$$n_\varepsilon := \sup \{(\delta(\varepsilon/2))^{-4}, M^2/\varepsilon^2\}. \tag{7}$$

The estimate (7) gives information on how large we must take n in order for B_n to approximate f within ε.

The Weierstrass Approximation Theorem 4.7.12 can be derived from the Bernšteĭn Approximation Theorem 4.7.13 by a change of variable. Specifically, we replace $f : [a,b] \to \boldsymbol{R}$ by a function $F : [0,1] \to \boldsymbol{R}$, defined by

$$F(t) := f(a + (b - a)t) \quad \text{for} \quad t \in [0,1].$$

The function F can be approximated by Bernšteĭn polynomials for F on the interval $[0,1]$, which can then yield polynomials on $[a,b]$ that approximate f.

Exercises for Section 4.7

1. Show that the function $f(x) := 1/x$ for $x \in [1,\infty)$ is uniformly continuous on $[1,\infty)$.
2. Show that the following functions are not uniformly continuous on the given domains.
   (a) $f(x) := 1/x$, $\quad D(f) := \{x \in \boldsymbol{R} : 0 < x < 1\}$;
   (b) $g(x) := x^2$, $\quad D(g) := \boldsymbol{R}$;
   (c) $h(x) := 1/x^2$, $\quad D(h) := \{x \in \boldsymbol{R} : x > 0\}$;
   (d) $k(x) := \sin(1/x)$, $\quad D(k) := \{x \in \boldsymbol{R} : x > 0\}$.
3. Show that the function $f(x) := 1/(1 + x^2)$ for $x \in \boldsymbol{R}$ is uniformly continuous on $\boldsymbol{R}$.
4. Show that if f and g are uniformly continuous on $\boldsymbol{R}$ to $\boldsymbol{R}$, then $f + g$ is uniformly continuous on $\boldsymbol{R}$.
5. Show that if f and g are uniformly continuous from $\boldsymbol{R}$ to $\boldsymbol{R}$ and if they are *both* bounded on $\boldsymbol{R}$, then their product fg is uniformly continuous on $\boldsymbol{R}$.
6. If $f(x) := x$ and $g(x) := \sin x$, show that both f and g are uniformly continuous on $\boldsymbol{R}$, but that their product fg is not uniformly continuous on $\boldsymbol{R}$.
7. Let $A \subseteq \boldsymbol{R}$ and let $f : A \to \boldsymbol{R}$ be uniformly continuous on A. Prove that if (x_n) is a Cauchy sequence in A, then $(f(x_n))$ is a Cauchy sequence in $\boldsymbol{R}$.
8. Let $f : (0,1] \to \boldsymbol{R}$ be uniformly continuous on $(0,1]$ and let $L := \lim (f(1/n))$. Prove that if (x_n) is any sequence in $(0,1]$ such that $\lim (x_n) = 0$, then $\lim (f(x_n)) = L$. Deduce that if f is defined at 0 by $f(0) := L$, then f becomes continuous at 0.
9. Let f be uniformly continuous on (a,b). Show that f can be defined at a and b in such a way that the extended function is continuous on $[a,b]$.
10. Let $A \subseteq \boldsymbol{R}$ and let $f : A \to \boldsymbol{R}$ be uniformly continuous on A. Show that if A is a bounded set, then f is bounded on A.
11. Let A, B be subsets of $\boldsymbol{R}$ and let f be defined on $A \cup B$ to $\boldsymbol{R}$. Suppose that f is uniformly continuous on the set A and also uniformly continuous on the set B. Prove that if f is continuous on $A \cup B$, then it is uniformly continuous on $A \cup B$.
12. Let $A \subseteq \boldsymbol{R}$ and suppose that $f : A \to \boldsymbol{R}$ has the following property: for each $\varepsilon > 0$ there exists a function $g_\varepsilon : A \to \boldsymbol{R}$ such that g_ε is uniformly continuous on A and $|f(x) - g_\varepsilon(x)| < \varepsilon$ for all $x \in A$. Prove that f is uniformly continuous on A.

13. A function $f : \boldsymbol{R} \to \boldsymbol{R}$ is said to be **periodic** on $\boldsymbol{R}$ if there exists a number $p > 0$ such that $f(x + p) = f(x)$ for all $x \in \boldsymbol{R}$. Prove that a continuous periodic function on $\boldsymbol{R}$ is bounded and uniformly continuous on $\boldsymbol{R}$.
14. If $f_0(x) := 1$ for $x \in [0,1]$, calculate the first few Bernšteĭn polynomials for f_0. Show that they coincide with f_0. [*Hint*: The Binominal Theorem asserts that

$$(a + b)^n = \sum_{k=0}^{n} \binom{n}{k} a^k b^{n-k}].$$

15. If $f_1(x) := x$ for $x \in [0,1]$, calculate the first few Bernšteĭn polynomials for f_1. Show that they coincide with f_1.
16. If $f_2(x) := x^2$ for $x \in (0,1)$, calculate the first few Bernšteĭn polynomials for f_2. Show that $B_n(x) = (1 - 1/n)x^2 + (1/n)x$.
17. Using the result of the preceding exercise for f_2, how large must n be so that the nth Bernšteĭn polynomial B_n for f_2 satisfies $|f(x) - B_n(x)| \leqslant 0.001$ for all $x \in [0,1]$?

SECTION 4.8 Monotone and Inverse Functions

Recall that if $A \subseteq \boldsymbol{R}$, then a function $f : A \to \boldsymbol{R}$ is said to be **increasing on** A if whenever $x_1, x_2 \in A$ and $x_1 \leqslant x_2$, then $f(x_1) \leqslant f(x_2)$. The function f is said to be **strictly increasing on** A if whenever $x_1, x_2 \in A$ and $x_1 < x_2$, then $f(x_1) < f(x_2)$. Similarly, $g : A \to \boldsymbol{R}$ is said to be **decreasing on** A if whenever $x_1, x_2 \in A$ and $x_1 \leqslant x_2$ then $g(x_1) \geqslant g(x_2)$. The function g is said to be **strictly decreasing on** A if whenever $x_1, x_2 \in A$ and $x_1 < x_2$ then $g(x_1) > g(x_2)$.

If a function is either increasing or decreasing on A, we say that it is **monotone** on A. If f is either strictly increasing or strictly decreasing on A, we say that f is **strictly monotone** on A.

We note that if $f : A \to \boldsymbol{R}$ is increasing on A then $g := -f$ is decreasing on A; similarly if $\varphi : A \to \boldsymbol{R}$ is decreasing on A then $\psi := -\varphi$ is increasing on A.

In this section, we shall be concerned with monotone functions that are defined on an interval $I \subseteq \boldsymbol{R}$. We shall discuss increasing functions explicitly, but it is clear that there are corresponding results for decreasing functions. These results can either be obtained directly from the results for increasing functions or proved by similar arguments.

Monotone functions are not necessarily continuous. For example, if $f(x) := 0$ for $x \in [0, 1]$ and $f(x) := 1$ for $x \in (1,2]$, then f is increasing on $[0,2]$, but fails to be continuous at $x = 1$. However, the next result shows that a monotone function always has one-sided limits (see Definition 4.3.1) in $\boldsymbol{R}$ at every point that is not an end point of its domain.

4.8.1 Theorem. *Let $I \subseteq \boldsymbol{R}$ be an interval and let $f : I \to \boldsymbol{R}$ be increasing on I. Suppose that $c \in I$ is not an end point of I. Then*

(i) $\lim_{x \to c-} f = \sup \{f(x) : x \in I, x < c\}$,

(ii) $\lim_{x \to c+} f = \inf \{f(x) : x \in I, x > c\}$.

Proof. First note that if $x \in I$ and $x < c$, then $f(x) \leqslant f(c)$. Hence the set $\{f(x) : x \in I, x < c\}$, which is non-void because c is not an end point of I, is bounded above by $f(c)$. Thus the indicated supremum exists; we denote it by L.

If $\varepsilon > 0$ is given, then $L - \varepsilon$ is not an upper bound of this set. Hence there exists $y_\varepsilon \in I$, $y_\varepsilon < c$ such that $L - \varepsilon < f(y_\varepsilon) \leq L$. Since f is increasing, we deduce that if $\delta(\varepsilon) := c - y_\varepsilon$ and if $0 < c - y < \delta(\varepsilon)$, then $y_\varepsilon < y < c$ so that

$$L - \varepsilon < f(y_\varepsilon) \leq f(y) \leq L.$$

Therefore $|f(y) - L| < \varepsilon$ when $0 < c - y < \delta(\varepsilon)$. Since $\varepsilon > 0$ is arbitrary we infer that (i) holds.

The proof of (ii) is similar. Q.E.D.

The next result gives criteria for the continuity of an increasing function f at a point c that is not an end point of the interval on which f is defined.

4.8.2 Corollary. *Let $I \subseteq \mathbf{R}$ be an interval and let $f : I \to \mathbf{R}$ be increasing on I. Suppose that $c \in I$ is not an end point of I. Then the following statements are equivalent.*

(a) *f is continuous at c.*

(b) $\lim_{x \to c-} f = f(c) = \lim_{x \to c+}$ f.

(c) $\sup \{f(x) : x \in I, x < c\} = f(c) = \inf \{f(x) : x \in I, x > c\}$.

This follows easily from Theorems 4.8.1 and 4.3.3. We leave the details to the reader.

Let I be an interval and let $f : I \to \mathbf{R}$ be an increasing function. If a is the left end point of I, it is an exercise to show that f is continuous at a if and only if

$$f(a) = \inf \{f(c) : x \in I, a < x\}$$

or if and only if $f(a) = \lim_{x \to c+} f$.

Similar conditions apply at a right end point, and for decreasing functions.

If $f : I \to \mathbf{R}$ is increasing on I and if c is not an end point of I, we define (as in Section 4.4) the **jump of f at** c to be $j_f(c) := \lim_{x \to c+} f - \lim_{x \to c-} f$. (See Figure 4.8.1.) It follows from Theorem 4.8.1 that

$$j_f(c) = \inf \{f(x) : x \in I, x > c\} - \sup \{f(x) : x \in I, x < c\},$$

for an increasing function. If the left end point a of I belongs to I, we define the **jump of f at** a to be $j_f(a) := \lim_{x \to a+} f - f(a)$. If the right end point b of I belongs to I, we define the **jump of f at** b to be $j_f(b) := f(b) - \lim_{x \to c-} f$.

4.8.3 Theorem. *Let $I \subseteq \mathbf{R}$ be an interval and let $f : I \to \mathbf{R}$ be increasing on I. If $c \in I$, then f is continuous at c if and only if $j_f(c) = 0$.*

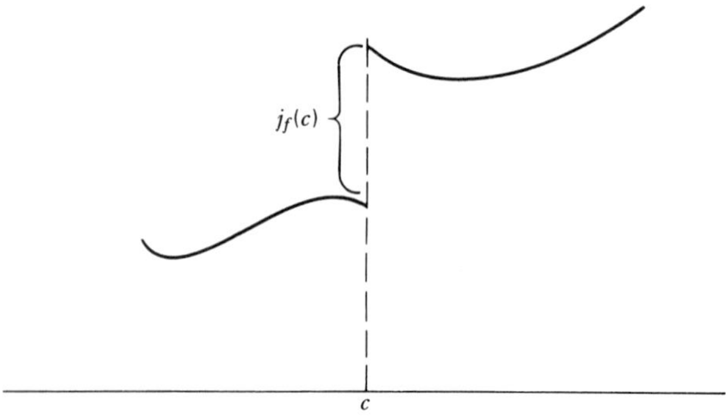

FIGURE 4.8.1 The jump of *f* at c.

Proof. If c is not an end point, this follows immediately from Corollary 4.8.2. If $c \in I$ is the left end point of I, then f is continuous at c if and only if $f(c) = \lim_{x \to c+} f$, which is equivalent to $j_f(c) = 0$. Similar remarks apply to the case of a right end point. Q.E.D.

We now show that there can be at most a countable set of points at which a monotone function is discontinuous.

4.8.4 Theorem. *Let $I \subseteq \mathbf{R}$ be an interval and let $f : I \to \mathbf{R}$ be monotone on I. Then the set of points $D \subseteq I$ at which f is discontinuous is a countable set.*

Proof. We shall suppose that f is increasing on I. It follows from Theorem 4.8.3 that $D = \{x \in I : j_f(x) \neq 0\}$. We shall consider the case that $I := [a,b]$ is a closed bounded interval, leaving the case of an arbitrary interval to the reader.

We first note that since f is increasing, then $j_f(c) \geqslant 0$ for all $c \in I$. Moreover, if $a \leqslant x_1 \leqslant \ldots \leqslant x_n \leqslant b$, then (why?) we have

$$f(a) \leqslant f(a) + j_f(x_1) + \cdots + j_f(x_n) \leqslant f(b),$$

whence it follows that

$$j_f(x_1) + \cdots + j_f(x_n) \leqslant f(b) - f(a).$$

(See Figure 4.8.2.) Consequently there can be at most k points in $I = [a,b]$ where $j_f(x) \geqslant (f(b) - f(a))/k$. We conclude that there is at most one point $x \in I$ where $j_f(x) = f(b) - f(a)$; there are at most two points in I where $j_f(x) \geqslant (f(b) - f(a))/2$; at most three points in I where $j_f(x) \geqslant (f(b) - f(a))/3$, and so on. Therefore there is at most a countable set of points x where $j_f(x) > 0$. But since every point in D must be included in this set, we deduce that D is a countable set. Q.E.D.

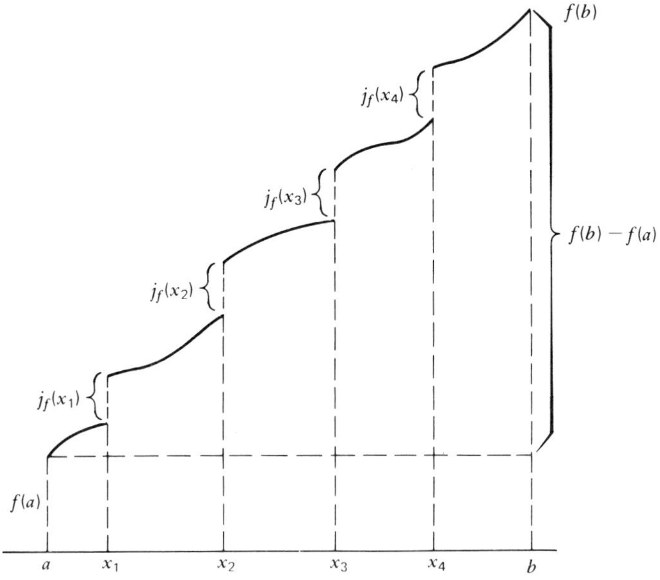

FIGURE 4.8.2 $j_f(x_1) + \cdots + j_f(x_n) \le f(b) - f(a)$.

Theorem 4.8.4 has some useful applications. For example, it was seen in Exercise 4.5.12 that if $h : \mathbf{R} \to \mathbf{R}$ satisfies the identity

$$(*) \qquad h(x + y) = h(x) + h(y) \qquad \text{for all} \quad x, y \in \mathbf{R},$$

and if h is continuous at a single point x_0, then h is continuous at *every* point of $\mathbf{R}$. It follows that if h is a monotone function satisfying $(*)$, then h must be continuous on $\mathbf{R}$. [It follows from this that $h(x) = Cx$ for all $x \in \mathbf{R}$, where $C := h(1)$.]

Inverse Functions

We shall now consider the existence of inverses for functions that are continuous on an interval $I \subseteq \mathbf{R}$. We recall (see Section 1.2) that a function $f : I \to \mathbf{R}$ has an inverse function if and only if f is injective (= one-one); that is, $x, y \in I$ and $x \neq y$ imply that $f(x) \neq f(y)$. We note that a strictly monotone function is injective and so has an inverse. In the next theorem, we show that if $f : I \to \mathbf{R}$ is a strictly monotone *continuous* function, then f has an inverse function g on $J := f(I)$ that is strictly monotone and continuous on J. In particular, if f is strictly increasing then so is g, and if f is strictly decreasing then so is g.

4.8.5 Continuous Inverse Theorem. *Let $I \subseteq \mathbf{R}$ be an interval and let $f : I \to \mathbf{R}$ be strictly monotone and continuous on I. Then the function g inverse to f is strictly monotone and continuous on $J := f(I)$.*

Proof. We consider the case that f is strictly increasing, leaving the case that f is strictly decreasing to the reader.

Since f is continuous and I is an interval, it follows from the Preservation of Intervals Theorem 4.6.10 that $J = f(I)$ is an interval. Moreover, since f is strictly increasing on I, it is injective on I; therefore the function $g : J \to \boldsymbol{R}$ inverse to f exists. We claim that g is strictly increasing. Indeed, if $y_1, y_2 \in J$ with $y_1 < y_2$, then $y_1 = f(x_1)$ and $y_2 = f(x_2)$ for some $x_1, x_2 \in I$. We must have $x_1 < x_2$; otherwise $x_1 \geqslant x_2$, which implies that $y_1 = f(x_1) \geqslant f(x_2) = y_2$, contrary to the hypothesis that $y_1 < y_2$. Therefore we have

$$g(y_1) = x_1 < x_2 = g(y_2).$$

Since y_1 and y_2 are arbitrary elements of J with $y_1 < y_2$, we conclude that g is strictly increasing on J.

It remains to show that g is continuous on J. However, this is a consequence of the fact that $g(J) = I$ is an interval. Indeed, if g is discontinuous at a point $c \in J$, then the jump of g at c is non-zero so that

$$\lim_{y \to c-} g < \lim_{y \to c+} g.$$

If we choose any number $x \neq g(c)$ satisfying $\lim_{y \to c-} g < x < \lim_{y \to c+} g$, then x has the property that $x \neq g(y)$ for any $y \in J$. (See Figure 4.8.3.) Hence $x \notin I$, which contradicts the fact that I is an interval. Therefore we conclude that g is continuous on J.

Q.E.D.

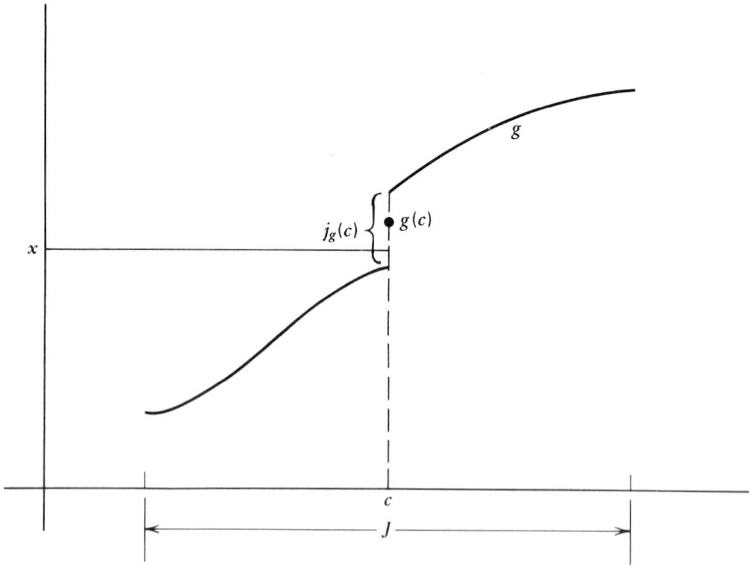

FIGURE 4.8.3 $g(y) \neq x$ **for** $y \in J$.

The hypothesis that f is strictly monotone, made in the Continuous Inverse Theorem 4.8.5, is a strong one. Hence it is natural to ask if it is necessary. We now show that it is, and that an injective continuous function on an interval must be strictly monotone.

4.8.6 Theorem. *Let $I \subseteq \mathbf{R}$ be an interval and let $f : I \to \mathbf{R}$ be an injective continuous function on I. Then f is strictly monotone on I.*

Proof. We shall treat the case where $I = [a,b]$, leaving it to the reader to prove the result for an arbitrary interval. Since f is injective on $[a,b]$, we must have either $f(a) < f(b)$ or $f(a) > f(b)$. We shall suppose that $f(a) < f(b)$ and prove that f is strictly increasing on $[a,b]$. [A similar argument shows that if $f(a) > f(b)$, then f is strictly decreasing on $[a,b]$.]

Let $x \in [a,b]$; we shall show that $f(a) < f(x) < f(b)$ by eliminating the other possibilities. If $f(x) < f(a) < f(b)$, then Bolzano's Intermediate Value Theorem 4.6.6 applied to the interval $[x,b]$ implies the existence of a number $a' \in (x,b)$ such that $f(a) = f(a')$, contrary to the hypothesis that f is injective. (See Figure 4.8.4.) Similarly, if $f(a) < f(b) < f(x)$, there exists a number $b' \in (a,x)$ such that $f(b) = f(b')$, contrary to the hypothesis that f is injective. Therefore $f(a) < f(x) < f(b)$ for any $x \in (a,b)$.

Now let $y \in (a,b)$ be such that $x < y$; by the preceding paragraph we infer that $f(a) < f(y) < f(b)$. If $f(a) < f(y) < f(x)$, there exists $y' \in (a,x)$ such that

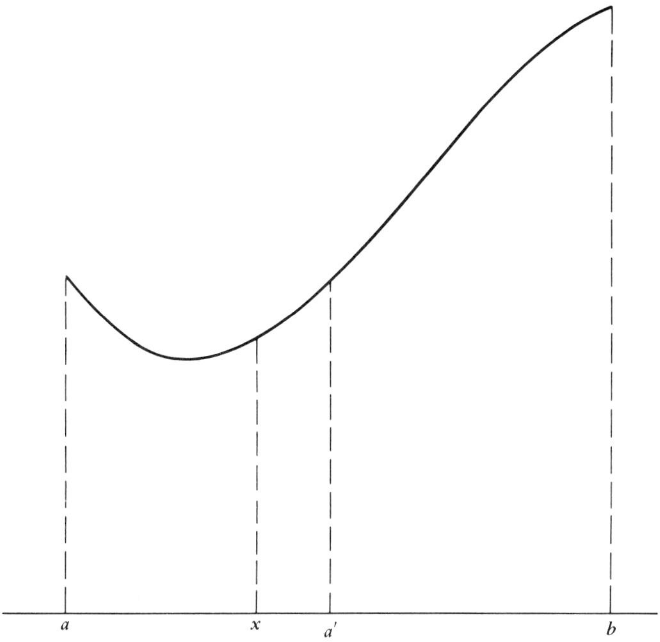

FIGURE 4.8.4 There exists a' such that $f(a) = f(a')$.

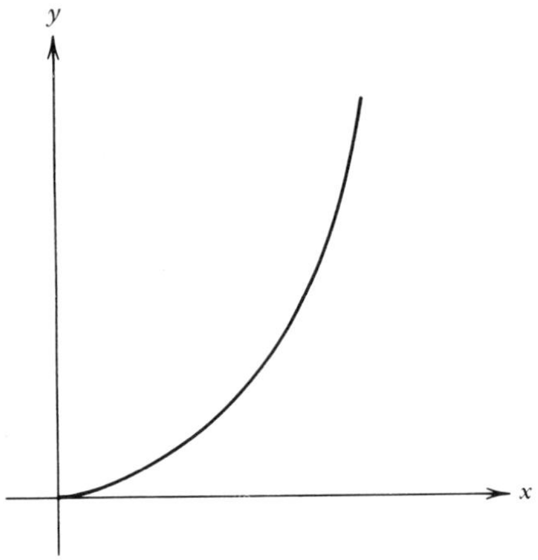

FIGURE 4.8.5 Graph of $f(x) = x^n$ ($x \geq 0$, n even).

$f(y') = f(y)$, a contradiction. Therefore we must have $f(x) < f(y)$ for $x < y$. Since x and y are arbitrary elements in (a,b) with $x < y$, we conclude that f is strictly increasing on $[a,b]$. Q.E.D.

The *n*th Root Function

We shall apply the Continuous Inverse Theorem 4.8.5 to the nth power function. We need to distinguish two cases: (i) n even, and (ii) n odd.

(i) n even. In order to obtain a function that is strictly monotone, we restrict our attention to the interval $I := [0,\infty)$. Thus, let $f(x) := x^n$ for $x \in I$. (See Figure 4.8.5.) We have seen (in Exercise 2.2.17) that if $0 \leq x < y$, then $f(x) = x^n < y^n = f(y)$; therefore f is strictly increasing on I. Moreover it follows from Example 4.5.4(a) that f is continuous on I. Therefore, by the Preservation of Intervals Theorem 4.6.10, $J := f(I)$ is an interval. We shall show that $J = [0,\infty)$. Let $y \geq 0$ be arbitrary; by the Archimedean Property, there exists $k \in \mathbf{N}$ such that $0 \leq y < k$. Since

$$f(0) = 0 \leq y < k \leq k^n = f(k),$$

it follows from Bolzano's Intermediate Value Theorem 4.6.6 that $y \in J$. Since $y \geq 0$ is arbitrary, we deduce that $J = [0,\infty)$.

We conclude from the Continuous Inverse Theorem 4.8.5 that the function g that is inverse to $f(x) = x^n$ on $I = [0,\infty)$ is strictly increasing and continuous on $J = [0,\infty)$. We usually write

$$g(x) = x^{1/n} \qquad \text{or} \qquad g(x) = \sqrt[n]{x}$$

for $x \geq 0$ (n even), and call $x^{1/n} = \sqrt[n]{x}$ the ***n*th root of** $x \geq 0$ (n even). The function g is called the ***n*th root function** (n even). (See Figure 4.8.6.)

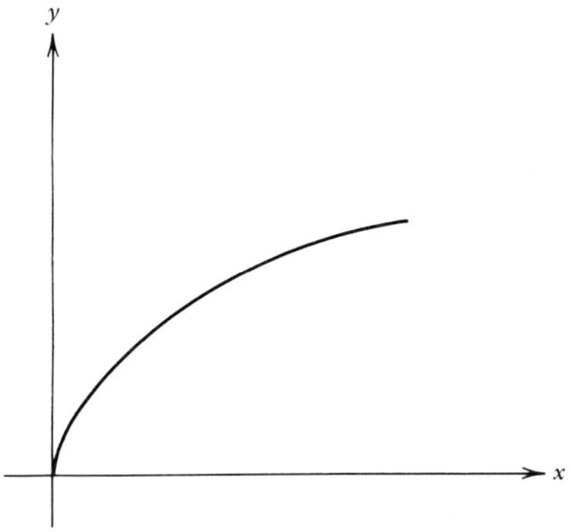

FIGURE 4.8.6 Graph of $g(x) = x^{1/n}$ ($x \geqslant 0$, n even).

Since g is inverse to f, we have

$$g(f(x)) = x \qquad \text{and} \qquad f(g(x)) = x \qquad \text{for all} \quad x \in [0,\infty).$$

We can write these equations in the following form:

$$(x^n)^{1/n} = x \qquad \text{and} \qquad (x^{1/n})^n = x$$

for all $x \in [0,\infty)$ and n even.

(ii) n odd. In this case we let $F(x) := x^n$ for all $x \in \boldsymbol{R}$; by 4.5.4(a), F is continuous on $\boldsymbol{R}$. We leave it to the reader to show that F is strictly increasing on $\boldsymbol{R}$ and that $F(\boldsymbol{R}) = \boldsymbol{R}$. (See Figure 4.8.7 on the next page.)

It follows from the Continuous Inverse Theorem 4.8.5 that the function G that is inverse to $F(x) = x^n$ for $x \in \boldsymbol{R}$, is strictly increasing and continuous on $\boldsymbol{R}$. We usually write

$$G(x) = x^{1/n} \qquad \text{or} \qquad G(x) = \sqrt[n]{x} \qquad \text{for} \quad x \in \boldsymbol{R}, \quad n \text{ odd},$$

and call $x^{1/n}$ the **nth root of $x \in \mathbf{R}$**. The function G is called the **nth root function** (n odd). (See Figure 4.8.8.) Here we have

$$(x^n)^{1/n} = x \qquad \text{and} \qquad (x^{1/n})^n = x$$

for all $x \in \boldsymbol{R}$ and n odd.

Rational Powers

Now that the nth root functions have been defined for $n \in \boldsymbol{N}$, it is easy to define rational powers.

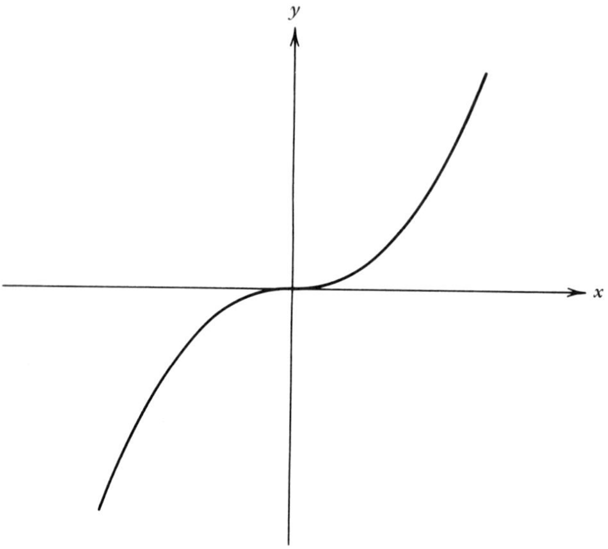

FIGURE 4.8.7 Graph of $F(x) = x^n$ ($x \in R$, n odd).

4.8.7 Definition. (i) If $m, n \in \mathbf{N}$ and $x \geqslant 0$, we define $x^{m/n} := (x^{1/n})^m$. (ii) If $m, n \in \mathbf{N}$ and $x > 0$, we define $x^{-m/n} := (x^{1/n})^{-m}$.

Hence we have defined x^r when r is a rational number and $x > 0$. The graphs of $x \mapsto x^r$ depend on whether $r > 1$, $r = 1$, $0 < r < 1$, $r = 0$, or $r < 0$. (See

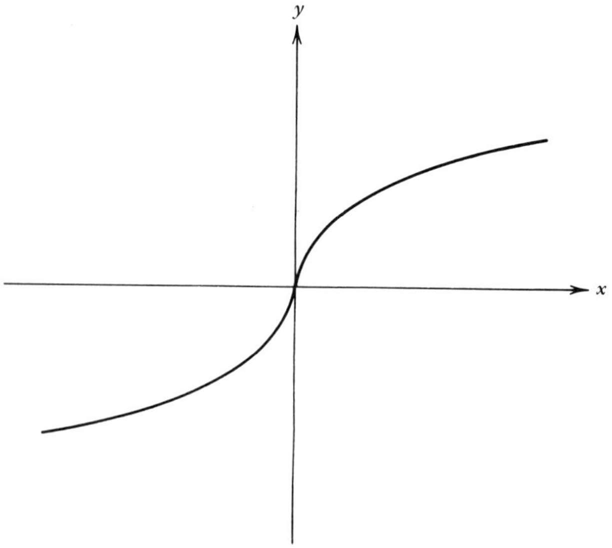

FIGURE 4.8.8 Graph of $G(x) = x^{1/n}$ ($x \in R$, n odd).

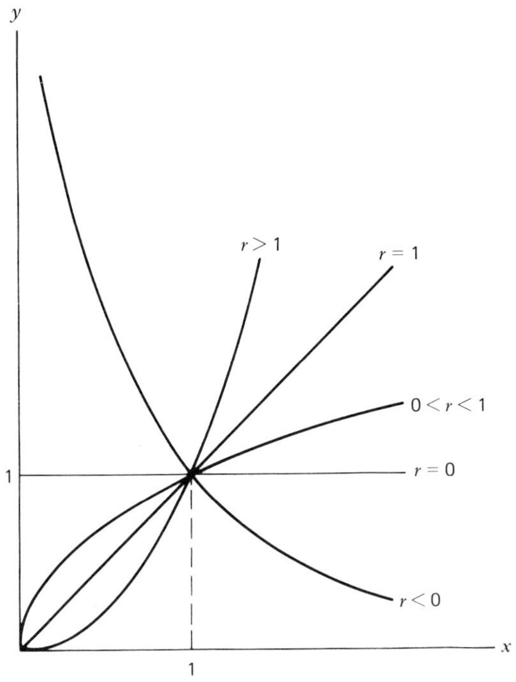

FIGURE 4.8.9 Graphs of $x \mapsto x^r$ $(x \geqslant 0)$.

Figure 4.8.9.) Since a rational number $r \in \boldsymbol{Q}$ can be written in the form $r = m/n$ with $m \in \mathbf{Z}$, $n \in \mathbf{N}$, in many ways, it should be shown that Definition 4.8.7 is not ambiguous. That is, if $r = m/n = p/q$ with $m, p \in \mathbf{Z}$ and $n, q \in \mathbf{N}$ and if $x > 0$, then $(x^{1/n})^m = (x^{1/q})^p$. We leave it as an exercise to the reader to establish this relation.

4.8.8 Theorem. *If $m \in \mathbf{Z}$, $n \in \mathbf{N}$, and $x > 0$, then $x^{m/n} = (x^m)^{1/n}$.*

Proof. If $x > 0$ and $m, n \in \mathbf{Z}$, then $(x^m)^n = x^{mn} = (x^n)^m$. Now let $y := x^{m/n} = (x^{1/n})^m > 0$ so that $y^n = ((x^{1/n})^m)^n = ((x^{1/n})^n)^m = x^m$. Therefore it follows that $y = (x^m)^{1/n}$. Q.E.D.

The reader should also show, as an exercise, that if $x > 0$ and $r, s \in \boldsymbol{Q}$, then

$$x^r x^s = x^{r+s} = x^s x^r \qquad \text{and} \qquad (x^r)^s = x^{rs} = (x^s)^r.$$

Exercises for Section 4.8

1. If $I := [a,b]$ is an interval and $f : I \to \boldsymbol{R}$ is an increasing function, then the point a [respectively, b] is an absolute minimum [respectively, maximum] point for f on I. If f is strictly increasing, then a is the only absolute minimum point for f on I.

2. If f and g are increasing functions on an interval $I \subseteq \mathbf{R}$, show that $f + g$ is an increasing function on I. If f is also strictly increasing on I, then $f + g$ is strictly increasing on I.
3. Show that both $f(x) := x$ and $g(x) := x - 1$ are strictly increasing on $I := [0,1]$, but that their product fg is not increasing on I.
4. Show that if f and g are positive increasing functions on an interval I, then their product fg is increasing on I.
5. Show that if $I := [a,b]$ and $f : I \to \mathbf{R}$ is increasing on I, then f is continuous at a if and only if $f(a) = \inf \{f(x) : x \in (a,b]\}$.
6. Let $I \subseteq \mathbf{R}$ be an interval and let $f : I \to \mathbf{R}$ be increasing on I. Suppose that $c \in I$ is not an end point of I. Show that f is continuous at c if and only if there exists a sequence (x_n) in I such that $x_n < c$ for $n = 1, 3, 5, \ldots;$ $x_n > c$ for $n = 2, 4, 6, \ldots;$ and such that $c = \lim (x_n)$ and $f(c) = \lim (f(x_n))$.
7. Let $I \subseteq \mathbf{R}$ be an interval and let $f : I \to \mathbf{R}$ be increasing on I. If c is not an end point of I, show that the jump $j_f(c)$ of f at c is given by $\inf \{f(y) - f(x) : x < c < y,\ x, y \in I\}$.
8. Let f, g be increasing on an interval $I \subseteq \mathbf{R}$ and let $f(x) > g(x)$ for all $x \in I$. If $y \in f(I) \cap g(I)$, show that $f^{-1}(y) < g^{-1}(y)$. [*Hint*: First interpret this statement geometrically.]
9. Let $I := [0,1]$ and let $f : I \to \mathbf{R}$ be defined by $f(x) := x$ for x rational, and $f(x) := 1 - x$ for x irrational. Show that f is injective on I and that $f(f(x)) = x$ for all $x \in I$. (Hence f is its own inverse function!) Show that f is continuous only at the point $x = 1/2$.
10. Let $I := [a,b]$ and let $f : I \to \mathbf{R}$ be continuous on I. If f has an absolute maximum [respectively, minimum] at an interior point c of I, show that f is not injective on I.
11. Let $f(x) := x$ for $x \in [0,1]$, and $f(x) := 1 + x$ for $x \in (1,2]$. Show that f and f^{-1} are strictly increasing. Are f and f^{-1} continuous at every point?
12. Let $f : \mathbf{R} \to \mathbf{R}$ be a continuous function that does not take on any of its values twice. Show that f must be strictly monotone. [*Hint*: Use Theorem 4.8.6.]
13. Let $h : [0,1] \to \mathbf{R}$ be a function that takes on each of its values exactly twice. Show that h cannot be continuous at every point. [*Hint*: If $c_1 < c_2$ are the points where h attains its supremum, show that $c_1 = 0$, $c_2 = 1$. Now examine the points where h attains its infimum.]
14. Let $x \in \mathbf{R}$, $x > 0$. Show that if $m, p \in \mathbf{Z}$, $n, q \in \mathbf{N}$, and $mq = np$, then $(x^{1/n})^m = (x^{1/q})^p$.
15. If $x \in \mathbf{R}$, $x > 0$, and if $r, s \in \mathbf{Q}$, show that $x^r x^s = x^{r+s} = x^s x^r$ and $(x^r)^s = x^{rs} = (x^s)^r$.

SECTION 4.9 Compact Sets†

In advanced analysis and topology, the notion of a "compact" set is of enormous importance. This is less true for sets in $\mathbf{R}$ because of the Heine-Borel Theorem, which gives a very simple characterization of compact sets in $\mathbf{R}$. The purpose of this section is to acquaint the reader with this notion and to give alternate proofs

† This section can be omitted in a first reading.

of a few results that we have already established. Explicit mention of compact sets will not be made later in this book, but the reader will find many uses of compactness in *The Elements of Real Analysis* (= *ERA*).

4.9.1 Definition. A set $K \subseteq \boldsymbol{R}$ is said to be **compact** if, whenever it is contained in the union of a collection $\mathcal{G} = \{G_\alpha\}$ of open sets in $\boldsymbol{R}$, then it is contained in the union of some *finite* number of sets in $\mathcal{G}$.

We note that, in order to apply the definition to prove that a set K is compact, we need to examine an *arbitrary* collection $\mathcal{G}$ of open sets whose union contains K, and show that K is contained in the union of some finite number of sets in $\mathcal{G}$. On the other hand, to show that a set H is not compact, it is sufficient to exhibit a *specific* collection $\mathcal{G}$ of open sets whose union contains H, but such that the union of any finite number of sets in $\mathcal{G}$ does not contain H.

4.9.2 Examples. (a) Let $K := \{x_1, \ldots, x_n\}$ be a finite subset of $\boldsymbol{R}$. It is clear that if $\mathcal{G} = \{G_\alpha\}$ is a collection of open sets in $\boldsymbol{R}$ and if every point of K is contained in some set of $\mathcal{G}$, then at most n carefully chosen sets in $\mathcal{G}$ will contain K in their union. Thus K is compact.

(b) Let $H := [0,\infty)$ and let $G_n := (-1,n)$ for $n \in \boldsymbol{N}$. It is evident that $H \subseteq \bigcup_{n=1}^{\infty} G_n$. However, if $\{G_{n_1}, \ldots, G_{n_k}\}$ is any finite subcollection of $\mathcal{G} := \{G_n\}$, and if $M := \sup\{n_1, \ldots, n_k\}$, then

$$G_{n_1} \cup \cdots \cup G_{n_k} = G_M.$$

However, $M + 1 \in H$, but $M + 1 \notin G_M$. Thus, no finite subcollection of $\mathcal{G}$ will have its union containing all of H. Therefore H is not compact.

(c) Let $J := (0,1)$ and for $n \in \{3,4, \ldots\}$ let $G_n := (1/n, 1 - 1/n)$. It is an exercise to show that

$$J \subseteq \bigcup_{n=3}^{\infty} G_n.$$

However, if $\{G_{n_1}, \ldots, G_{n_p}\}$ is any finite subcollection of $\mathcal{G} := \{G_n\}$ and if $P := \sup\{n_1, \ldots, n_p\}$, then

$$G_{n_1} \cup \cdots \cup G_{n_p} = G_P.$$

However $1/P \in J$ but $1/P \notin G_P$. Therefore J is not compact.

It can be shown directly (see *ERA*, pp. 73–74) that the interval $[0,1]$ is compact. However, the argument is somewhat delicate. Instead of going through it, we shall now establish the basic result concerning compact subsets of $\boldsymbol{R}$.

4.9.3 Heine-Borel Theorem. *A subset K of $\boldsymbol{R}$ is compact if and only if it is closed and bounded.*

Proof. ($\Rightarrow$). Let $K \subseteq \boldsymbol{R}$ be compact. We shall first show that K is bounded in $\boldsymbol{R}$. In fact for each $m \in \boldsymbol{N}$, let $H_m := (-m,m)$ so that H_m is open. Clearly $K \subseteq \bigcup_{m=1}^{\infty} H_m$; since K is compact, there exists $M \in \boldsymbol{N}$ such that

$$K \subseteq \bigcup_{m=1}^{M} H_m = H_M.$$

Therefore $x \in (-M,M)$ for all $x \in K$, showing that K is bounded.

We now show that K is closed, by showing that its complement $\mathscr{C}(K)$ is open. To do so, let $u \in \mathscr{C}(K)$ be arbitrary and for each $n \in \boldsymbol{N}$, we let $G_n := \{y \in \boldsymbol{R} : |y - u| > 1/n\}$. It is an exercise to show that each set G_n is open and that

$$\boldsymbol{R} \setminus \{u\} = \bigcup_{n=1}^{\infty} G_n.$$

Since $u \notin K$, we have $K \subseteq \bigcup_{n=1}^{\infty} G_n$. Since K is compact, there exist $M \in \boldsymbol{N}$ such that

$$K \subseteq \bigcup_{n=1}^{M} G_n = G_M.$$

Now it follows from this that $K \cap (u - 1/M, u + 1/M) = \emptyset$, so that $(u - 1/M, u + 1/M) \subseteq \mathscr{C}(K)$. But since u was an arbitrary point in $\mathscr{C}(K)$, we infer that $\mathscr{C}(K)$ is open.

($\Leftarrow$) Now suppose that $K \subseteq \boldsymbol{R}$ is closed and bounded and that $\mathscr{G} = \{G_\alpha\}$ is a collection of open sets with $K \subseteq \bigcup G_\alpha$. We wish to show that K is contained in the union of some finite subcollection from $\mathscr{G}$. The proof is by contradiction; we assume that:

(*) K is not contained in the union of any finite number of sets in $\mathscr{G}$.

By hypothesis, K is bounded, so there exists $r > 0$ such that $K \subseteq [-r,r]$. We let $I_1 := [-r,r]$ and bisect I_1 into two closed subintervals $I_1' := [-r,0]$ and $I_1'' := [0,r]$. At least one of the two subsets $K \cap I_1'$ and $K \cap I_1''$ must be non-void and have the property that it is not contained in the union of any finite number of sets in $\mathscr{G}$. [For if both of the sets $K \cap I_1'$ and $K \cap I_1''$ are contained in the union of some finite number of sets in $\mathscr{G}$, then $K = (K \cap I_1') \cup (K \cap I_1'')$ is contained in the union of some finite number of sets in $\mathscr{G}$, contrary to the assumption (*).] If $K \cap I_1'$ is not contained in the union of some finite number of sets in $\mathscr{G}$, we let $I_2 := I_1'$; otherwise $K \cap I_1''$ has this property and we let $I_2 := I_1''$.

We now bisect I_2 into two closed subintervals I_2' and I_2''. If $K \cap I_2'$ is non-void and is not contained in the union of some finite number of sets in $\mathscr{G}$, we let $I_3 := I_2'$; otherwise $K \cap I_2''$ has this property and we let $I_3 := I_2''$.

Continuing this process, we obtain a nested sequence of intervals (I_n). By the Nested Intervals Property 2.5.1, there is a point z that belongs to all of the I_n, $n \in \mathbf{N}$. Since each interval I_n contains points in K, the point z is a cluster point of K. Moreover, since K is assumed to be closed, it follows from Theorem 2.6.7 that $z \in K$. Therefore there exists a set G_λ in $\mathcal{G}$ with $z \in G_\lambda$. Since G_λ is open, there exists $\varepsilon > 0$ such that

$$(z - \varepsilon, z + \varepsilon) \subseteq G_\lambda.$$

On the other hand, since the intervals I_n are obtained by repeated bisections of $I_1 = [-r,r]$, the length of I_n is $r/2^{n-2}$. It follows that if n is so large that $r/2^{n-2} < \varepsilon$, then $I_n \subseteq (z - \varepsilon, z + \varepsilon) \subseteq G_\lambda$. But this means that if n is such that $r/2^{n-2} < \varepsilon$, then $K \cap I_n$ is contained in the *single* set G_λ in $\mathcal{G}$, contrary to our construction of I_n. This contradiction shows that the assumption (*) that the closed bounded set K requires an infinite number of sets in $\mathcal{G}$ to enclose it is untenable. We conclude that K is compact. Q.E.D.

Remark. It was seen in Example 4.9.2(b) that the closed set $H := [0,\infty)$ is *not* compact; note that H is not bounded. It was also seen in Example 4.9.2(c) that the bounded set $J := (0,1)$ is *not* compact; note that J is not closed. Thus, we cannot drop either hypothesis of the Heine-Borel Theorem.

The fundamental connection between compactness and continuity is that the continuous image of a compact set is compact, as we now prove.

4.9.4 Preservation of Compactness. *If K is a compact subset of $\mathbf{R}$ and if $f : K \to \mathbf{R}$ is continuous on K, then the set $f(K)$ is compact.*

Proof. Let $\mathcal{G} = \{G_\alpha\}$ be any collection of open sets in $\mathbf{R}$ whose union contains the set $f(K)$. Then, since $f(K) \subseteq \bigcup G_\alpha$, it follows that $K \subseteq \bigcup f^{-1}(G_\alpha)$. It may happen that the sets $f^{-1}(G_\alpha)$ are not open sets; however, for each α we can find an open set H_α such that $H_\alpha \cap K = f^{-1}(G_\alpha)$ and thereby obtain a collection $\{H_\alpha\}$ of open set whose union contains K.

For a given set $f^{-1}(G_\alpha)$ we obtain the corresponding set H_α as follows. For x in K such that $x \in f^{-1}(G_\alpha)$, we have $f(x) \in G_\alpha$. Then G_α is a neighborhood of $f(x)$, and since f is continuous, there exists an open neighborhood U_x of x such that $f(U_x \cap K) \subseteq G_\alpha$. (See Definition 4.4.1.) We now define $H_\alpha := \bigcup \{U_x : x \in f^{-1}(G_\alpha)\}$. Since the union of any collection of open sets is open by Theorem 2.6.3(a), the set H_α is an open set. Also it is readily verified that $H_\alpha \cap K = f^{-1}(G_\alpha)$, so that we have

$$\bigcup H_\alpha \supseteq \bigcup f^{-1}(G_\alpha) \supseteq K.$$

Therefore $\{H_\alpha\}$ is a collection of open sets whose union contains K.

We now use the compactness of K to obtain a finite subcollection $\{H_{\alpha_1}, \ldots, H_{\alpha_n}\}$ of $\{H_\alpha\}$ whose union contains K. Then we have

$$\bigcup_{i=1}^{n} f^{-1}(G_{\alpha_i}) = \bigcup_{i=1}^{n} H_{\alpha_i} \cap K \supseteq K$$

from which it follows that $\bigcup_{i=1}^{n} G_{\alpha_i} \supseteq f(K)$. Hence we have found a finite subcollection $\{G_{\alpha_1}, \ldots, G_{\alpha_n}\}$ of $\mathcal{G} = \{G_\alpha\}$ whose union contains $f(K)$.

We have shown that if $\mathcal{G} = \{G_\alpha\}$ is any collection of open sets whose union contains $f(K)$, then there is a finite subcollection of sets in $\mathcal{G}$ whose union contains $f(K)$. Therefore $f(K)$ is compact. Q.E.D.

4.9.5 Some Applications. We shall now show how to apply the notion of compactness (and the Heine-Borel Theorem) to obtain alternative proofs of some important results that we have proved earlier by using the Bolzano-Weierstrass Theorem. In fact, these theorems remain true if the intervals are replaced by arbitrary compact sets in $\boldsymbol{R}$.

(A) The Boundedness Theorem 4.6.2 is an immediate consequence of Theorem 4.9.4 and the Heine-Borel Theorem 4.9.3. Indeed, if $K \subseteq \boldsymbol{R}$ is compact and if $f : K \to \boldsymbol{R}$ is continuous on K, then $f(K)$ is compact and hence bounded.

(B) The Maximum-Minimum Theorem 4.6.4 also is an easy consequence of Theorem 4.9.4 and the Heine-Borel Theorem. As before we find that $f(K)$ is compact and hence bounded in $\boldsymbol{R}$, so that $s^* := \sup f(K)$ exists. If $f(K)$ is a finite set, then $s^* \in f(K)$. If $f(K)$ is an infinite set, then s^* is a cluster point of $f(K)$ [see Example 2.5.3(e)]. Since $f(K)$ is a closed set, by the Heine-Borel Theorem, it follows from Theorem 2.6.7 that $s^* \in f(K)$. We conclude that $s^* = f(x^*)$ for some $x^* \in K$.

(C) We conclude this section by giving a proof of the Uniform Continuity Theorem 4.7.3, based on the notion of compactness. To do so, let $K \subseteq \boldsymbol{R}$ be compact and let $f : K \to \boldsymbol{R}$ be continuous on K. Then given $\varepsilon > 0$ and $u \in K$, there is a number $\delta_u := \delta(\frac{1}{2}\varepsilon, u) > 0$ such that if $x \in K$ and $|x - u| < \delta_u$, then $|f(x) - f(u)| < \frac{1}{2}\varepsilon$. For each $u \in K$, let $G_u := (u - \frac{1}{2}\delta_u, u + \frac{1}{2}\delta_u)$ so that G_u is open; we consider the collection $\mathcal{G} := \{G_u : u \in K\}$. Since $u \in G_u$ for $u \in K$, it is trivial that $K \subseteq \bigcup_{u \in I} G_u$. Since K is compact, there are a finite number of sets, say $G_{u_1}, \ldots, G_{u_M}$ whose union contains K. We now define

$$\delta(\varepsilon) := \tfrac{1}{2} \inf \{\delta_{u_1}, \ldots, \delta_{u_M}\},$$

so that $\delta(\varepsilon) > 0$. Now if $x, u \in K$ and $|x - u| < \delta(\varepsilon)$, then there exists some u_k with $k = 1, \ldots, M$ such that $x \in G_{u_k}$; therefore $|x - u_k| < \frac{1}{2}\delta_{u_k}$. Since we have $\delta(\varepsilon) < \frac{1}{2}\delta_{u_k}$ it follows that

$$|u - u_k| \le |u - x| + |x - u_k| < \delta_{u_k}.$$

But since $\delta_{u_k} = \delta(\frac{1}{2}\varepsilon, u_k)$ it follows that both

$$|f(x) - f(u_k)| < \tfrac{1}{2}\varepsilon \qquad \text{and} \qquad |f(u) - f(u_k)| < \tfrac{1}{2}\varepsilon.$$

Therefore we have $|f(x) - f(u)| < \varepsilon$.

We have shown that if $\varepsilon > 0$, then there exist $\delta(\varepsilon) > 0$ such that if x, u are any points in K with $|x - u| < \delta(\varepsilon)$, then $|f(x) - f(u)| < \varepsilon$. Since $\varepsilon > 0$ is arbitrary, this shows that f is uniformly continuous on K, as asserted.

Exercises for Section 4.9

1. Exhibit a collection of open sets whose union contains the interval $(1,2]$ such that no finite subcollection has that property.
2. Prove, using Definition 4.9.1, that if K is compact in $\boldsymbol{R}$ and F is a closed subset of K, then F is compact.
3. Prove, using Definition 4.9.1, that if K_1 and K_2 are compact sets in $\boldsymbol{R}$, then their union $K_1 \cup K_2$ is compact in $\boldsymbol{R}$.
4. Find an infinite collection $\{K_n : n \in \boldsymbol{N}\}$ of compact sets in $\boldsymbol{R}$ such that the union $\bigcup_{n=1}^{\infty} K_n$ is not compact.
5. Prove that the intersection of an arbitrary collection of compact sets in $\boldsymbol{R}$ is compact.
6. Let $\{K_n : n \in \boldsymbol{N}\}$ be a sequence of non-empty compact sets in $\boldsymbol{R}$ such that $K_1 \supseteq K_2 \supseteq \cdots \supseteq K_n \supseteq \cdots$. Prove that there exists at least one point $x \in \boldsymbol{R}$ such that $x \in K_n$ for all $n \in \boldsymbol{N}$; that is, the intersection $\bigcap_{n=1}^{\infty} K_n$ is not empty.
7. Let K be a compact set in $\boldsymbol{R}$. Show that $\inf K$ and $\sup K$ exist and belong to K.
8. Prove that if K is a compact set in $\boldsymbol{R}$, then every sequence in K has a subsequence that converges to a point of K.
9. Let K be compact in $\boldsymbol{R}$ and let $c \in \boldsymbol{R}$. Prove that there exists a point a in K such that $|c - a| = \inf\{|c - x| : x \in K\}$.
10. Let K be compact in $\boldsymbol{R}$ and let $c \in \boldsymbol{R}$. Prove that there exists a point b in K such that $|c - b| = \sup\{|c - x| : x \in \boldsymbol{K}\}$.
11. Use the notion of compactness to give an alternative proof of Exercise 4.6.16.

CHAPTER FIVE

DIFFERENTIATION

With the creation of analytic geometry by René Descartes (1596–1650) and Pierre de Fermat (1601–1665) in the 1630s, many geometric problems were recast in terms of algebraic expressions and new classes of curves were determined by algebraic equations rather than geometric conditions. The concept of the derivative evolved in this new context out of attempts to solve two problems that at first seemed to be unrelated. One was the classical problem of determining the line tangent to a curve at a given point. The other was the problem of finding the maximum and minimum values of a function. The connection between tangent lines to curves and the velocity of a moving particle was discovered by Isaac Newton (1642–1727) in the late 1660s. His theory of fluxions, with appropriate changes in terminology and notation, would be familiar to any student of modern differential calculus. The vital observation, made by Newton and, independently, by Gottfried Leibniz (1646–1716), that area problems could be solved by reversing the differentiation problem led to a coherent theory that became known as differential and integral calculus.

In this chapter we shall develop the theory of differentiation. Integration theory, including the fundamental theorem that relates differentiation and integration, will be the subject of the next chapter. We shall assume that the reader is already familiar with the geometrical and physical interpretations of the derivative of a function as described in introductory calculus courses. Consequently, we shall concentrate on the mathematical aspects of the derivative and not go into its applications in geometry, physics, economics, and so on.

The first section is devoted to a presentation of the basic results concerning the differentiation of functions. In Section 5.2 we discuss the fundamental Mean Value Theorem and some of its applications. In Section 5.3 the important L'Hospital Rules are presented for the calculation of certain types of "indeterminant" limits. In Section 5.4 we give a brief discussion of Taylor's Theorem and a few of its applications—for example, to convex functions and to Newton's Method for the location of roots.

SECTION 5.1 The Derivative

In this section we shall present some of the elementary properties of the derivative. We begin with the definition of the derivative of a function.

5.1.1 Definition. Let $I \subseteq \boldsymbol{R}$ be an interval, let $f : I \to \boldsymbol{R}$, and let $c \in I$. We say that a real number L is the **derivative of** f **at** c if for any given number $\varepsilon > 0$ there exists a number $\delta(\varepsilon) > 0$ such that if $x \in I$ and $0 < |x - c| < \delta(\varepsilon)$, then

$$(1) \qquad \left| \frac{f(x) - f(c)}{x - c} - L \right| < \varepsilon.$$

In this case we say that f is *differentiable at* c, and we write $f'(c)$ for L.

In other words, the derivative of f at c is given by the limit

$$(2) \qquad f'(c) = \lim_{x \to c} \frac{f(x) - f(c)}{x - c}$$

provided this limit exists. (We allow the possibility that c may be the end point of an interval.)

Note. It is possible to define the derivative of a function having a domain more general than an interval (since the point c need only be an element of the domain and also a cluster point of the domain) but the significance of the concept is most naturally apparent for functions defined on intervals. Consequently we shall limit our attention to such functions.

Whenever the derivative of $f : I \to \boldsymbol{R}$ exists at a point $c \in I$, its value is denoted by $f'(c)$. In this way we obtain a function f' whose domain is a subset of the domain of f. In working with the function f', it is convenient to regard it also as a function of x. For example, if $f(x) := x^2$ for $x \in \boldsymbol{R}$, then at any c in $\boldsymbol{R}$ we have

$$f'(c) = \lim_{x \to c} \frac{f(x) - f(c)}{x - c} = \lim_{x \to c} \frac{x^2 - c^2}{x - c} = \lim_{x \to c} (x + c) = 2c.$$

Thus, in this case, the function f' is defined on all of $\boldsymbol{R}$ and $f'(x) = 2x$ for $x \in \boldsymbol{R}$.

We now show that continuity of f at a point c is a necessary (but not sufficient) condition for the existence of the derivative at c.

5.1.2 Theorem. *If $f : I \to \boldsymbol{R}$ has a derivative at $c \in I$, then f is continuous at c.*

Proof. For all $x \in I$, $x \neq c$, we have

$$f(x) - f(c) = \left(\frac{f(x) - f(c)}{x - c} \right) (x - c).$$

Since $f'(c)$ exists, we may apply Theorem 4.2.4 concerning the limit of a product to conclude that

$$\begin{aligned}\lim_{x\to c}\big(f(x) - f(c)\big) &= \left(\lim_{x\to c}\frac{f(x) - f(c)}{x - c}\right)\lim_{x\to c}(x - c)\big)\\ &= f'(c)\cdot 0 = 0.\end{aligned}$$

Therefore, $\lim_{x\to c} f(x) = f(c)$, so that f is continuous at c. Q.E.D.

The continuity of $f: I \to \boldsymbol{R}$ at a point does not assure the existence of the derivative at that point. For example, if $f(x) := |x|$ for $x \in \boldsymbol{R}$, then for $x \neq 0$ we have $(f(x) - f(0))/(x - 0) = |x|/x$ which is equal to 1 if $x > 0$ and equal to -1 if $x < 0$. Thus the limit at 0 does not exist [see Example 4.1.7(b)] and therefore the function is not differentiable at 0. Hence, continuity at a point c is *not* a sufficient condition for the derivative to exist at c.

Remark. By taking simple algebraic combinations of functions of the form $x \mapsto |x - c|$, it is not difficult to construct continuous functions that do not have a derivative at a finite (or even a countable) number of points. In 1872, Karl Weierstrass astounded the mathematical world by giving an example of a function that is *continuous at every point but whose derivative does not exist anywhere.* Such a function defied geometric intuition about curves and tangent lines, and consequently spurred much deeper investigations into the concepts of real analysis. (It can be shown that the function f defined by the series

$$f(x) := \sum_{n=0}^{\infty} \frac{1}{2^n}\cos(3^n x)$$

has the stated property. However, we shall not go through the proof, but refer the reader to the books of Titchmarsh and Boas listed in the references at the end of this book.)

There are a number of basic properties of the derivative that are very useful in the calculation of the derivatives of various combinations of functions. We now provide the justification of some of these properties, which will be familiar to the reader from earlier courses.

5.1.3 Theorem. *Let $I \subseteq \boldsymbol{R}$ be an interval, let $c \in I$, and let $f: I \to \boldsymbol{R}$ and $g: I \to \boldsymbol{R}$ be functions that are differentiable at c. Then:*

(a) *If $\alpha \in \boldsymbol{R}$, then the function αf is differentiable at c, and*

(3) $$(\alpha f)'(c) = \alpha f'(c).$$

(b) *The function $f + g$ is differentiable at c, and*

(4) $$(f + g)'(c) = f'(c) + g'(c).$$

(c) (*Product Rule*) *The function fg is differentiable at c, and*

$$(fg)'(c) = f'(c)g(c) + f(c)g'(c). \tag{5}$$

(d) (*Quotient Rule*) *If $g(c) \neq 0$, then the function f/g is differentiable at c, and*

$$\left(\frac{f}{g}\right)'(c) = \frac{f'(c)g(c) - f(c)g'(c)}{g(c)^2}. \tag{6}$$

Proof. We shall prove (c) and (d), leaving (a) and (b) as exercises for the reader.

(c) Let $p := fg$; then for $x \in I$, $x \neq c$, we have

$$\begin{aligned}
\frac{p(x) - p(c)}{x - c} &= \frac{f(x)g(x) - f(c)g(c)}{x - c} \\
&= \frac{f(x)g(x) - f(c)g(x) + f(c)g(x) - f(c)g(c)}{x - c} \\
&= \frac{f(x) - f(c)}{x - c} \cdot g(x) + f(c) \cdot \frac{g(x) - g(c)}{x - c}.
\end{aligned}$$

Since g is continuous at c, by Theorem 5.1.2, then $\lim_{x \to c} g(x) = g(c)$. Since f and g are differentiable at c, we deduce from Theorem 4.2.4 on properties of limits that

$$\lim_{x \to c} \frac{p(x) - p(c)}{x - c} = f'(c)g(c) + f(c)g'(c).$$

Hence $p = fg$ is differentiable at c and (5) holds.

(d) Let $q := f/g$. Since g is differentiable at c, it is continuous at that point (by Theorem 5.1.2). Therefore, since $g(c) \neq 0$, we know from Theorem 4.2.11 that there exists an interval $J \subseteq I$ with $c \in J$ such that $g(x) \neq 0$ for all $x \in J$. For $x \in J$, $x \neq c$, we have

$$\begin{aligned}
\frac{q(x) - q(c)}{x - c} &= \frac{f(x)/g(x) - f(c)/g(c)}{x - c} \\
&= \frac{f(x)g(c) - f(c)g(x)}{g(x)g(c)(x - c)} \\
&= \frac{f(x)g(c) - f(c)g(c) + f(c)g(c) - f(c)g(x)}{g(x)g(c)(x - c)} \\
&= \frac{1}{g(x)g(c)} \left[\frac{f(x) - f(c)}{x - c} \cdot g(c) - f(c) \cdot \frac{g(x) - g(c)}{x - c} \right].
\end{aligned}$$

Using the continuity of g at c and the differentiability of f and g at c, we deduce that

$$q'(c) = \lim_{x \to c} \frac{q(x) - q(c)}{x - c} = \frac{f'(c)g(c) - f(c)g'(c)}{g(c)^2}.$$

Thus, $q = f/g$ is differentiable at c and equation (6) holds. Q.E.D.

Mathematical induction may be utilized to obtain the following extensions of the differentiation rules.

5.1.4 Corollary. *If $f_1, f_2, \ldots, f_n$ are functions on an interval I to $\mathbf{R}$ that are differentiable at $c \in I$, then:*

(a) *The function $f_1 + f_2 + \cdots + f_n$ is differentiable at c and*

(7) $$(f_1 + f_2 + \cdots + f_n)'(c) = f_1'(c) + f_2'(c) + \cdots + f_n'(c).$$

(b) *The function $f_1 f_2 \cdots f_n$ is differentiable at c, and*

(8) $$(f_1 f_2 \cdots f_n)'(c) = f_1'(c) f_2(c) \cdots f_n(c) + f_1(c) f_2'(c) \cdots f_n(c) + \cdots + f_1(c) f_2(c) \cdots f_n'(c).$$

5.1.5 Notation. If $I \subseteq \mathbf{R}$ is an interval and $f : I \to \mathbf{R}$, we have introduced the notation f' to denote the function whose domain is a subset of I and whose value at a point c is the derivative $f'(c)$ of f at c. There are other notations that are sometimes used for f'; for example, one sometimes writes Df for f'. Thus one can write formulas (4) and (5) in the form:

$$D(f + g) = Df + Dg, \qquad D(fg) = (Df)\cdot g + f\cdot(Dg).$$

When x is the "independent variable", it is common practice in elementary courses to write df/dx for f'. Thus formula (5) is sometimes written in the form

$$\frac{d}{dx}\left(f(x)g(x)\right) = \left(\frac{df}{dx}(x)\right)(g(x)) + f(x)\left(\frac{dg}{dx}(x)\right).$$

This last notation, due to Leibniz, has certain advantages. However, it also has certain disadvantages and must be used with some care.

The Chain Rule

If in Corollary 5.1.4(b) we have $f_1 = f_2 = \cdots = f_n = f$, then the result becomes

(9) $$(f^n)'(c) = n(f(c))^{n-1} f'(c).$$

This can also be derived from the following theorem on the differentiation of composite functions, known as the Chain Rule.

5.1.6 Chain Rule. *Let I, J be intervals in $\mathbf{R}$, let $g : I \to \mathbf{R}$ and $f : J \to \mathbf{R}$ be functions such that $f(J) \subseteq I$, and let $c \in J$. If f is differentiable at c and if g is differentiable at $f(c)$, then the composite function $g \circ f$ is differentiable at c and*

$$(10) \qquad (g \circ f)'(c) = g'(f(c))f'(c).$$

Proof. Let $d := f(c)$ and let G be defined on I by

$$G(y) := \frac{g(y) - g(d)}{y - d} \qquad \text{if } y \in I, \quad y \neq d,$$

$$:= g'(d) \qquad \text{if } y = d.$$

Since g is differentiable at d, we have $\lim_{y \to d} G(y) = g'(d) = G(d)$, so that G is continuous at d. Now since f is continuous at $c \in J$ (by Theorem 5.1.2) and $f(J) \subseteq I$, we conclude from Theorem 4.5.9 that $G \circ f$ is continuous at c so that

$$(11) \qquad \lim_{x \to c} G \circ f(x) = \lim_{y \to d} G(y) = g'(f(c)).$$

It follows from the definition of G that

$$g(y) - g(d) = G(y)(y - d)$$

for all $y \in I$. Hence, if $x \in J$ and we let $y = f(x)$, then we have

$$\begin{aligned} g \circ f(x) - g \circ f(c) &= g(f(x)) - g(f(c)) \\ &= G \circ f(x)\,(f(x) - f(c)). \end{aligned}$$

Therefore, if $x \in J$, $x \neq c$, then we have

$$\frac{g \circ f(x) - g \circ f(c)}{x - c} = G \circ f(x) \cdot \frac{f(x) - f(c)}{x - c}.$$

Consequently, by (11), we have

$$\lim_{x \to c} \frac{g \circ f(x) - g \circ f(c)}{x - c} = g'(f(c))f'(c).$$

We conclude that $g \circ f$ is differentiable at $c \in J$ and that equality (10) holds. Q.E.D.

If g is differentiable on I, if f is differentiable on J and if $f(J) \subseteq I$, then it follows from the Chain Rule that $(g \circ f)' = (g' \circ f) \cdot f'$, which can also be written in the form $D(g \circ f) = ((Dg) \circ f)\, Df$.

5.1.7 Examples. (a) If $f : I \to \boldsymbol{R}$ is differentiable on I and $g(y) := y^n$ for $y \in \boldsymbol{R}$ and $n \in \boldsymbol{N}$, then since $g'(y) = ny^{n-1}$, it follows from the Chain Rule 5.1.6 that

$$(g \circ f)'(x) = g'(f(x)) \cdot f'(x) \qquad \text{for} \quad x \in I.$$

Therefore we have $(f^n)'(x) = n(f(x))^{n-1} f'(x)$ for all $x \in I$ as was seen in (9).

(b) Suppose that $f : I \to \boldsymbol{R}$ is differentiable on I and that $f(x) \neq 0$ and $f'(x) \neq 0$ for $x \in I$. If $h(y) := 1/y$ for $y \neq 0$, then it is an exercise to show that $h'(y) = -1/y^2$ for $y \in \boldsymbol{R}$, $y \neq 0$. Therefore we have

$$\left(\frac{1}{f}\right)'(x) = (h \circ f)'(x) = h'(f(x))f'(x) = -\frac{f'(x)}{(f(x))^2}$$

for $x \in I$.

(c) It will be proved later that if $S(x) := \sin x$ and $C(x) := \cos x$ for all $x \in \boldsymbol{R}$, then

$$S'(x) = \cos x = C(x) \qquad \text{and} \qquad C'(x) = -\sin x = -S(x)$$

for all $x \in \boldsymbol{R}$. If we use these facts together with the definitions

$$\tan x := \frac{\sin x}{\cos x}, \qquad \sec x := \frac{1}{\cos x}$$

for $x \neq (2k + 1)\pi/2$, $k \in \boldsymbol{Z}$, and apply the Quotient Rule 5.1.3(d), we obtain

$$\begin{aligned} D \tan x &= \frac{(\cos x)(\cos x) - (\sin x)(-\sin x)}{(\cos x)^2} \\ &= \frac{1}{(\cos x)^2} = (\sec x)^2 \end{aligned}$$

and

$$D \sec x = \frac{0 - 1(-\sin x)}{(\cos x)^2} = \frac{\sin x}{(\cos x)^2} = (\sec x)(\tan x)$$

for $x \neq (2k + 1)\pi/2$, $k \in \boldsymbol{Z}$.

Similarly, since

$$\cot x := \frac{\cos x}{\sin x}, \qquad \operatorname{cosec} x := \frac{1}{\sin x}$$

for $x \neq k\pi$, $k \in \boldsymbol{Z}$, then we obtain

$$D \cot x = -(\operatorname{cosec} x)^2 \qquad \text{and} \qquad D \operatorname{cosec} x = -(\operatorname{cosec} x)(\cot x)$$

for $x \neq k\pi$, $k \in \mathbf{Z}$.

(d) Suppose that f is defined by

$$\begin{aligned} f(x) &:= x^2 \sin(1/x) && \text{for} \quad x \neq 0, \\ &:= 0 && \text{for} \quad x = 0. \end{aligned}$$

If we use the fact that $D \sin x = \cos x$ for all $x \in \mathbf{R}$ and apply the Product Rule 5.1.3(c) and Chain Rule 5.1.6, we obtain

$$f'(x) = 2x \sin \frac{1}{x} - \cos \frac{1}{x} \qquad \text{for} \quad x \neq 0.$$

If $x = 0$, none of the calculational rules may be applied. (Why?) Consequently, the derivative of f at $x = 0$ must be found by appealing to the definition of derivative. We find that

$$f'(0) = \lim_{x \to 0} \frac{f(x) - f(0)}{x - 0} = \lim_{x \to 0} \frac{x^2 \sin(1/x)}{x} = \lim_{x \to 0} x \sin \frac{1}{x} = 0.$$

Hence, the derivative f' of f exists at all $x \in \mathbf{R}$. However, the function f' does not have a limit at $x = 0$ (why?), and consequently f' is discontinuous at $x = 0$. Thus, a function f that is differentiable at every point of $\mathbf{R}$ need not have a continuous derivative f'.

Inverse Functions

We now relate the derivative of a function with the derivative of its inverse function, when this inverse function exists. In order to be able to use the Continuous Inverse Theorem 4.8.5, we shall limit our attention to a continuous strictly monotone function.

5.1.8 Theorem. *Let $I \subseteq \mathbf{R}$ be an interval and let $f : I \to \mathbf{R}$ be strictly monotone and continuous on I. Let $J := f(I)$ and let $g : J \to \mathbf{R}$ be the strictly monotone and continuous function inverse to f. If f is differentiable at $c \in I$ and $f'(c) \neq 0$, then g is differentiable at $d := f(c)$ and*

$$g'(d) = \frac{1}{f'(c)} = \frac{1}{f'(g(d))}. \tag{11}$$

Proof. For $y \in J$, $y \neq d$, define

$$H(y) := \frac{f(g(y)) - f(g(d))}{g(y) - g(d)}.$$

Since $g : J \to \mathbf{R}$ is strictly monotone, then $g(y) \neq g(d)$ *for* $y \in J$, $y \neq d$; therefore H is well-defined on J. Also since $y = f(g(y))$ and $d = f(g(d))$, then we have

$$H(y) = \frac{y - d}{g(y) - g(d)}$$

so that $H(y) \neq 0$ for $y \neq d$, $y \in J$.

We shall prove that $\lim_{y \to d} H(y) = f'(c)$. Indeed, if $\varepsilon > 0$ is given, since f is differentiable at $c = g(d)$, there exists $\delta > 0$ such that if $0 < |x - c| < \delta$, $x \in I$, then

$$\left| \frac{f(x) - f(c)}{x - c} - f'(c) \right| < \varepsilon .$$

But since g is continuous at $d = f(c)$, there exists $\gamma > 0$ such that if $|y - d| < \gamma$, $y \in J$, then $|g(y) - g(d)| < \delta$. Since g is injective and $c = g(d)$, we have $0 < |g(y) - c| < \delta$ if $0 < |y - d| < \gamma$, $y \in J$. Therefore it follows that

$$\left| \frac{f(g(y)) - f(g(d))}{g(y) - g(d)} - f'(c) \right| < \varepsilon$$

when $0 < |y - d| < \gamma$, $y \in J$. But since $\varepsilon > 0$ is arbitrary, this implies that $\lim_{y \to d} H(y) = f'(c)$.

However, it was noted above that $H(y) \neq 0$ for $y \neq d$, $y \in J$. Since we have

$$\frac{g(y) - g(d)}{y - d} = \frac{1}{H(y)}$$

for $y \neq d$, $y \in J$, we infer that

$$\lim_{y \to d} \frac{g(y) - g(d)}{y - d} = \lim_{y \to d} \frac{1}{H(y)} = \frac{1}{\lim_{y \to d} H(y)} = \frac{1}{f'(c)}.$$

Therefore $g'(d)$ exists and equals $1/f'(c)$. Q.E.D.

Note. The hypothesis, made in Theorem 5.1.8, that $f'(c) \neq 0$ is essential. In fact, if $f'(c) = 0$, then the inverse function g is *not* differentiable at $d = f(c)$. Indeed, if g were differentiable at d, since f is the function inverse to g, we can apply Theorem 5.1.8 to g to conclude that f is differentiable at $c = g(d)$ and that $1 = f'(c)g'(d) = 0$, which is a contradiction. Therefore g is not differentiable at d. The function $f(x) := x^3$, $x \in \mathbf{R}$, is such an example with $c = 0$.

5.1.9 Theorem. *Let $I \subseteq \mathbf{R}$ be an interval and let $f : I \to \mathbf{R}$ be strictly monotone on I. Let $J := f(I)$ and let $g : J \to \mathbf{R}$ be the function inverse to f. If*

f is differentiable on I and $f'(x) \neq 0$ for $x \in I$, then g is differentiable on J and

$$g' = \frac{1}{f' \circ g}. \tag{12}$$

Proof. If f is differentiable on I, it follows from Theorem 5.1.2 that f is continuous on I. Therefore by the Continuous Inverse Theorem 4.7.5, the inverse function g is continuous on J. Equation (12) now follows from Theorem 5.1.8. Q.E.D.

Remark. If $f : I \to \mathbf{R}$ and $g : J \to \mathbf{R}$ are the strictly monotone functions in Theorem 5.1.9, we have seen that if $f'(x) \neq 0$ for $x \in I$, then g is differentiable on J. If $x \in I$ and $y \in J$ are related by $y = f(x)$ (or $x = g(y)$), then equation (12) can be written in the form

$$g'(y) = \frac{1}{f' \circ g(y)} \qquad \text{for} \quad y \in J$$

or the form

$$g' \circ f(x) = \frac{1}{f'(x)} \qquad \text{for} \quad x \in I.$$

It can also be written in the form $g'(y) = 1/f'(x)$, *provided* that it is kept firmly in mind that x and y are related by $y = f(x)$ or $x = g(y)$.

5.1.10 Examples. (a) Let $n \in \mathbf{N}$ be even, let $I := [0,\infty)$, and let $f(x) := x^n$ for $x \in I$. It was seen at the end of Section 4.8 that f is strictly increasing and continuous on I, so that its inverse function $g(y) := y^{1/n}$ for $y \in J := [0,\infty)$ is also strictly increasing and continuous on J. Moreover, we have $f'(x) = nx^{n-1}$ for all $x \in I$. Hence it follows that if $y > 0$, then $g'(y)$ exists and

$$\begin{aligned} g'(y) &= \frac{1}{f'(g(y))} = \frac{1}{n(g(y))^{n-1}} \\ &= \frac{1}{n(y^{1/n})^{n-1}} = \frac{1}{ny^{(n-1)/n}}. \end{aligned}$$

Hence we deduce that

$$g'(y) = \frac{1}{n} y^{(1/n)-1} \qquad \text{for} \quad y > 0.$$

However, g is *not* differentiable at 0. (For a graph of f and g, see Figures 4.8.5 and 4.8.6 on pages 180 and 181.)

(b) Let $n \in \boldsymbol{N}$, $n \neq 1$, be odd, let $F(x) := x^n$ for $x \in \boldsymbol{R}$, and let $G(y) := y^{1/n}$ be its inverse function defined for all $y \in \boldsymbol{R}$. As in part (a) we find that G is differentiable for $y \neq 0$ and that $G'(y) = (1/n)y^{(1/n)-1}$ for $y \neq 0$. However, G is not differentiable at 0, even though G is differentiable for all $y \neq 0$. (For a graph of F and G, see Figures 4.8.7 and 4.8.8.)

(c) Let $r := m/n$ be a positive rational number, let $I := [0,\infty)$, and let $R(x) := x^r$ for $x \in I$. Then R is the composition of the functions $f(x) := x^m$ and $g(x) := x^{1/n}$, $x \in$ I. That is, $R(x) = f(g(x))$ for $x \in I$. If we apply the Chain Rule 5.1.6 and the results of (a) [or (b), depending on whether n is even or odd], then we obtain

$$\begin{aligned} R'(x) &= f'(g(x))g'(x) = m(x^{1/n})^{m-1} \cdot \frac{1}{n} x^{(1/n)-1} \\ &= \frac{m}{n} x^{(m/n)-1} = rx^{r-1} \end{aligned}$$

for all $x > 0$. If $r > 1$, then it is an exercise to show that the derivative also exists at $x = 0$ and $R'(0) = 0$. (For a graph of R see Figure 4.8.9.)

(d) The sine function sin is strictly increasing on the interval $I := [-\pi/2, \pi/2]$; therefore its inverse function, which we shall denote by Arcsin, exists on $J := [-1,1]$. That is, if $x \in [-\pi/2,\pi/2]$ and $y \in [-1,1]$ then $y = \sin x$ if and only if Arcsin $y = x$. It was asserted (without proof) in Example 5.1.7(c) that sin is differentiable on I and that $D \sin x = \cos x$ for $x \in I$. Since $\cos x \neq 0$ for x in $(-\pi/2,\pi/2)$ it follows from Theorem 5.1.8 that

$$\begin{aligned} D \text{ Arcsin } y &= \frac{1}{D \sin x} = \frac{1}{\cos x} \\ &= \frac{1}{\sqrt{1-(\sin x)^2}} = \frac{1}{\sqrt{1-y^2}} \end{aligned}$$

for all $y \in (-1, 1)$. The derivative of Arcsin does *not* exist at the points -1 and 1.

Exercises for Section 5.1

1. Use the definition to find the derivative of each of the following functions:
   (a) $f(x) := x^3$ for $x \in \boldsymbol{R}$;
   (b) $g(x) := 1/x$ for $x \in \boldsymbol{R}$, $x \neq 0$;
   (c) $h(x) := \sqrt{x}$ for $x > 0$;
   (d) $k(x) := 1/\sqrt{x}$ for $x > 0$.
2. Show that $f(x) := x^{1/3}$, $x \in \boldsymbol{R}$, is not differentiable at $x = 0$.
3. Prove Theorem 5.1.3(a), (b).
4. Let $f : \boldsymbol{R} \to \boldsymbol{R}$ be defined by $f(x) := x^2$ for x rational, $f(x) := 0$ for x irrational. Show that f is differentiable at $x = 0$, and find $f'(0)$.
5. Differentiate and simplify:
   (a) $f(x) := \dfrac{x}{1+x^2}$; (b) $g(x) := \sqrt{5 - 2x + x^2}$;
   (c) $h(x) := (\sin x^k)^m$ for $m, k \in \boldsymbol{N}$; (d) $k(x) := \tan(x^2)$ for $|x| < \pi/2$.

6. Let $n \in \boldsymbol{N}$ and let $f : \boldsymbol{R} \to \boldsymbol{R}$ be defined by $f(x) := x^n$ for $x \geqslant 0$ and $f(x) = 0$ for $x < 0$. For which values of n is f' continuous at 0? For which values of n is f' differentiable at 0?
7. Suppose that $f : \boldsymbol{R} \to \boldsymbol{R}$ is differentiable at c and that $f(c) = 0$. Show that $g(x) = |f(x)|$ is differentiable at c if and only if $f'(c) = 0$.
8. Determine where each of the following functions from $\boldsymbol{R}$ to $\boldsymbol{R}$ is differentiable and find the derivative:
(a) $f(x) := |x| + |x + 1|$; (b) $g(x) := 2x + |x|$;
(c) $h(x) := x|x|$; (d) $k(x) := |\sin x|$.
9. Prove that if $f : \boldsymbol{R} \to \boldsymbol{R}$ is an even function [that is, $f(-x) = f(x)$ for all $x \in \boldsymbol{R}$] and has a derivative at every point, then the derivative f' is an odd function [that is, $f'(-x) = -f'(x)$ for all $x \in \boldsymbol{R}$]. Also prove that if $g : \boldsymbol{R} \to \boldsymbol{R}$ is a differentiable odd function, then g' is an even function.
10. Let $g : \boldsymbol{R} \to \boldsymbol{R}$ be defined by $g(x) := x^2 \sin(1/x^2)$ for $x \neq 0$, and $g(0) := 0$. Show that g is differentiable for all $x \in \boldsymbol{R}$. Also show that the derivative g' is not bounded on the interval $[-1,1]$.
11. Assume that there exists a function $L : (0,\infty) \to \boldsymbol{R}$ such that $L'(x) = 1/x$ for $x > 0$. Calculate the derivatives of the following functions:
(a) $f(x) := L(2x + 3)$ for $x > 0$;
(b) $g(x) := (L(x^2))^3$ for $x > 0$;
(c) $h(x) := L(ax)$ for $a > 0$, $x > 0$;
(d) $k(x) := L(L(x))$ when $L(x) > 0$, $x > 0$.
12. If $r > 0$ is a rational number, let $f : \boldsymbol{R} \to \boldsymbol{R}$ Be defined by $f(x) := x^r \sin(1/x)$ for $x \neq 0$, and $f(0) := 0$. Determine those values of r for which $f'(0)$ exists.
13. If $f : \boldsymbol{R} \to \boldsymbol{R}$ is differentiable at $c \in \boldsymbol{R}$, show that

$$f'(c) = \lim \left(n\{f(c + 1/n) - f(c)\}\right).$$

However, show by example that the existence of the limit of this sequence does not imply the existence of $f'(c)$.
14. Given that the function $h(x) := x^3 + 2x + 1$ for $x \in \boldsymbol{R}$ has an inverse h^{-1} on $\boldsymbol{R}$, find the value of $(h^{-1})'(y)$ at the points corresponding to $x = 0, 1, -1$.
15. Given that the restriction of the cosine function cos to $I := [0,\pi]$ is strictly decreasing and that $\cos 0 = 1$, $\cos \pi = -1$, let $J := [-1,1]$, and let Arccos : $J \to \boldsymbol{R}$ be the function inverse to the restriction of cos to I. Show that Arccos is differentiable on $(-1,1)$ and that $D \text{ Arccos } y = (-1)(1 - y^2)^{1/2}$ for $y \in (-1,1)$. Show that Arccos is not differentiable at -1 and 1.
16. Given that the restriction of the tangent function tan to $I := (-\pi/2, \pi/2)$ is strictly increasing and that $\tan(I) = \boldsymbol{R}$, let Arctan : $\boldsymbol{R} \to \boldsymbol{R}$ be the function inverse to the restriction of tan to I. Show that Arctan is differentiable on $\boldsymbol{R}$ and that $D \text{ Arctan }(y) = (1 + y^2)^{-1}$ for $y \in \boldsymbol{R}$.

SECTION 5.2 The Mean Value Theorem

The Mean Value Theorem, which relates the values of a function to values of its derivative, is one of the most useful results in real analysis. In this section we shall establish this important theorem and sample some of its many consequences.

We begin by looking at the relationship between the relative extrema of a function and the values of its derivative. Recall that the function $f : I \to \mathbf{R}$ is said to have a **relative maximum** [respectively, **minimum**] at $c \in I$ if there exists a neighborhood U of c such that $f(x) \leqslant f(c)$ [respectively, $f(c) \leqslant f(x)$] for all x in $U \cap I$. We say that f has a **relative extremum** at $c \in I$ if it has either a relative maximum or a relative minimum at c.

The next result provides the theoretical justification for the familiar process of finding points at which f has relative extrema by examining the zeros of the derivatives. However, it must be realized that this procedure applies only to *interior* points of the interval. For example, if $f(x) := x$ on the interval $I := [0,1]$, then the endpoint $x = 0$ yields the unique relative minimum and the endpoint $x = 1$ yields the unique maximum of f on I, but neither value is a zero of the derivative of f.

5.2.1 Interior Extremum Theorem. *Let c be an interior point of the interval I at which $f : I \to \mathbf{R}$ has a relative extremum. If the derivative of f at c exists, then $f'(c) = 0$.*

Proof. We shall prove the result only for the case that f has a relative maximum at c; the proof for the case of a relative minimum is similar.

If $f'(c) > 0$, then by Theorem 4.2.11 there exists a neighborhood $U \subseteq I$ of c such that

$$\frac{f(x) - f(c)}{x - c} > 0 \qquad \text{for} \quad x \in U, x \neq c.$$

If $x \in U$ and $x > c$, then we have

$$f(x) - f(c) = (x - c) \cdot \frac{f(x) - f(c)}{x - c} > 0.$$

But this contradicts the hypothesis that f has a relative maximum at c. Thus we cannot have $f'(c) > 0$. Similarly (how?), we cannot have $f'(c) < 0$. Therefore we must have $f'(c) = 0$. Q.E.D.

5.2.2 Corollary. *Let $f : I \to \mathbf{R}$ be continuous on an interval I and suppose that f has a relative extremum at an interior point c of I. Then either the derivative of f at c does not exist, or it is equal to zero.*

We note that if $f(x) := |x|$ on $I := [-1,1]$, then f has an interior minimum at $x = 0$; however, the derivative of f fails to exist at $x = 0$.

5.2.3 Rolle's Theorem. *Suppose that f is continuous on a closed interval $I := [a,b]$, that the derivative f' exists at every point of the open interval (a,b), and that $f(a) = f(b) = 0$. Then there exists at least one point c in (a,b) such that $f'(c) = 0$. (See Figure 5.2.1.)*

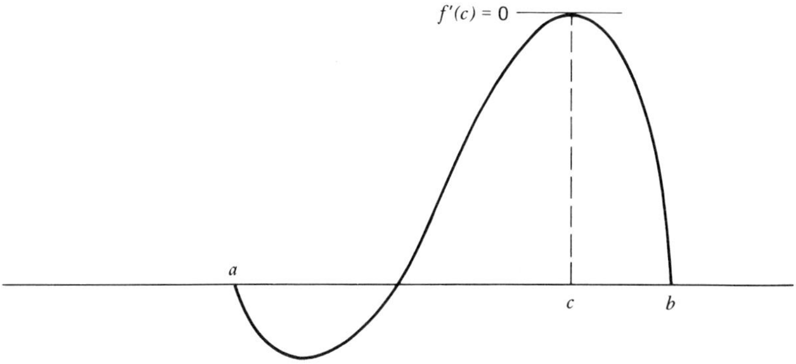

FIGURE 5.2.1 Rolle's Theorem.

Proof. If f vanishes identically on I, then any c in (a,b) will satisfy the conclusion of the theorem. Hence we suppose that f does not vanish identically; replacing f by $-f$ if necessary, we may suppose that f assumes some positive values. By the Maximum-Minimum Theorem 4.6.4, the function f attains the value $\sup \{f(x) : x \in I\} > 0$ at some point c in I. Since $f(a) = f(b) = 0$, the point c must lie in (a,b); therefore $f'(c)$ exists. Since f has a relative maximum at c, we conclude from the Interior Extremum Theorem 5.2.1 that $f'(c) = 0$. Q.E.D.

As a consequence of Rolle's Theorem, we obtain the fundamental Mean Value Theorem.

5.2.4 Mean Value Theorem. *Suppose that f is continuous on a closed interval $I := [a,b]$, and that f has a derivative in the open interval (a,b). Then there exists at least one point c in the open interval (a,b) such that*

$$f(b) - f(a) = f'(c)\,(b - a).$$

Proof. Consider the function φ defined on I by

$$\varphi(x) := f(x) - f(a) - \frac{f(b) - f(a)}{b - a}(x - a).$$

[The function φ is simply the difference of f and the function whose graph is the line segment joining the points $(a,f(a))$ and $(b,f(b))$; see Figure 5.2.2.] The hypotheses of Rolle's Theorem are satisfied by φ since φ is continuous on $[a,b]$, differentiable on (a,b), and $\varphi(a) = \varphi(b) = 0$. Therefore, there exists a point c in (a,b) such that

$$0 = \varphi'(c) = f'(c) - \frac{f(b) - f(a)}{b - a}.$$

Hence, $f(b) - f(a) = f'(c)\,(b - a)$. Q.E.D.

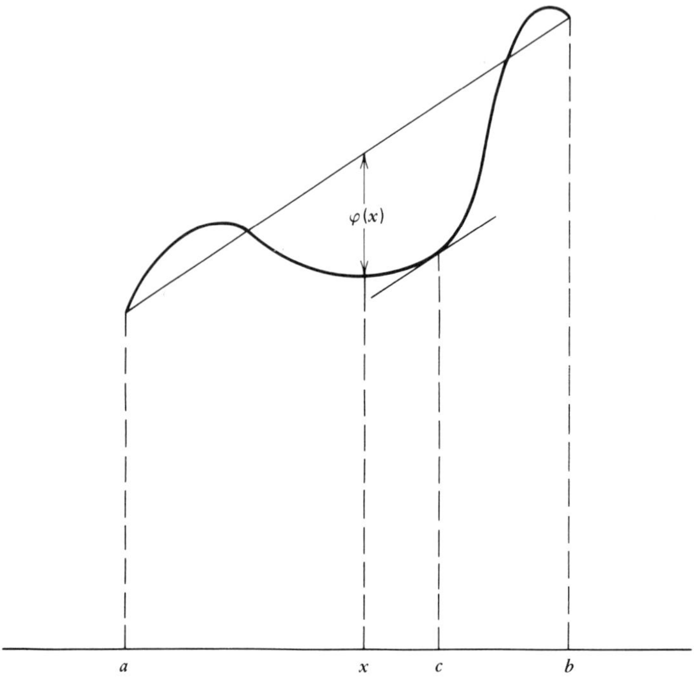

FIGURE 5.2.2 The Mean Value Theorem.

The geometric view of the Mean Value Theorem is that there is some point on the curve $y = f(x)$ at which the tangent line is parallel to the line segment through the points $(a,f(a))$ and $(b,f(b))$. Thus it is easy to remember the statement of the Mean Value Theorem by drawing appropriate diagrams. While this should not be discouraged, it tends to suggest that its importance is geometrical in nature, which is quite misleading. In fact the Mean Value Theorem is a wolf in sheep's clothing and is *the* Fundamental Theorem of Differential Calculus. In the remainder of this section, we shall present some of the consequences of this result. Other applications will be given later.

The Mean Value Theorem permits one to draw conclusions about the nature of a function f from information about its derivative f'. The following results are obtained in this manner.

5.2.5 Theorem. *Suppose that f is continuous on the closed interval $I := [a,b]$, that f is differentiable on the open interval (a,b), and that $f'(x) = 0$ for $x \in (a,b)$. Then f is constant on I.*

Proof. We shall show that $f(x) = f(a)$ for all $x \in I$. Indeed, if $x \in I$, $x > a$, is given, we apply the Mean Value Theorem to f on the closed interval $I_x := [a,x]$. We obtain a point c (depending on x) between a and x such that $f(x) - f(a) = f'(c)(x - a)$. Since $f'(c) = 0$ (by hypothesis), we deduce that $f(x) - f(a) = 0$. Hence, $f(x) = f(a)$ for any $x \in I$. Q.E.D.

5.2.6 Corollary. *Suppose that f and g are continuous on $I := [a,b]$, that they are differentiable on (a,b), and that $f'(x) = g'(x)$ for all $x \in (a,b)$. Then there exists a constant C such that $f = g + C$ on I.*

Recall that a function $f : I \to \mathbf{R}$ is said to be **increasing** on the interval I if whenever x_1, x_2 in I satisfy $x_1 < x_2$, then $f(x_1) \leqslant f(x_2)$. We say that f is **decreasing** on I if the function $-f$ is increasing on I.

5.2.7 Theorem. *Let $f : I \to \mathbf{R}$ be differentiable on the interval I. Then:*
(a) *f is increasing on I if and only if $f'(x) \geqslant 0$ for all $x \in I$.*
(b) *f is decreasing on I if and only if $f'(x) \leqslant 0$ for all $x \in I$.*

Proof. (a) Suppose that $f'(x) \geqslant 0$ for all $x \in I$. If x_1, x_2 in I satisfy $x_1 < x_2$, then we apply the Mean Value Theorem to f on the closed interval $J := [x_1, x_2]$ to obtain a point c in (x_1, x_2) such that

$$f(x_2) - f(x_1) = f'(c)\,(x_2 - x_1).$$

Since $f'(c) \geqslant 0$ and $x_2 - x_2 > 0$, it follows that $f(x_2) - f(x_1) \geqslant 0$. Hence, $f(x_1) \leqslant f(x_2)$ and, since $x_1 < x_2$ are arbitrary points in I, we conclude that f is increasing on I.

For the converse assertion, we suppose that f is differentiable and increasing on I. For any point c in I, if either $x > c$ or $x < c$ for $x \in I$ then we have that $(f(x) - f(c))/(x - c) \geqslant 0$. Hence, by Theorem 4.2.6 we conclude that

$$f'(c) = \lim_{x \to c} \frac{f(x) - f(c)}{x - c} \geqslant 0.$$

(b) The proof of part (b) is similar and will be omitted. Q.E.D.

A function f is said to be **strictly increasing** on an interval I if for any points x_1, x_2 in I such that $x_1 < x_2$, we have $f(x_1) < f(x_2)$. An argument along the same lines of the proof of Theorem 5.2.7 can be made to show that a function having a strictly positive derivative on an interval is strictly increasing there. (See Exercise 5.2.13.) However, the converse assertion fails to be true, since a strictly increasing differentiable function may have a derivative that vanishes at certain points. For example, the function $f : \mathbf{R} \to \mathbf{R}$ defined by $f(x) := x^3$ is strictly increasing on $\mathbf{R}$, but $f'(0) = 0$. The situation for strictly decreasing functions is similar.

Remark. It is reasonable to define a function to be **increasing at a point** if there is a neighborhood of the point on which the function is increasing. One might suppose that, if the derivative is strictly positive at a point, then the function is increasing at this point. However, this supposition is false; indeed, the differentiable function defined by

$$g(x) := x + 2x^2 \sin(1/x) \qquad \text{if} \quad x \neq 0,$$
$$:= 0 \qquad \text{if} \quad x = 0,$$

is such that $g'(0) = 1$, yet it can be shown that g is not increasing in any neighborhood of $x = 0$. (See Exercise 5.2.10.)

We next verify a sufficient condition for a function to have a relative extremum at an interior point of an interval. This condition is often referred to as the "First Derivative Test".

5.2.8 First Derivative Test for Extrema. *Let f be continous on the interval $I := [a,b]$ and let c be an interior point of I. Assume that f is differentiable on (a,c) and (c,b). Then:*

(a) *If there exists a neighborhood $(c - \delta, c + \delta) \subseteq I$ such that $f'(x) \geqslant 0$ for $c - \delta < x < c$ and $f'(x) \leqslant 0$ for $c < x < c + \delta$, then f has a relative maximum at c.*

(b) *If there exists a neighborhood $(c - \delta, c + \delta) \subseteq I$ such that $f'(x) \leqslant 0$ for $c - \delta < x < c$ and $f'(x) \geqslant 0$ for $c < x < c + \delta$, then f has a relative minimum at c.*

Proof. (a) If $x \in (c - \delta, c)$, then it follows from the Mean Value Theorem that there exists a point $c_x \in (x, c)$ such that $f(c) - f(x) = (c - x) f'(c_x)$. Since $f'(c_x) \geqslant 0$ we infer that $f(x) \leqslant f(c)$ for $x \in (c - \delta, c)$. Similarly, it follows (how?) that $f(x) \leqslant f(c)$ for $x \in (c, c + \delta)$. Therefore $f(x) \leqslant f(c)$ for all $x \in (c - \delta, c + \delta)$ so that f has a relative maximum at c.

(b) The proof is similar. Q.E.D.

Remark. The converse of the First Derivative Test 5.2.8 is *not* true. For example, there exists a differentiable function $f : \boldsymbol{R} \to \boldsymbol{R}$ with absolute minimum at $x = 0$ but such that f' takes on both positive and negative values on both sides of (and arbitrarily close to) $x = 0$. (See Exercise 5.2.9.)

Further Applications of the Mean Value Theorem

We shall continue giving other types of applications of the Mean Value Theorem; in doing so we shall draw more freely than before on the past experience of the reader and his or her knowledge concerning the derivatives of certain well-known functions.

5.2.9 Examples. (a) Rolle's Theorem can be used for the location of roots of a function. For, if a function g can be identified as the derivative of a function f, then between any two roots of f there is at least one root of g. For example, let $g(x) := \cos x$, then g is known to be the derivative of $f(x) := \sin x$. Hence, between any two roots of $\sin x$ there is at least one root of $\cos x$. On the other hand, $g'(x) = -\sin x = -f(x)$, so another application of Rolle's Theorem tells us that between any two roots of cos there is at least one root of sin. Therefore, we conclude that the roots of sin and cos *interlace each other*. This conclusion is probably not news to the reader; however, the same type of argument can be applied to the *Bessel functions* J_n of order $n = 0, 1, 2, \ldots$ by using the relations

$$[x^n J_n(x)]' = x^n J_{n-1}(x), \qquad [x^{-n} J_n(x)]' = -x^{-n} J_{n+1}(x) \qquad \text{for} \quad x > 0.$$

The details of this argument should be supplied by the reader.

(b) We can apply the Mean Value Theorem for approximate calculations and to obtain error estimates. For example, suppose it is desired to evaluate $\sqrt{105}$. We employ the Mean Value Theorem with $f(x) := \sqrt{x}$, $a = 100$, $b = 105$ to obtain

$$\sqrt{105} - \sqrt{100} = \frac{5}{2\sqrt{c}},$$

for some number c with $100 < c < 105$. Since $10 < \sqrt{c} < \sqrt{105} < \sqrt{121} = 11$, we can assert that

$$\frac{5}{2(11)} < \sqrt{105} - 10 < \frac{5}{2(10)},$$

whence it follows that $10.22 < \sqrt{105} < 10.25$. This estimate may not be as sharp as desired. It is clear that the estimate $\sqrt{c} < \sqrt{105} < \sqrt{121}$ was wasteful and can be improved by making use of our conclusion that $\sqrt{105} < 10.25$. Thus, $\sqrt{c} < 10.25$ and we easily determine that

$$0.243 < \frac{5}{2(10.25)} < \sqrt{105} - 10.$$

Our improved estimate is $10.243 < \sqrt{105} < 10.250$.

Inequalities

One very important use of the Mean Value Theorem is to obtain certain inequalities. Whenever information concerning the range of the derivative of a function is available, this information can be used to deduce certain properties of the function itself. The following examples illustrate the valuable role that the Mean Value Theorem plays in this respect.

5.2.10 Examples. (a) The exponential function $f(x) := e^x$ has the derivative $f'(x) = e^x$ for all $x \in \boldsymbol{R}$. Thus $f'(x) > 1$ for $x > 0$, and $f'(x) < 1$ for $x < 0$. From these relationships, we shall derive the inequality

$$(*) \qquad e^x \geq 1 + x, \qquad x \in \boldsymbol{R}$$

with equality occurring if and only if $x = 0$.

If $x = 0$, we have equality with both sides equal to 1. If $x > 0$, we apply the Mean Value Theorem to the function f on the interval $[0,x]$. Then for some c with $0 < c < x$ we have

$$e^x - e^0 = e^c (x - 0).$$

Since $e^0 = 1$ and $e^c > 1$, this becomes $e^x - 1 > x$ so that we have $e^x > 1 + x$ for $x > 0$. A similar argument establishes the same strict inequality for $x < 0$. Thus the inequality $(*)$ holds for all x, and equality occurs only if $x = 0$.

(b) The function $g(x) := \sin x$ has the derivative $g'(x) = \cos x$ for all $x \in \boldsymbol{R}$. On the basis of the fact that $-1 \leqslant \cos x \leqslant 1$ for all $x \in \boldsymbol{R}$, we shall show that

$$(**) \qquad -x \leqslant \sin x \leqslant x \qquad \text{for all} \quad x \geqslant 0.$$

Indeed, if we apply the Mean Value Theorem to g on the interval $[0,x]$, where $x > 0$, we obtain

$$\sin x - \sin 0 = (\cos c)(x - 0)$$

for some c between 0 and x. Since $\sin 0 = 0$ and $-1 \leqslant \cos c \leqslant 1$, this yields $-x \leqslant \sin x \leqslant x$. Since equality holds at $x = 0$, the inequality $(**)$ is established.

(c) (Bernoulli's inequality) If $\alpha > 1$, then

$$(\dagger) \qquad (1 + x)^\alpha \geqslant 1 + \alpha x \qquad \text{for all} \quad x > -1,$$

with equality if and only if $x = 0$.

This inequality was established earlier, in Example 2.2.13(c), for positive integer values of α by using mathematical induction. We now derive the more general version by employing the Mean Value Theorem.

If $h(x) := (1 + x)^\alpha$, then $h'(x) = \alpha(1 + x)^{\alpha-1}$ for all $x > -1$. [For rational α, this derivative was established in Example 5.1.10(c). The extension to irrational α will be discussed in Section 7.3.] If $x > 0$, we infer from the Mean Value Theorem applied to h on the interval $[0,x]$ that there exists c satisfying $0 < c < x$ such that $h(x) - h(0) = h'(c)(x - 0)$. Thus, we have

$$(1 + x)^\alpha - 1 = \alpha(1 + c)^{\alpha-1}x.$$

Since $c > 0$ and $\alpha - 1 > 0$, it follows that $(1 + c)^{\alpha-1} > 1$ and hence that $(1 + x)^\alpha > 1 + \alpha x$. If $-1 < x < 0$ a similar use of the Mean Value Theorem on the interval $[x,0]$ leads to the same strict inequality. Since the case $x = 0$ results in equality, we conclude that $(\dagger)$ is valid for all $x > -1$ with equality if and only if $x = 0$.

(d) Let α be a real number satisfying $0 < \alpha < 1$ and let $g(x) = \alpha x - x^\alpha$ for $x \geqslant 0$. Then $g'(x) = \alpha(1 - x^{\alpha-1})$, so that $g'(x) < 0$ for $0 < x < 1$ and $g'(x) > 0$

for $x > 1$. Consequently, if $x \geqslant 0$, then $g(x) \geqslant g(1)$ and $g(x) = g(1)$ if and only if $x = 1$. Therefore, if $x \geqslant 0$ and $0 < \alpha < 1$, then we have

$$x^{\alpha} \leqslant \alpha x + (1 - \alpha).$$

If $a > 0$ and $b > 0$ and if we let $x = a/b$ and multiply by b, we obtain the inequality

$$(\dagger\dagger) \qquad a^{\alpha} b^{1-\alpha} \leqslant \alpha a + (1 - \alpha) b,$$

where equality holds if and only if $a = b$. (This inequality is often the starting point in establishing the important Hölder Inequality.)

The Intermediate Value Property of Derivatives

We conclude this section with an interesting result, often referred to as Darboux's Theorem. It states that if a function f is differentiable at every point of an interval I, then the function f' has the Intermediate Value Property. This means that if f' takes on values A and B, then it also takes on all values between A and B. The reader will recognize this property as one of the important consequences of continuity as established in Theorem 4.6.6. It is remarkable that derivatives, which need not be continuous functions, also possess this property.

5.2.11 Lemma. *Let $I \subseteq \mathbf{R}$ be an interval, let $f : I \to \mathbf{R}$, let $c \in I$, and assume that f has a derivative at c. Then:*

(a) *If $f'(c) > 0$, then there is a number $\delta > 0$ such that $f(x) > f(c)$ for $x \in I$ such that $c < x < c + \delta$.*

(b) *If $f'(c) < 0$, then there is a number $\delta > 0$ such that $f(x) > f(c)$ for $x \in I$ such that $c - \delta < x < c$.*

Proof. (a) Since

$$\lim_{x \to c} \frac{f(x) - f(c)}{x - c} = f'(c) > 0,$$

it follows from Theorem 4.2.11 that there is a number $\delta > 0$ such that if $x \in I$ and $0 < |x - c| < \delta$, then

$$\frac{f(x) - f(c)}{x - c} > 0.$$

If $x \in I$ also satisfies $x > c$, then we have

$$f(x) - f(c) = (x - c) \cdot \frac{f(x) - f(c)}{x - c} > 0.$$

Hence, if $x \in I$ and $c < x < c + \delta$, then $f(x) > f(c)$.

The proof of (b) is similar. Q.E.D.

5.2.12 Darboux's Theorem. *If f is differentiable on $I = [a,b]$ and if k is a number between $f'(a)$ and $f'(b)$, then there is at least one point c in (a,b) such that $f'(c) = k$.*

Proof. Suppose that $f'(a) < k < f'(b)$. We define g on I by $g(x) := k(x - a) - f(x)$ for $x \in I$. Since g is continuous, it attains a maximum value on I. Since $g'(a) = k - f'(a) > 0$, it follows from Lemma 5.2.11(a) that the maximum of g does not occur at $x = a$. Similarly, since $g'(b) = k - f'(b) < 0$, it follows from Lemma 5.2.11(b) that the maximum does not occur at $x = b$. Therefore, g attains its maximum at some c in (a,b). Then from Theorem 5.2.1 we have $0 = g'(c) = k - f'(c)$. Hence, $f'(c) = k$. Q.E.D.

5.2.13. Example. The function $g : [-1,1] \to \boldsymbol{R}$ defined by

$$\begin{aligned} g(x) &:= 1 && \text{for } x > 0, \\ &:= 0 && \text{for } x = 0, \\ &:= -1 && \text{for } x < 0, \end{aligned}$$

(which is a restriction of the signum function) clearly fails to satisfy the intermediate value property on the interval $[-1,1]$. Therefore, by Darboux's Theorem, there does not exist a function f such that $f'(x) = g(x)$ for all $x \in [-1,1]$. In other words, g is **not** the derivative on $[-1,1]$ of any function on this interval.

Exercises for Section 5.2

1. For each of the following functions on $\boldsymbol{R}$ to $\boldsymbol{R}$, find points of relative extrema, the intervals on which the function is increasing, and those on which it is decreasing:
   (a) $f(x) := x^2 - 3x + 5$; (b) $g(x) := 3x - 4x^2$;
   (c) $h(x) := x^3 - 3x - 4$; (d) $k(x) := x^4 + 2x^2 - 4$.
2. Find the points of relative extrema, the intervals on which the following functions are increasing, and those on which they are decreasing:
   (a) $f(x) := x + 1/x$ for $x \neq 0$;
   (b) $g(x) := x/(x^2 + 1)$ for $x \in \boldsymbol{R}$;
   (c) $h(x) := \sqrt{x} - 2\sqrt{x + 2}$ for $x > 0$;
   (d) $k(x) := 2x + 1/x^2$ for $x \neq 0$.
3. Find the points of relative extrema of the following functions on the specified domain:
   (a) $f(x) := |x^2 - 1|$ for $-4 \leqslant x \leqslant 4$;
   (b) $g(x) := 1 - (x - 1)^{2/3}$ for $0 \leqslant x \leqslant 2$;
   (c) $h(x) := x|x^2 - 12|$ for $-2 \leqslant x \leqslant 3$;
   (d) $k(x) := x(x - 8)^{1/3}$ for $0 \leqslant x \leqslant 9$.

4. Let $a_1, a_2, \ldots, a_n$ be real numbers and let f be defined on $\mathbf{R}$ by

$$f(x) := \sum_{i=1}^{n} (a_i - x)^2, \qquad x \in \mathbf{R}.$$

Find the unique point of relative minimum for f.

5. Let $a > b > 0$ and let $n \in \mathbf{N}$ satisfy $n \geqslant 2$. Prove that $a^{1/n} - b^{1/n} < (a - b)^{1/n}$. [*Hint*: Show that $f(x) := x^{1/n} - (x-1)^{1/n}$ is decreasing for $x \geqslant 1$, and evaluate f at 1 and a/b.]
6. Use the Mean Value Theorem to prove that $|\sin x - \sin y| \leqslant |x - y|$ for all x, y in $\mathbf{R}$.
7. Use the Mean Value Theorem to prove that $(x-1)/x < \log x < x - 1$ for $x > 1$. [Use the fact that $D \log x = 1/x$ for $x > 0$.]
8. Let $f : [a,b] \to \mathbf{R}$ be continuous on $[a,b]$ and differentiable in (a,b). Show that if $\lim\limits_{x \to a} f'(x) = A$, then $f'(a)$ exists and equals A. [*Hint*: Use the definition of $f'(a)$ and the Mean Value Theorem.]
9. Let $f : \mathbf{R} \to \mathbf{R}$ be defined by $f(x) := 2x^4 + x^4 \sin(1/x)$ for $x \neq 0$ and $f(0) := 0$. Show that f has an absolute minimum at $x = 0$, but that its derivative has strictly positive and strictly negative values in every neighbohrood of 0.
10. Let $g : \mathbf{R} \to \mathbf{R}$ be defined by $g(x) := x + 2x^2 \sin(1/x)$ for $x \neq 0$ and $g(0) := 0$. Show that $g'(0) = 1$, but in every neighborhood of 0 the derivative $g'(x)$ takes on both strictly positive and strictly negative values. Thus g is not monotonic in any neighborhood of 0.
11. Give an example of a uniformly continuous function on $[0,1]$ that is differentiable on $(0,1)$ but whose derivative is not bounded on $(0,1)$.
12. If $h(x) := 0$ for $x < 0$ and $h(x) := 1$ for $x \geqslant 0$, prove there does not exist a function $f : \mathbf{R} \to \mathbf{R}$ such that $f'(x) = h(x)$ for all $x \in \mathbf{R}$. Give examples of two functions, not differing by a constant, whose derivatives equal $h(x)$ for all $x \neq 0$.
13. Let I be an interval and let $f : I \to \mathbf{R}$ be differentiable on I. Show that if f' is strictly positive on I, then f is strictly increasing on I.
14. Let I be an interval and let $f : I \to \mathbf{R}$ be differentiable on I. Show that if the derivative f' is never 0 on I, then either $f'(x) > 0$ for all $x \in I$ or $f'(x) < 0$ for all $x \in I$.
15. Let I be an interval. Prove that if f is differentiable on I and if the derivative f' is bounded on I, then f satisfies a Lipschitz condition on I. [Recall that a function $f : I \to \mathbf{R}$ is said to satisfy a **Lipschitz condition** on I if there exists a constant K such that $|f(x) - f(y)| \leqslant K|x - y|$ for all x, y in I.]
16. Let $f : [0,\infty) \to \mathbf{R}$ be differentiable on $(0,\infty)$ and assume that $f'(x) \to b$ as $x \to \infty$.
    (a) Show that for any $h > 0$, we have $\lim\limits_{x \to \infty} ((f(x + h) - f(x))/h = b$.
    (b) Show that if $f(x) \to a$ as $x \to \infty$, then $b = 0$.
    (c) Show that $\lim\limits_{x \to \infty} f(x)/x = b$.
17. Let f, g be differentiable on $\mathbf{R}$ and suppose that $f(0) = g(0)$ and $f'(x) \leqslant g'(x)$ for all $x \geqslant 0$. Show that $f(x) \leqslant g(x)$ for all $x \geqslant 0$.
18. Let $I := [a,b]$ and let $f : I \to \mathbf{R}$ be differentiable at $c \in I$. Show that for every $\varepsilon > 0$ there exists $\delta > 0$ such that if $0 < |x - y| < \delta$ and $a \leqslant x \leqslant c \leqslant y \leqslant b$, then

$$\left| \frac{f(x) - f(y)}{x - y} - f'(c) \right| < \varepsilon.$$

19. A differentiable function $f : I \to \mathbf{R}$ is said to be **uniformly differentiable** on $I := [a,b]$ if for every $\varepsilon > 0$ there exists $\delta > 0$ such that if $0 < |x - y| < \delta$ and $x, y \in I$, then

$$\left| \frac{f(x) - f(y)}{x - y} - f'(x) \right| < \varepsilon .$$

Show that if f is uniformly differentiable on I, then f' is continuous on I.

20. Suppose that $f : [0,2] \to \mathbf{R}$ is continuous on $[0,2]$ and differentiable on $(0,2)$, and that $f(0) = 0$, $f(1) = 1$, $f(2) = 1$.
(a) Show that there exists $c_1 \in (0,1)$ such that $f'(c_1) = 1$.
(b) Show that there exists $c_2 \in (1,2)$ such that $f'(c_2) = 0$.
(c) Show that there exists $c \in (0,2)$ such that $f'(c) = \frac{1}{3}$.

SECTION 5.3 L'Hospital's Rules

The Marquis Guillame François L'Hospital (1661–1704) was the author of the first calculus book, *L'Analyse des infiniment petits*, published in 1696. He studied the then new differential calculus from Johann Bernoulli (1667–1748), first when Bernoulli visited L'Hospital's country estate and subsequently through a series of letters. The book was the result of L'Hospital's studies. The limit theorem that became known as L'Hospital's Rule first appeared in this book, though in fact it was discovered by Bernoulli.

The initial theorem was refined and extended, and the various results are collectively referred to as L'Hospital's Rules. In this section we establish the most basic of these results and indicate how others can be derived.

Indeterminate Forms

In the preceding chapters we have often been concerned with methods of evaluating limits. It was shown in Theorem 4.2.4(b) that if $A := \lim_{x \to c} f(x)$ and $B := \lim_{x \to c} g(x)$, and if $B \neq 0$, then

$$\lim_{x \to c} \frac{f(x)}{g(x)} = \frac{A}{B}.$$

However, if $B = 0$, then no conclusion was deduced. It will be seen in Exercise 5.3.2 that if $B = 0$ and $A \neq 0$, then the limit is infinite (when it exists).

The case $A = 0$, $B = 0$ has not been covered previously. In this case, the limit of the quotient f/g is said to be "indeterminate". We will see that in this case the limit may not exist or may be any real value, depending on the particular functions f and g. The symbolism $0/0$ is used to refer to this situation. For example, if α is any real number, and if we define $f(x) := \alpha x$ and $g(x) := x$, then

$$\lim_{x\to 0} \frac{f(x)}{g(x)} = \lim_{x\to 0} \frac{\alpha x}{x} = \lim_{x\to 0} \alpha = \alpha.$$

Thus the indeterminate form $0/0$ can lead to any real number α as a limit.

Other indeterminate forms are represented by the symbols ∞/∞, $0\cdot\infty$, 0^0, 1^∞, ∞^0, and $\infty - \infty$. These notations correspond to the indicated limiting behavior and juxtaposition of the functions f and g. Our attention will be focused on the indeterminate forms $0/0$ and ∞/∞. The other indeterminate cases are usually reduced to the form $0/0$ or ∞/∞ by taking logarithms, exponentials, or algebraic manipulations.

L'Hospital's Rule: The Case 0/0

To show that the use of differentiation in this context is a natural and not surprising development, we first establish an elementary result that is based simply on the definition of the derivative.

5.3.1 Theorem. *Let f and g be defined on $[a,b]$, let $f(a) = g(a) = 0$, and let $g(x) \neq 0$ for $a < x < b$. If f and g are differentiable at a and if $g'(a) \neq 0$, then the limit of f/g at a exists and is equal to $f'(a)/g'(a)$. Thus*

$$\lim_{x\to a+} \frac{f(x)}{g(x)} = \frac{f'(a)}{g'(a)}.$$

Proof. Since $f(a) = g(a) = 0$, we can write the quotient $f(x)/g(x)$ for $a < x < b$ as follows:

$$\frac{f(x)}{g(x)} = \frac{f(x) - f(a)}{g(x) - g(a)} = \frac{\dfrac{f(x) - f(a)}{x - a}}{\dfrac{g(x) - g(a)}{x - a}}.$$

Applying Theorem 4.2.4(b), we thus obtain

$$\lim_{x\to a+} \frac{f(x)}{g(x)} = \frac{\lim\limits_{x\to a+} \dfrac{f(x) - f(a)}{x - a}}{\lim\limits_{x\to a+} \dfrac{g(x) - g(a)}{x - a}} = \frac{f'(a)}{g'(a)}.$$

Q.E.D.

Warning. The hypothesis that $f(a) = g(a) = 0$ is essential here. For example, if $f(x) := x + 17$ and $g(x) := 2x + 3$ for $x \in \boldsymbol{R}$, then

$$\lim_{x\to 0} \frac{f(x)}{g(x)} = \frac{17}{3}, \quad \text{while} \quad \frac{f'(0)}{g'(0)} = \frac{1}{2}.$$

The preceding result enables us to deal with limits such as

$$\lim_{x \to 0} \frac{x^2 + x}{\sin 2x} = \frac{2 \cdot 0 + 1}{2 \cos 0} = \frac{1}{2}.$$

To handle limits where f and g are not differentiable at the point a, we need a more general version of the Mean Value Theorem due to Cauchy.

5.3.2 Cauchy Mean Value Theorem. *Let f and g be continuous on $[a,b]$ and differentiable on (a,b), and assume that $g'(x) \neq 0$ for all x in (a,b). Then there exists c in (a,b) such that*

$$\frac{f(b) - f(a)}{g(b) - g(a)} = \frac{f'(c)}{g'(c)}.$$

Proof. As in the proof of the Mean Value Theorem, we introduce a function to which Rolle's Theorem will apply. First we note that since $g'(x) \neq 0$ for x in (a,b), it follows from Rolle's Theorem that $g(a) \neq g(b)$. For x in $[a,b]$, we now define

$$h(x) := \frac{f(b) - f(a)}{g(b) - g(a)} \left(g(x) - g(a)\right) - \left(f(x) - f(a)\right).$$

Then h is continuous on $[a,b]$, differentiable on (a,b), and $h(a) = h(b) = 0$. Therefore, it follows from Rolle's Theorem 5.2.3 that there exists a point c in (a,b) such that

$$0 = h'(c) = \frac{f(b) - f(a)}{g(b) - g(a)} g'(c) - f'(c).$$

Since $g'(c) \neq 0$, we obtain the desired result by dividing by $g'(c)$. Q.E.D.

Remark. The preceding theorem has a geometric interpretation that is similar to that of the Mean Value Theorem 4.2.4. The functions f and g can be viewed as determining a curve in the plane by means of the parametric equations $x = f(t)$, $y = g(t)$ where $a \leq t \leq b$. Then the conclusion of the theorem is that there exists a point $(f(c), g(c))$ on the curve for some c in (a,b) such that the slope $g'(c)/f'(c)$ of the line tangent to the curve at that point is equal to the slope of the line segment joining the end points of the curve. Note that if $g(x) := x$, then the Cauchy Mean Value Theorem reduces to the Mean Value Theorem 4.2.4.

We shall next establish the main result of this section, which we refer to as L'Hospital's Rule. The reader should observe that, in contrast to Theorem 5.3.1, the following result does not assume the differentiability of the functions at the point a. This result asserts that the limiting behavior of f/g at a is the same as the limiting behavior of f'/g' at a, including the case in which the limit is infinite.

5.3.3 L'Hospital's Rule. *Suppose that f and g are continuous on* $[a,b]$, *differentiable on* (a,b), *that* $f(a) = g(a) = 0$, *and that* $g(x) \neq 0$ *and* $g'(x) \neq 0$ for $a < x < b$. *Then:*

(a) *if* $\lim_{x \to a+} \dfrac{f'(x)}{g'(x)} = L$ *for* $L \in \mathbf{R}$, *then* $\lim_{x \to a+} \dfrac{f(x)}{g(x)} = L$;

(b) *if* $\lim_{x \to a+} \dfrac{f'(x)}{g'(x)} = \infty$ [*respectively*, $-\infty$], *then* $\lim_{x \to a+} \dfrac{f(x)}{g(x)} = \infty$ [*resp.*, $-\infty$].

Proof. (a) If $\varepsilon > 0$ is given, then there exists $\delta > 0$ such that if $a < x < a + \delta$ then

$$\left| \frac{f'(x)}{g'(x)} - L \right| < \varepsilon .$$

For each x that satisfies $a < x < a + \delta$ we obtain (by the Cauchy Mean Value Theorem 5.3.2) a point c_x such that $a < c_x < x$ and

$$\frac{f(x)}{g(x)} = \frac{f'(c_x)}{g'(c_x)}.$$

Since c_x satisfies $a < c_x < a + \delta$, the preceding inequality implies that

$$\left| \frac{f(x)}{g(x)} - L \right| = \left| \frac{f'(c_x)}{g'(c_x)} - L \right| < \varepsilon .$$

Since this is valid for all x with $a < x < a + \delta$, we conclude that $\lim_{x \to a+} \dfrac{f(x)}{g(x)} = L.$

(b) We shall consider the case $+\infty$ only. If any $K > 0$ is given, then there exists $\delta > 0$ such that if $a < x < a + \delta$ then $f'(x)/g'(x) > K$. For each such x, we may apply the Cauchy Mean Value Theorem 5.3.2 to obtain c_x such that $a < c_x < x < a + \delta$ and

$$\frac{f(x)}{g(x)} = \frac{f'(c_x)}{g'(c_x)} > K.$$

Since K is arbitrary, we conclude that $\lim_{x \to a+} f(x)/g(x) = +\infty$. Q.E.D.

The preceding result was established for right-hand limits, but it is clear that combining the results for left and right limits produces the corresponding result for two-sided limits.

We shall now extend the results to the case of limits at infinity; we shall consider only the case ∞.

5.3.4 Theorem. *Suppose that f and g are continuous and differentiable on $[b,\infty)$, that $\lim_{x\to\infty} f(x) = \lim_{x\to\infty} g(x) = 0$, and that $g(x) \neq 0$ and $g'(x) \neq 0$ for $x > b$. Then*

$$\lim_{x\to\infty} \frac{f(x)}{g(x)} = \lim_{x\to\infty} \frac{f'(x)}{g'(x)}.$$

Proof. The change of variables $t := 1/x$ enables us to consider the functions F and G defined on the interval $[0, 1/b]$ by

$$\begin{aligned} F(t) &:= f(1/t), && \text{if } 0 < t \leq 1/b, \\ &:= 0, && \text{if } t = 0; \end{aligned}$$

and

$$\begin{aligned} G(t) &:= g(1/t), && \text{if } 0 < t \leq 1/b, \\ &:= 0, && \text{if } t = 0. \end{aligned}$$

Note that $\lim_{t\to 0+} F(t) = \lim_{x\to+\infty} f(x)$ and $\lim_{t\to 0+} G(t) = \lim_{x\to+\infty} g(x)$. The hypotheses of Theorem 5.3.3 then apply to F and G. For $0 < t < 1/b$, the Chain Rule 5.1.6 implies that $F'(t) = (-1/t^2)\, f'(1/t)$ and $G'(t) = (-1/t^2)g'(1/t)$. Hence, applying Theorem 5.3.3, we conclude that

$$\lim_{x\to+\infty} \frac{f(x)}{g(x)} = \lim_{t\to 0+} \frac{F(t)}{G(t)} = \lim_{t\to 0+} \frac{f'(1/t)}{g'(1/t)} = \lim_{x\to+\infty} \frac{f'(x)}{g'(x)}.$$

Q.E.D.

5.3.5 Examples. (a) We have

$$\lim_{x\to 0+} \frac{\sin x}{\sqrt{x}} = \lim_{x\to 0+} \frac{\cos x}{1/2\sqrt{x}} = \lim_{x\to 0+} 2\sqrt{x} \cos x = 0.$$

Observe that the denominator is not differentiable at $x = 0$ so that Theorem 5.3.1 can not be applied.

(b) We have $\lim_{x\to 0} \dfrac{(1 - \cos x)}{x^2} = \lim_{x\to 0} \dfrac{\sin x}{2x}$.

The quotient in the second limit is again indeterminate in the form 0/0. However, the hypotheses of L'Hospital's Rule are again satisfied so that a second application of L'Hospital's Rule is permissible. Hence, we obtain

$$\lim_{x\to 0} \frac{1 - \cos x}{x^2} = \lim_{x\to 0} \frac{\sin x}{2x} = \lim_{x\to 0} \frac{\cos x}{2} = \frac{1}{2}.$$

(c) We have $\lim_{x\to 0} (e^x - 1)/x = \lim_{x\to 0} e^x/1 = 1$.

Similiarly, we have

$$\lim_{x\to 0} \frac{e^x - 1 - x}{x^2} = \lim_{x\to 0} \frac{e^x - 1}{2x} = \frac{1}{2}.$$

(d) We have $\lim_{x\to 1} (\log x)/(x-1) = \lim_{x\to 1} (1/x)/1 = 1.$

The Case ∞/∞

In the following result we deal with the indeterminate form ∞/∞.

5.3.6 Theorem. *Suppose that f and g are differentiable on (a,b), that $\lim_{x\to a+} f(x) = \infty$ and $\lim_{x\to a+} g(x) = \infty$, and that $g(x) \neq 0$ and $g'(x) \neq 0$ for $a < x < b$. Then:*

(a) *if* $\lim_{x\to a+} \dfrac{f'(x)}{g'(x)} = L$ *for* $L \in \mathbf{R}$, *then* $\lim_{x\to a+} \dfrac{f(x)}{g(x)} = L$;

(b) *if* $\lim_{x\to a+} \dfrac{f'(x)}{g'(x)} = \infty$ *[respectively,* $-\infty$*], then* $\lim_{x\to a+} \dfrac{f(x)}{g(x)} = \infty$ *[resp.,* $-\infty$*].*

Proof. (a) By hypothesis, if $0 < \varepsilon \leq \frac{1}{2}$ is given, then there exists $\delta > 0$ such that if $a < x < a + \delta$, then

$$\left| \frac{f'(x)}{g'(x)} - L \right| < \varepsilon.$$

We choose c_1 in $(a, a + \delta)$ and, because f has an infinite right-hand limit at a, we may choose c_2 in (a,c_1) such that $f(x) \neq f(c_1)$ for $a < x < c_2$. We now define the function F on (a,c_2) by

$$F(x) := \frac{1 - f(c_1)/f(x)}{1 - g(c_1)/g(x)} \qquad \text{for} \quad a < x < c_2.$$

Note that since $g'(x) \neq 0$ for $a < x < b$, we have $g(x) \neq g(c_1)$ for $a < x < c_2$. From the hypotheses, we see that $\lim_{x\to a+} F(x) = 1$. Therefore, there exists c_3 with $a < c_3 < c_2$ such that $|F(x) - 1| < \varepsilon$ whenever $a < x < c_3$. Hence, if $a < x < c_3$, we have

$$\frac{1}{|F(x)|} < \frac{1}{1 - \varepsilon} < 2.$$

We observe that

$$\frac{f(x)}{g(x)} = \frac{f(x)}{g(x)} \cdot \frac{F(x)}{F(x)} = \frac{f(x) - f(c_1)}{g(x) - g(c_1)} \cdot \frac{1}{F(x)}.$$

Then by applying the Cauchy Mean Value Theorem 5.3.2, there exists ξ in (x,c_1) such that

$$\frac{f(x)}{g(x)} = \frac{f'(\xi)}{g'(\xi)} \cdot \frac{1}{F(x)}.$$

Since $a < x < c_3 < c_2 < c_1 < a + \delta$ it follows that

$$\begin{aligned}
\left|\frac{f(x)}{g(x)} - L\right| &= \left|\frac{f'(\xi)}{g'(\xi)} \cdot \frac{1}{F(x)} - L\right| \\
&= \left|\frac{f'(\xi)}{g'(\xi)} - L\,F(x)\right| \cdot |F(x)|^{-1} \\
&\leqslant \left\{\left|\frac{f'(\xi)}{g'(\xi)} - L\right| + |L - L\,F(x)|\right\} |F(x)|^{-1} \\
&\leqslant (\varepsilon + |L|\varepsilon)2 = \{2(1 + |L|)\}\varepsilon.
\end{aligned}$$

Since $\varepsilon > 0$ is arbitrary, we conclude that $\lim_{x \to a+} f(x)/g(x) = L$.

(b) We leave the proof as an exercise. Q.E.D.

There is a theorem analogous to Theorem 5.3.6 that is valid under the same hypotheses for the calculation of limits as $x \to \infty$ (or as $x \to -\infty$). This result is established from Theorem 5.3.6 in the same way that Theorem 5.3.4 was derived from Theorem 5.3.3. We shall make free use of this result, leaving the details to the reader.

5.3.7 Examples. (a) Let $I = (0,\infty)$ and consider $\lim_{x\to\infty} (\log x)/x$.

If we apply the modification of Theorem 5.3.6, we obtain $\lim_{x\to\infty} (\log x)/x = \lim_{x\to\infty} (1/x)/1 = 0$.

(b) Let $I = \boldsymbol{R}$ and consider $\lim_{x\to\infty} x^2/e^x$.

We have $\lim_{x\to\infty} x^2/e^x = \lim_{x\to\infty} 2x/e^x = \lim_{x\to\infty} 2/e^x = 0$.

(c) Let $I = (0,\pi)$ and consider $\lim_{x\to 0+} (\log \sin x)/\log x$.

If we apply Theorem 5.3.6, we obtain

$$\lim_{x\to 0+} \frac{\log \sin x}{\log x} = \lim_{x\to 0+} \frac{\dfrac{\cos x}{\sin x}}{1/x} = \lim_{x\to 0+} \left(\frac{x}{\sin x}\right) \cos x.$$

Since $\lim_{x\to 0+} x/(\sin x) = 1$ and $\lim_{x\to 0+} \cos x = 1$, we conclude that the limit under consideration equals 1.

Other Indeterminate Forms

Indeterminate forms such as $\infty - \infty$, $0 \cdot \infty$, 1^∞, 0^0, ∞^0 can be reduced to the previously considered cases by algebraic manipulations and the use of the logarithmic and exponential functions. Instead of formulating these variations as theorems, we illustrate the pertinent techniques by means of examples.

5.3.8 Examples. (a) Let $I := (0, \pi/2)$ and consider

$$\lim_{x \to 0+} \left(\frac{1}{x} - \frac{1}{\sin x} \right),$$

which has the indeterminate form $\infty - \infty$. We have

$$\begin{aligned} \lim_{x \to 0+} \left(\frac{1}{x} - \frac{1}{\sin x} \right) &= \lim_{x \to 0+} \frac{\sin x - x}{x \sin x} \\ &= \lim_{x \to 0+} \frac{\cos x - 1}{\sin x + x \cos x} = \lim_{x \to 0+} \frac{-\sin x}{2 \cos x - x \sin x} \\ &= \frac{0}{2} = 0. \end{aligned}$$

(b) Let $I := (0, \infty)$ and consider $\lim\limits_{x \to 0+} x \log x$, which has the indeterminate form $0 \cdot (-\infty)$. We have

$$\lim_{x \to 0+} x \log x = \lim_{x \to 0+} \frac{\log x}{1/x} = \lim_{x \to 0+} \frac{1/x}{-1/x^2} = \lim_{x \to 0+} (-x) = 0.$$

(c) Let $I := (0, \infty)$ and consider $\lim\limits_{x \to 0+} x^x$, which has the indeterminate form 0^0. We recall from calculus (see Section 7.3) that $x^x = e^{x \log x}$. It follows from part (b) and the continuity of the function $y \mapsto e^y$ at $y = 0$ that $\lim\limits_{x \to 0+} x^x = e^0 = 1$.

(d) Let $I := (1, \infty)$ and consider $\lim\limits_{x \to \infty} (1 + 1/x)^x$, which has the indeterminate form 1^∞. We note that

$$(*) \qquad (1 + 1/x)^x = e^{x \log (1 + 1/x)}.$$

Moreover, we have

$$\begin{aligned} \lim_{x \to \infty} x \log (1 + 1/x) &= \lim_{x \to \infty} \frac{\log (1 + 1/x)}{1/x} \\ &= \lim_{x \to \infty} \frac{(1 + 1/x)^{-1}(-x^{-2})}{-x^{-2}} = \lim_{x \to \infty} \frac{1}{1 + 1/x} = 1. \end{aligned}$$

Since $y \mapsto e^y$ is continuous at $y = 1$, we infer that $\lim\limits_{x \to \infty} (1 + 1/x)^x = e$.

(e) Let $I := (0,\infty)$ and consider $\lim_{x\to 0+} (1 + 1/x)^x$, which has the indeterminate form ∞^0. In view of formula (*), we consider

$$\lim_{x\to 0+} x \log (1 + 1/x) = \lim_{x\to 0+} \frac{\log (1 + 1/x)}{1/x} = \lim_{x\to 0+} \frac{1}{1 + 1/x} = 0.$$

Therefore we have $\lim_{x\to 0+} (1 + 1/x)^x = e^0 = 1$.

Exercises for Section 5.3

1. Suppose that f and g are continuous on $[a,b]$, differentiable on (a,b), that $c \in [a,b]$ and that $g(x) \neq 0$ for $x \in [a,b]$, $x \neq c$. Let $A := \lim_{x\to c} f$ and $B := \lim_{x\to c} g$. If $B = 0$, and if $\lim_{x\to c} f(x)/g(x)$ exists in $\mathbf{R}$, show that we must have $A = 0$. [*Hint:* $f(x) = \{f(x)/g(x)\}g(x)$.]
2. In addition to the suppositions of the preceding exercise, let $g(x) > 0$ for $x \in [a,b]$, $x \neq c$. If $A > 0$ and $B = 0$, prove that we must have $\lim_{x\to c} f(x)/g(x) = \infty$. If $A < 0$ and $B = 0$, prove that we must have $\lim_{x\to c} f(x)/g(x) = -\infty$.
3. Let $f(x) := x^2 \sin (1/x)$ for $0 < x \leq 1$ and $f(0) := 0$, and let $g(x) := x^2$ for $x \in [0,1]$. Then both f and g are differentiable on $[0,1]$ and $g(x) > 0$ for $x \neq 0$. Show that $\lim_{x\to 0} f(x) = 0 = \lim_{x\to 0} g(x)$ and that $\lim_{x\to 0} f(x)/g(x)$ does not exist.
4. Let $f(x) := x^2$ for x rational, let $f(x) := 0$ for x irrational, and let $g(x) := \sin x$ for $x \in \mathbf{R}$. Use Theorem 5.3.1 to show that $\lim_{x\to 0} f(x)/g(x) = 0$. Explain why Theorem 5.3.3 cannot be used.
5. Let $f(x) := x^2 \sin(1/x)$ for $x \neq 0$, let $f(0) := 0$, and let $g(x) := \sin x$ for $x \in \mathbf{R}$. Show that $\lim_{x\to 0} f(x)/g(x) = 0$ but that $\lim_{x\to 0} f'(x)/g'(x)$ does not exist.
6. Evaluate the following limits, where the domain of the quotient is as indicated.

   (a) $\displaystyle\lim_{x\to 0+} \frac{\log (x+1)}{\sin x}$ $(0, \pi/2)$; (b) $\displaystyle\lim_{x\to 0+} \frac{\tan x}{x}$ $(0, \pi/2)$;

   (c) $\displaystyle\lim_{x\to 0+} \frac{\log \cos x}{x}$ $(0, \pi/2)$; (d) $\displaystyle\lim_{x\to 0+} \frac{\tan x - x}{x^3}$ $(0, \pi/2)$.

7. Evaluate the following limits:

   (a) $\displaystyle\lim_{x\to 0} \frac{\text{Arctan } x}{x}$ $(-\infty,\infty)$; (b) $\displaystyle\lim_{x\to 0} \frac{1}{x(\log x)^2}$ $(0,1)$;

   (c) $\displaystyle\lim_{x\to 0+} x^3 \log x$ $(0,\infty)$; (d) $\displaystyle\lim_{x\to \infty} \frac{x^3}{e^x}$ $(0,\infty)$.

8. Evaluate the following limits:

   (a) $\displaystyle\lim_{x\to \infty} \frac{\log x}{x^2}$ $(0,\infty)$; (b) $\displaystyle\lim_{x\to \infty} \frac{\log x}{\sqrt{x}}$ $(0,\infty)$;

   (c) $\displaystyle\lim_{x\to 0} x \log \sin x$ $(0,\pi)$; (d) $\displaystyle\lim_{x\to \infty} \frac{x + \log x}{x \log x}$ $(0,\infty)$.

9. Evaluate the following limits:

(a) $\lim_{x \to 0+} x^{2x}$ $(0, \infty)$;

(b) $\lim_{x \to 0} (1 + 3/x)^x$ $(0, \infty)$;

(c) $\lim_{x \to \infty} (1 + 3/x)^x$ $(0, \infty)$;

(d) $\lim_{x \to 0+} \left(\frac{1}{x} - \frac{1}{\text{Arctan } x}\right)$ $(0, \infty)$.

10. Evaluate the following limits:

(a) $\lim_{x \to \infty} x^{1/x}$ $(0, \infty)$;

(b) $\lim_{x \to 0+} (\sin x)^x$ $(0, \pi)$;

(c) $\lim_{x \to 0+} x^{\sin x}$ $(0, \infty)$;

(d) $\lim_{x \to \pi/2-} (\sec x - \tan x)$ $(0, \pi/2)$.

11. Let f be differentiable on $(0,\infty)$ and suppose that $\lim_{x \to \infty} (f(x) + f'(x)) = L$. Show that $\lim_{x \to \infty} f(x) = L$ and $\lim_{x \to \infty} f'(x) = 0$. [*Hint:* $f(x) = e^x f(x)/e^x$.]

SECTION 5.4 Taylor's Theorem

A very useful technique in the analysis of real functions is the approximation of functions by polynomials. In this section we shall prove a fundamental theorem in this area which goes back to Brook Taylor (1685–1731), although the remainder term was not provided until much later by Joseph-Louis Lagrange (1736–1813). Taylor's Theorem is a powerful result that has many applications. We shall illustrate the versatility of Taylor's Theorem by briefly discussing some of its applications to numerical estimation, inequalities, extreme values of a function, and convex functions.

Taylor's Theorem can be regarded as an extension of the Mean Value Theorem to "higher order" derivatives. Whereas the Mean Value Theorem relates the values of a function and its derivative, Taylor's Theorem provides a relation between the values of a function and its higher order derivatives.

Derivatives of order greater than one are obtained by a natural extension of the differentiation process. If the derivative $f'(x)$ of a function f exists at every point x in an interval I containing a point c, then we can consider the existence of the derivative of the function f' at the point c. In case f' has a derivative at the point c, we refer to the resulting number as the **second derivative** of f at c, and we denote this number by $f''(c)$ or by $f^{(2)}(c)$. In similar fashion we define the third derivative $f'''(c) = f^{(3)}(c), \ldots$, and the ***n*th derivative** $f^{(n)}(c)$, whenever these derivatives exist. It is noted that the existence of the nth derivative at c presumes the existence of the $(n-1)$st derivative in an interval containing c, but we do allow the possibility that c might be an end point of such an interval.

If a function f has an nth derivative at a point x_0, it is not difficult to construct an nth degree polynomial P_n such that $P_n(x_0) = f(x_0)$ and $P_n^{(k)}(x_0) = f^{(k)}(x_0)$ for $k = 1, 2, \ldots, n$. In fact, the polynomial

$$P_n(x) := f(x_0) + f'(x_0)(x - x_0) + \frac{f''(x_0)}{2!}(x - x_0)^2 + \cdots + \frac{f^{(n)}(x_0)}{n!}(x - x_0)^n, \tag{1}$$

has the property that it and its derivatives up to order n agree with the function f and its derivatives up to order n, at the specified point x_0. This polynomial P_n is called the nth **Taylor polynomial for f at x_0**. It is natural to expect this polynomial to provide a reasonable approximation to f for points near x_0, but to gauge the quality of the approximation, it is necessary to have information concerning the remainder $R_n := f - P_n$. The following fundamental result provides such information.

5.4.1 Taylor's Theorem. *Let $n \in \mathbf{N}$, let $I := [a,b]$, and let $f : I \to \mathbf{R}$ be such that f and its derivatives f', f'', $\ldots$, $f^{(n)}$ are continuous on I and that $f^{(n+1)}$ exists on (a,b). If $x_0 \in I$, then for any x in I there exists a point c between x and x_0 such that*

$$f(x) = f(x_0) + f'(x_0)(x - x_0) + \frac{f''(x_0)}{2!}(x - x_0)^2 + \cdots + \frac{f^{(n)}(x_0)}{n!}(x - x_0)^n + \frac{f^{(n+1)}(c)}{(n+1)!}(x - x_0)^{n+1}. \tag{2}$$

Proof. Let x_0 and x be given and let J denote the closed interval with end points x_0 and x. We define the function F on J by

$$F(t) := f(x) - f(t) - (x - t)f'(t) - \cdots - \frac{(x - t)^n}{n!} f^{(n)}(t)$$

for $t \in J$. Then an easy calculation shows that we have

$$F'(t) = -\frac{(x - t)^n}{n!} f^{(n+1)}(t).$$

If we define G on J by

$$G(t) := F(t) - \left(\frac{x - t}{x - x_0}\right)^{n+1} F(x_0)$$

for $t \in J$, then $G(x_0) = G(x) = 0$. An application of Rolle's Theorem 5.2.3 yields a point c between x and x_0 such that

$$0 = G'(c) = F'(c) + (n + 1)\frac{(x - c)^n}{(x - x_0)^{n+1}} F(x_0).$$

Hence, we obtain

$$F(x_0) = -\frac{1}{n+1}\frac{(x-x_0)^{n+1}}{(x-c)^n}F'(c)$$

$$= \frac{1}{n+1}\frac{(x-x_0)^{n+1}}{(x-c)^n}\frac{(x-c)^n}{n!}f^{(n+1)}(c) = \frac{f^{(n+1)}(c)}{(n+1)!}(x-x_0)^{n+1},$$

which is the stated result. Q.E.D.

We shall use the notation P_n for the nth Taylor polynomial (1) of f, and R_n for the remainder. Thus we may write the conclusion of Taylor's Theorem as $f(x) = P_n(x) + R_n(x)$ where R_n is given by

(3)
$$R_n(x) := \frac{f^{(n+1)}(c)}{(n+1)!}(x-x_0)^{n+1}$$

for some point c between x and x_0. This formula for R_n is referred to as the **Lagrange form** (or the **derivative form**) of the remainder. Other expressions for R_n are known; one is in terms of integration and will be given later. (See Theorem 6.3.14.)

Applications of Taylor's Theorem

The remainder term R_n in Taylor's Theorem can be used to estimate the error in approximating a function by its Taylor polynomial P_n. If the number n is prescribed, then the question of the accuracy of the approximation arises. On the other hand, if a certain accuracy is specified, then the question of finding a suitable value of n is germane. The following examples illustrate how one responds to these questions.

5.4.2 Examples. (a) Use Taylor's Theorem with $n = 2$ to approximate $\sqrt[3]{1+x}$, $x > -1$.

We take the function $f(x) := (1+x)^{1/3}$, the point $x_0 = 0$, and $n = 2$. Since $f'(x) = \frac{1}{3}(1+x)^{-2/3}$ and $f''(x) = \frac{1}{3}(-\frac{2}{3})(1+x)^{-5/3}$, we have $f'(0) = \frac{1}{3}$ and $f''(0) = -2/9$. Thus we obtain

$$f(x) = P_2(x) + R_2(x) = 1 + \tfrac{1}{3}x - \tfrac{1}{9}x^2 + R_2(x),$$

where $R_2(x) = \dfrac{f'''(c)}{3!}x^3 = \dfrac{5}{81}(1+c)^{-8/3}x^3$ for some point c between 0 and x. If we let $x = 0.3$, we get the approximation $P_2(0.3) = 1.09$ for $\sqrt[3]{1.3}$. Moreover, since $c > 0$ in this case, then $(1+c)^{-8/3} < 1$ and so the error is at most

$$R_2(0.3) \leq \frac{5}{81}\left(\frac{3}{10}\right)^3 = \frac{1}{600} < 0.17 \times 10^{-2}.$$

Hence, we have $|\sqrt[3]{1.3} - 1.09| < 0.5 \times 10^{-2}$, so that two decimal place accuracy is assured.

(b) Approximate the number e with error less than 10^{-5}.

We shall consider the function $g(x) := e^x$ and take $x_0 = 0$ and $x = 1$ in Taylor's Theorem. We need to determine n so that $|R_n(1)| < 10^{-5}$. To do so, we shall use the fact that $g'(x) = e^x$ and the initial bound of $e^x \leq 3$ for $0 \leq x \leq 1$.

Since $g'(x) = e^x$, it follows that $g^{(k)}(x) = e^x$ for all $k \in \mathbf{N}$, and therefore $g^{(k)}(0) = 1$ for all $k \in \mathbf{N}$. Consequently the nth Taylor polynomial is given by

$$P_n(x) := 1 + x + \frac{x^2}{2!} + \cdots + \frac{x^n}{n!}$$

and the remainder for $x = 1$ is given by $R_n(1) = e^c/(n+1)!$ for some c satisfying $0 < c < 1$. Since $e^c < 3$, we seek a value of n such that $3/(n+1)! < 10^{-5}$. A calculation reveals that $9! = 362{,}880 > 3 \times 10^5$ so that the value $n = 8$ will provide the desired accuracy; moreover, since $8! = 40{,}320$, no smaller value of n will be certain to suffice. Thus, we obtain

$$e \approx P_8(1) = 1 + 1 + \frac{1}{2!} + \cdots + \frac{1}{8!} = 2.71828$$

with error less than 10^{-5}.

Taylor's Theorem is also used to derive inequalities.

5.4.3 Examples. (a) $1 - \frac{1}{2}x^2 \leq \cos x$ for all $x \in \mathbf{R}$.

Using $f(x) := \cos x$ and $x_0 = 0$ in Taylor's Theorem, we obtain

$$\cos x = 1 - \tfrac{1}{2}x^2 + R_2(x),$$

where for some c between 0 and x we have

$$R_2(x) = \frac{f'''(c)}{3!}x^3 = \frac{\sin c}{6}x^3.$$

If $0 \leq x \leq \pi$, then $0 \leq c < \pi$; since c and x^3 are both positive, we have $R_2(x) \geq 0$. Also, if $-\pi \leq x \leq 0$, then $-\pi \leq c \leq 0$; since $\sin c$ and x^3 are both negative, we again have $R_2(x) \geq 0$. Therefore, we see that $1 - \frac{1}{2}x^2 \leq \cos x$ for $|x| \leq \pi$. If $|x| \geq \pi$, then $1 - \frac{1}{2}x^2 < -3 \leq \cos x$ and the inequality is trivially valid. Hence, the inequality holds for all $x \in \mathbf{R}$.

(b) For any $k \in \mathbf{N}$, and for all $x > 0$, we have

$$x - \frac{1}{2}x^2 + \cdots - \frac{1}{2k}x^{2k} < \log(1+x) < x - \frac{1}{2}x^2 + \cdots + \frac{1}{2k+1}x^{2k+1}.$$

Using the fact that the derivative of $\log(1+x)$ is $1/(1+x)$ for $x > 0$, we see that the nth Taylor polynomial for $\log(1+x)$ with $x_0 = 0$ is

$$P_n(x) = x - \frac{1}{2}x^2 + \cdots + (-1)^{n-1}\frac{1}{n}x^n$$

and the remainder is given by

$$R_n(x) = \frac{(-1)^n c^{n+1}}{n+1}x^{n+1}$$

for some c satisfying $0 < c < x$. Thus for any $x > 0$, if $n = 2k$ is even, then we have $R_{2k}(x) > 0$; and if $n = 2k+1$ is odd, then we have $R_{2k+1}(x) < 0$. The stated inequality then follows immediately.

Relative Extrema

It was established in Theorem 5.2.1 that if a function $f : I \to \mathbf{R}$ is differentiable at a point c interior to the interval I, then a necessary condition for f to have a relative extremum at c is that $f'(c) = 0$. One way to determine whether f has a relative maximum or relative minimum [or neither] at c, is to use the First Derivative Test 5.2.8. Higher order derivatives, if they exist, can also be used in this determination, as we now show.

5.4.4 Theorem. *Let I be an interval, let x_0 be an interior point of I, and let $n \geqslant 2$. Suppose that the derivatives $f', f'', \ldots, f^{(n)}$ exist and are continuous in a neighborhood of x_0 and that $f'(x_0) = \ldots = f^{(n-1)}(x_0) = 0$, but $f^{(n)}(x_0) \neq 0$.*

(i) *If n is even and $f^{(n)}(x_0) > 0$, then f has a relative minimum at x_0.*

(ii) *If n is even and $f^{(n)}(x_0) < 0$, then f has a relative maximum at x_0.*

(iii) *If n is odd, then f has neither a relative minimum nor relative maximum at x_0.*

Proof. Applying Taylor's Theorem at x_0, we find that for $x \in I$ we have

$$f(x) = P_{n-1}(x) + R_{n-1}(x) = f(x_0) + \frac{f^{(n)}(c)}{n!}(x - x_0)^n,$$

where c is some point between x_0 and x. Since $f^{(n)}$ is continuous, if $f^{(n)}(x_0) \neq 0$, then there exists an interval U containing x_0 such that $f^{(n)}(x)$ will have the same sign as $f^{(n)}(x_0)$ for $x \in U$. If $x \in U$, then the point c also belongs to U and consequently $f^{(n)}(c)$ and $f^{(n)}(x_0)$ will have the same sign.

(i) If n is even and $f^{(n)}(x_0) > 0$, then for $x \in U$ we have $f^{(n)}(c) > 0$ and $(x - x_0)^n \geqslant 0$ so that $R_{n-1}(x) \geqslant 0$. Hence, $f(x) \geqslant f(x_0)$ for $x \in U$, and therefore f has a relative minimum at x_0.

(ii) If n is even and $f^{(n)}(x_0) < 0$, then it follows that $R_{n-1}(x) \leqslant 0$ for $x \in U$, so that $f(x) \leqslant f(x_0)$ for $x \in U$. Therefore, f has a relative maximum at x_0.

(iii) If n is odd, then $(x - x_0)^n$ is strictly positive if $x > x_0$ and strictly negative if $x < x_0$. Consequently, if $x \in U$, then $R_{n-1}(x)$ will have opposite signs to the left and to the right of x_0. Therefore, f has neither a relative minimum nor a relative maximum at x_0. Q.E.D.

Convex Functions

The notion of convexity plays an important role in a number of areas, particularly in the modern theory of optimization. We shall briefly look at convex functions of one real variable and their relation to differentiation. The basic results, when appropriately modified, can be extended to higher dimensional spaces.

5.4.5 Definition. Let $I \subseteq \boldsymbol{R}$ be an interval. A function $f : I \to \boldsymbol{R}$ is said to be **convex** on I if for any t satisfying $0 \leqslant t \leqslant 1$ and any points x_1, x_2 in I, we have

$$f((1-t)x_1 + tx_2) \leqslant (1-t)f(x_1) + tf(x_2).$$

Note that if $x_1 < x_2$, then as t ranges from 0 to 1, the point $(1-t)x_1 + tx_2$ traverses the interval from x_1 to x_2. Thus if f is convex on I and if $x_1, x_2 \in I$, then the chord joining any two points $(x_1, f(x_1))$ and $(x_2, f(x_2))$ on the graph of f lies above the graph of f. (See Figure 5.4.1.)

A convex function need not be differentiable, as the example $f(x) := |x|$, $x \in \boldsymbol{R}$, reveals. However, it can be shown that if I is an open interval and if $f : I \to \boldsymbol{R}$ is convex on I, then the left and right derivatives of f exist at every point of I. As a consequence, it follows that a convex function on an open interval is necessarily continuous. We shall not verify the preceding assertions, nor shall we develop many other interesting properties of convex functions. Rather, we shall restrict ourselves to establishing the connection between a convex function f and its second derivative f'', assuming that f'' exists.

5.4.6 Theorem. *Let I be an open interval and suppose that $f : I \to \boldsymbol{R}$ has a second derivative on I. Then f is a convex function on I if and only if $f''(x) \geqslant 0$ for all $x \in I$.*

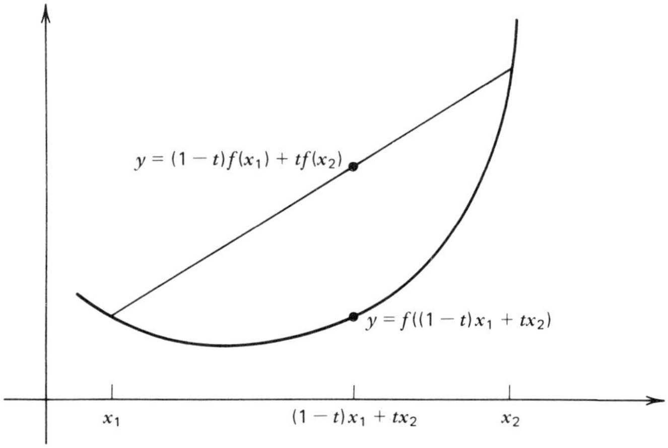

FIGURE 5.4.1 A convex function.

Proof. To prove the necessity of the condition, we shall make use of the fact that the second derivative is given by the limit

$$(*) \qquad f''(a) = \lim_{h \to 0} \frac{f(a+h) - 2f(a) + f(a-h)}{h^2}$$

for each $a \in I$. (See Exercise 5.4.16.) Given $a \in I$, let h be such that $a + h$ and $a - h$ belong to I. Then $a = \frac{1}{2}((a+h) + (a-h))$, and since f is convex on I, we have

$$f(a) = f(\tfrac{1}{2}(a+h) + \tfrac{1}{2}(a-h)) \leq \tfrac{1}{2} f(a+h) + \tfrac{1}{2} f(a-h).$$

Therefore, we have $f(a+h) - 2f(a) + f(a-h) \geq 0$. Since $h^2 > 0$ for all $h \neq 0$, we see that the limit in $(*)$ must be positive. Hence, we obtain $f''(a) \geq 0$ for any $a \in I$.

To prove the sufficiency of the condition we shall use Taylor's Theorem. Let x_1, x_2 be any two points of I, let $0 < t < 1$, and let $x_0 = (1-t)x_1 + tx_2$. Applying Taylor's Theorem to f at x_0 we obtain a point c_1 between x_0 and x_1 such that

$$f(x_1) = f(x_0) + f'(x_0)(x_1 - x_0) + \tfrac{1}{2} f''(c_1)(x_1 - x_0)^2,$$

and a point c_2 between x_0 and x_2 such that

$$f(x_2) = f(x_0) + f'(x_0)(x_2 - x_0) + \tfrac{1}{2} f''(c_2)(x_2 - x_0)^2.$$

If f'' is positive on I, then the term

$$R := \frac{(1-t)}{2} f''(c_1)(x_1 - x_0)^2 + \frac{t}{2} f''(c_2)(x_2 - x_0)^2$$

is positive. Thus we obtain

$$\begin{aligned}(1-t)f(x_1) + tf(x_2) &= f(x_0) + f'(x_0)((1-t)x_1 + tx_2 - x_0) \\ &\quad + \frac{(1-t)}{2} f''(c_1)(x_1 - x_0)^2 + \frac{t}{2} f''(c_2)(x_2 - x_0)^2 \\ &= f(x_0) + R \\ &\geqslant f(x_0) = f((1-t)x_1 + tx_2).\end{aligned}$$

Hence, we see that f is a convex function on I. Q.E.D.

Newton's Method

It is often desirable to estimate a solution of an equation with a high degree of accuracy. The method of interval bisection, used in the proof of the Location of Roots Theorem 4.6.5, provides one estimation procedure, but it has the disadvantage of converging to a solution rather slowly. A method that often results in much more rapid convergence is based on the geometric idea of successively approximating a curve by tangent lines. The method is named after its discoverer, Isaac Newton.

Let f be a differentiable function that has a zero at r and let x_1 be an initial estimate of r. The line tangent to the graph at $(x_1, f(x_1))$ has the equation $y = f(x_1) + f'(x_1)(x - x_1)$, and crosses the x-axis at the point

$$x_2 := x_1 - \frac{f(x_1)}{f'(x_1)}.$$

(See Figure 5.4.2.) If we replace x_1 by the second estimate x_2, then we obtain a point x_3, and so on. At the nth iteration we get the point x_{n+1} from the point x_n by the formula

$$x_{n+1} := x_n - \frac{f(x_n)}{f'(x_n)}.$$

Under suitable hypotheses, the sequence (x_n) will converge rapidly to a root of the equation $f(x) = 0$, as we now show. The key tool in establishing the rapid rate of convergence is Taylor's Theorem.

5.4.7 Newton's Method. *Let $I := [a,b]$ and let $f : I \to \mathbf{R}$ be twice differentiable on I. Suppose that $f(a)f(b) < 0$ and that there are constants m,M*

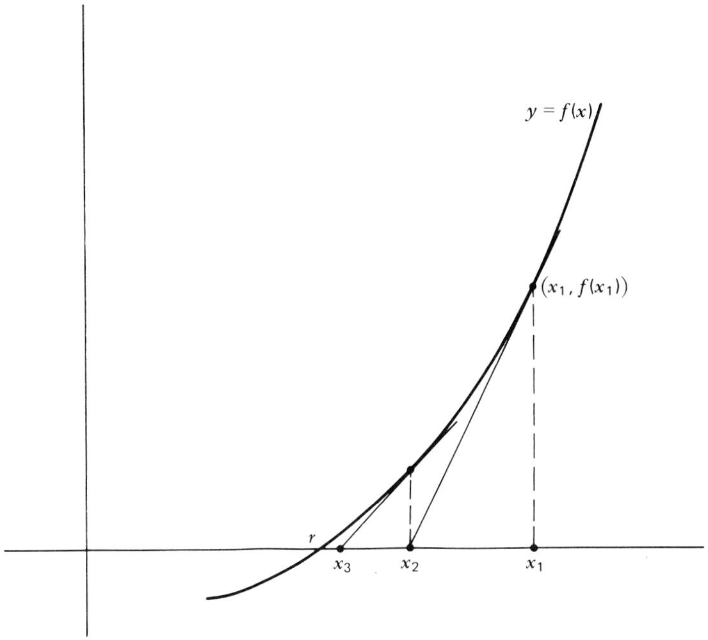

FIGURE 5.4.2 Newton's Method.

such that $|f'(x)| \geqslant m > 0$ *and* $|f''(x)| \leqslant M$ *for all* $x \in I$ *and let* $K := M/2m$. *Then there exists a subinterval* I^* *containing a zero* r *of* f *such that for any* $x_1 \in I^*$ *the sequence* (x_n) *defined by*

$$(*) \qquad x_{n+1} := x_n - \frac{f(x_n)}{f'(x_n)} \quad \text{for} \quad n \in \mathbf{N},$$

belongs to I^* *and* (x_n) *converges to* r. *Moreover*

$$(**) \qquad |x_{n+1} - r| \leqslant K|x_n - r|^2 \quad \text{for} \quad n \in \mathbf{N}.$$

Proof. Since $f(a)f(b) < 0$, the numbers $f(a)$ and $f(b)$ have opposite signs; hence by Theorem 4.6.5 there exists $r \in I$ such that $f(r) = 0$. Since f' is never zero on I, it follows from Rolle's Theorem 5.2.3 that f does not vanish at any other point of I.

We now let $x' \in I$ be arbitrary; by Taylor's Theorem 5.4.1 there exists a point c' between x' and r such that

$$0 = f(r) = f(x') + f'(x')(r-x') + \tfrac{1}{2}f''(c')(r-x')^2,$$

from which it follows that

$$-f(x') = f'(x')(r-x') + \tfrac{1}{2}f''(c')(r-x')^2.$$

If x'' is the number defined from x' by "the Newton procedure":

$$x'' := x' - \frac{f(x')}{f'(x')},$$

then an elementary calculation shows that

$$x'' = x' + (r-x') + \frac{1}{2}\frac{f''(c')}{f'(x')}(r-x')^2,$$

whence it follows that

$$x'' - r = \frac{1}{2}\frac{f''(c')}{f'(x')}(x' - r)^2.$$

Since $c' \in I$, the assumed bounds on f' and f'' hold and, setting $K := M/2m$, we obtain the inequality

$$(\dagger) \qquad |x'' - r| \leq K|x' - r|^2.$$

We now choose $\delta > 0$ so small that $\delta < 1/K$ and that the interval $I^* := [r-\delta, r+\delta]$ is contained in I. If $x_n \in I^*$, then $|x_n - r| \leq \delta$ and it follows from $(\dagger)$ that $|x_{n+1} - r| \leq K|x_n - r|^2 \leq K\delta^2 < \delta$; hence $x_n \in I^*$ implies that $x_{n+1} \in I^*$. Therefore if $x_1 \in I^*$, we infer that $x_n \in I^*$ for all $n \in \mathbf{N}$. Also if $x_1 \in I^*$, then an elementary induction argument using $(\dagger)$ shows that $|x_{n+1} - r| < (K\delta)^n|x_1 - r|$ for $n \in \mathbf{N}$. But since $K\delta < 1$ this proves that $\lim x_n = r$. Q.E.D.

5.4.8 Example. We shall illustrate Newton's Method by using it to approximate $\sqrt{2}$. If we let $f(x) := x^2 - 2$ for $x \in \boldsymbol{R}$, then we seek the positive root of the equation $f(x) = 0$. Since $f'(x) = 2x$, the iteration formula is

$$\begin{aligned} x_{n+1} &= x_n - \frac{f(x_n)}{f'(x_n)} \\ &= x_n - \frac{x_n^2 - 2}{2x_n} = \frac{1}{2}\left(x_n + \frac{2}{x_n}\right). \end{aligned}$$

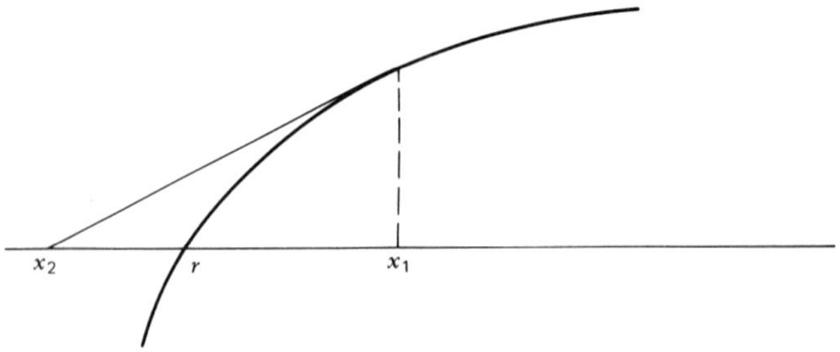

FIGURE 5.4.3 *f* not defined at x_2.

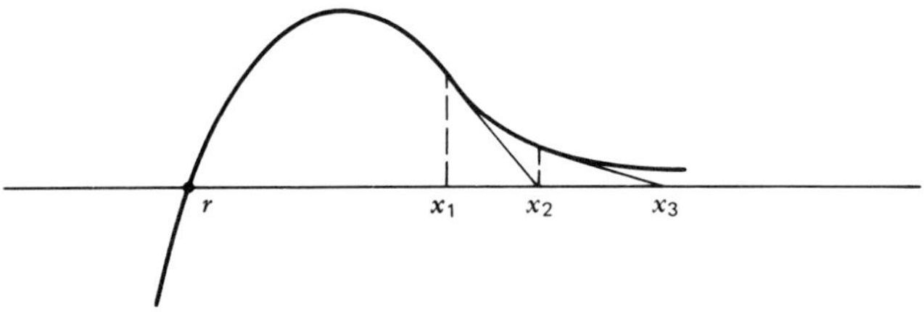

FIGURE 5.4.4 $x_n \to \infty$.

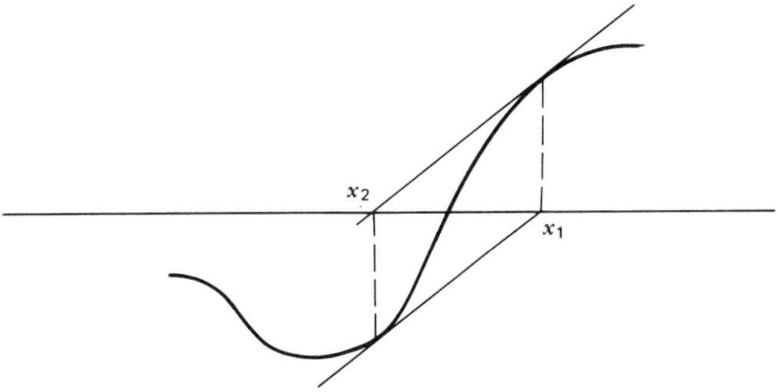

FIGURE 5.4.5 (x_n) **oscillates between** x_1 **and** x_2.

If we take $x_1 := 1$ as our initial estimate, we obtain the successive values $x_2 = 3/2 = 1.5$, $x_3 = 17/12 = 1.41666\ldots$, $x_4 = 577/408 = 1.414215\ldots$, and $x_5 = 665857/470832 = 1.414213562374\ldots$, which is correct to eleven places.

Remarks. (1) If we let $e_n := x_n - r$ be the error in approximating r, then inequality $(**)$ can be written in the form $|Ke_{n+1}| \leqslant |Ke_n|^2$. Consequently, if $|Ke_n| < 10^{-m}$ then $|Ke_{n+1}| < 10^{-2m}$ so that the number of significant digits in Ke_n has been doubled. Because of this doubling, the sequence generated by Newton's Method is said to converge "quadratically".

(2) In practice, when Newton's Method is programmed for a computer, one often makes an initial guess x_1 and lets the computer run. If x_1 is poorly chosen, or if the root is too near the end point of I, the procedure may not converge to a zero of f. A variety of possible difficulties are illustrated in Figure 5.4.3 through 5.4.5. One familiar strategy is to use the bisection method to arrive at a fairly close estimate of the root and then to switch to Newton's Method for the coup de grâce.

Exercises for Section 5.4

1. Let $f(x) := \cos ax$ for $x \in \boldsymbol{R}$ where $a \neq 0$. Find $f^{(n)}(x)$ for $n \in \boldsymbol{N}$, $x \in \boldsymbol{R}$.
2. Let $g(x) := |x^3|$ for $x \in \boldsymbol{R}$. Find $g'(x)$ and $g''(x)$ for $x \in \boldsymbol{R}$, and $g'''(x)$ for $x \neq 0$. Show that $g'''(0)$ does not exist.

3. Use induction to prove Leibniz's rule for the nth derivative of a product

$$(fg)^{(n)}(x) = \sum_{k=0}^{n} \binom{n}{k} f^{(n-k)}(x)g^{(k)}(x).$$

4. Show that if $x > 0$, then

$$1 + \frac{x}{2} - \frac{x^2}{8} \leqslant \sqrt{1+x} \leqslant 1 + \frac{x}{2}.$$

5. Use the preceding exercise to approximate $\sqrt{1.2}$ and $\sqrt{2}$. What is the best accuracy you can be sure of, using this inequality?
6. Use Taylor's Theorem with $n = 2$ to obtain more accurate approximations for $\sqrt{1.2}$ and $\sqrt{2}$.
7. If $x > 0$ show that

$$\left|(1+x)^{1/3} - \left(1 + \frac{x}{3} - \frac{x^2}{9}\right)\right| \leqslant \frac{5x^3}{81}.$$

Use this inequality to approximate $\sqrt[3]{1.2}$ and $\sqrt[3]{2}$.
8. If $f(x) := e^x$, show that the remainder term in Taylor's Theorem converges to zero as $n \to \infty$, for each fixed x_0 and x. [*Hint*: See Theorem 3.2.11.]
9. If $g(x) :=$ sin x, show that the remainder term in Taylor's Theorem converges to zero as $n \to \infty$, for each fixed x_0 and x.
10. Let $h(x) := e^{-1/x^2}$ for $x \neq 0$ and $h(0) := 0$. Show that $h^{(n)}(0) = 0$ for all $n \in \mathbf{N}$. Conclude that the remainder term in Taylor's Theorem for $x_0 = 0$ does *not* converge to zero as $n \to \infty$ for $x \neq 0$. [*Hint*: By L'Hospital's Rule, $\lim_{x\to 0} h(x)/x^k = 0$ for any $k \in \mathbf{N}$. Use Exercise 3 to calculate $f^{(n)}(x)$ for $x \neq 0$.]
11. If $x \in [0,1]$ and $n \in \mathbf{N}$, show that

$$\left|\log(1+x) - \left(x - \frac{x^2}{2} + \frac{x^3}{3} + \cdots + (-1)^{n-1}\frac{x^n}{n}\right)\right| < \frac{x^{n+1}}{n+1}.$$

Use this to approximate log 1.5 with an error less than 0.01. Less than 0.001.
12. We wish to approximate sin by a polynomial on $[-1,1]$ so that the error is less than 0.001. Show that we have

$$\left|\sin x - \left(x - \frac{x^3}{6} + \frac{x^5}{120}\right)\right| < \frac{1}{5040} \quad \text{for} \quad |x| \leqslant 1.$$

13. Calculate e correct to 7 decimal places.
14. Determine whether or not $x = 0$ is a point of relative extremum of the following functions:
(a) $f(x) := x^3 + 2$; (b) $g(x) := \sin x - x$;
(c) $h(x) := \sin x + \frac{1}{6}x^3$; (d) $k(x) := \cos x - 1 + \frac{1}{2}x^2$
15. Let f be continuous on $[a,b]$ and assume the second derivative f'' exists on (a,b). Suppose that the graph of f and the line segment joining the points $(a,f(a))$ and $(b,f(b))$ intersect at a point $(x_0,f(x_0))$ where $a < x_0 < b$. Show that there exists a point $c \in (a,b)$ such that $f''(c) = 0$.

16. Let $I \subseteq \boldsymbol{R}$ be an open interval, let $f : I \to \boldsymbol{R}$ be differentiable on I, and suppose $f''(a)$ exists at $a \in I$. Show that

$$f''(a) = \lim_{h \to 0} \frac{f(a+h) - 2f(a) + f(a-h)}{h^2}.$$

Give an example where this limit exists, but the function does not have a second derivative at a.

17. Suppose that $I \subseteq \boldsymbol{R}$ is an open interval and that $f''(x) \geqslant 0$ for all $x \in I$. If $c \in I$, show that the part of the graph of f on I is never below the tangent line to the graph at $(c, f(c))$.

18. Let $I \subseteq \boldsymbol{R}$ be an interval and let $c \in I$. Suppose that f and g are defined on I and that the derivatives $f^{(n)}$, $g^{(n)}$ exist and are continuous on I. If $f^{(k)}(c) = 0$ and $g^{(k)}(c) = 0$ for $k = 0, 1, \ldots, n-1$, but $g^{(n)}(c) \neq 0$, show that

$$\lim_{x \to c} \frac{f(x)}{g(x)} = \frac{f^{(n)}(c)}{g^{(n)}(c)}.$$

19. Show that the function $f(x) := x^3 - 2x - 5$ has a zero r in the interval $I := [2, 2.2]$. If $x_1 := 2$ and if we define the sequence (x_n) using the Newton procedure, show that $|x_{n+1} - r| \leqslant (0.7)|x_n - r|^2$. Show that x_4 is accurate to within 6 decimal places.

20. Approximate the real zeros of $g(x) := x^4 - x - 3$.

21. Approximate the real zeros of $h(x) := x^3 - x - 1$. Apply Newton's Method starting with the initial choices (a) $x_1 := 2$, (b) $x_1 := 0$, (c) $x_1 = -2$. Explain what happens.

22. The equation $\log x = x - 2$ has two solutions Approximate them using Newton's Method. What happens if $x_1 := \frac{1}{2}$ is the initial point?

23. The function $f(x) := 8x^3 - 8x^2 + 1$ has two zeros in $[0,1]$. Approximate them, using Newton's Method, with the starting points (a) $x_1 := \frac{1}{8}$, and (b) $x_1 := \frac{1}{4}$. Explain what happens.

24. Approximate the solution of the equation $x = \cos x$, accurate to within six decimals.

25. Let $I := [a,b]$ and let $f : I \to \boldsymbol{R}$ be differentiable on I. Suppose that $f(a) < 0 < f(b)$ and that there exist m, M such that $0 < m < f'(x) \leqslant M$ for $x \in I$. Let $x_1 \in I$ be arbitrary and define $x_{n+1} := x_n - f(x_n)/M$ for $n \in \boldsymbol{N}$. Show that the sequence (x_n) is well defined and converges to the unique zero $r \in I$ of f and that

$$|x_{n+1} - r| \leqslant (1 - m/M)^n |x_1 - r| \leqslant (1 - m/M)^n |f(x_1)|/m.$$

[*Hint*: If $\varphi(x) := x - f(x)/M$, show that $0 \leqslant \varphi'(x) \leqslant 1 - m/M < 1$ and $\varphi(I) \subseteq I$.]

26. Apply the result in the preceding exercise to some of the functions considered in Exercises 19–24.

CHAPTER SIX

THE RIEMANN INTEGRAL

We have already mentioned the developments, during the 1630s, by Fermat and Descartes leading to analytic geometry and the theory of the derivative. However, the subject we know as calculus did not begin to take shape until the late 1660s when Isaac Newton (1642–1727) created his theory of "fluxions" and invented the method of "inverse tangents" to find areas under curves. The reversal of the process for finding tangent lines to find areas was also discovered in the 1680s by Gottfried Leibniz (1646–1716), who was unaware of Newton's unpublished work and who arrived at the discovery by a very different route. Leibniz introduced the terminology "calculus differentialis" and "calculus integralis", since finding tangent lines involved differences and finding areas involved summations. Thus, they had discovered that integration, being a process of summation, was inverse to the operation of differentiation.

During a century and a half of development and refinement of techniques, calculus consisted of these paired operations and their applications, primarily to physical problems. In the 1850s, Bernhard Riemann (1826–1866) adopted a new and different viewpoint. He separated the concept of integration from its companion, differentiation, and examined the motivating summation and limit process of finding areas by itself. He broadened the scope by considering all functions on an interval for which this process of "integration" could be defined: the class of "integrable" functions. The Fundamental Theorem of Calculus became a result that held only for a restricted set of integrable functions. The viewpoint of Riemann led others to invent other integration theories, the most significant being Lebesgue's theory of integration.

In this chapter we shall begin by defining the concept of Riemann integrability of functions on an interval by means of upper and lower sums. In Section 6.2, we shall discuss the basic properties of the integral and the class of integrable functions on an interval. The Fundamental Theorem of Calculus is the subject of Section 6.3. Other approaches to the Riemann integral are discussed in Section 6.4 and their equivalence is established. A second version of the Fundamental Theorem is also included in this section, as is a brief introduction to "improper integrals". In Section 6.5 we discuss several methods of approximating integrals, a subject that has become increasingly important during this era of high-speed computers.

SECTION 6.1 Riemann Integrability

In this section we shall define the upper integral and the lower integral of an arbitrary *bounded function* on a *closed, bounded interval.* Such a function is then said to be Riemann integrable if its upper integral and lower integral are equal, and the *Riemann integral* of the function is defined to be this common value. We shall also consider several elementary examples and establish a criterion for integrability that is reminiscent of the Cauchy criterion for the convergence of a sequence. Since we assume that the reader is acquainted—at least informally—with the integral from a calculus course, we shall not provide an extensive motivation of it or discuss the many important interpretations of it that are used in applications.

Upper and Lower Sums

To discuss the concept of Riemann integral, we must first introduce some terminology and notation.

If $I := [a,b]$ is a closed, bounded interval in $\boldsymbol{R}$, then a **partition of** I is a finite, ordered set $P := (x_0, x_1, \ldots, x_n)$ of points in I such that

$$a = x_0 < x_1 < x_2 < \cdots < x_n = b.$$

(See Figure 6.1.1.) The points of the partition P can be used to divide I into non-overlapping subintervals:

$$[x_0,x_1],\ [x_1,x_2],\ \ldots,\ [x_{n-1},x_n].$$

Let $f : I \to \boldsymbol{R}$ be a bounded function on I and let $P = (x_0,x_1, \ldots, x_n)$ be a partition of I. For $k = 1,2, \ldots, n$ we let

$$m_k := \inf\{f(x) : x \in [x_{k-1},x_k]\}, \qquad M_k := \sup\{f(x) : x \in [x_{k-1},x_k]\}.$$

The **lower sum** of f corresponding to the partition P is defined to be

$$L(P;f) := \sum_{k=1}^{n} m_k(x_k - x_{k-1}),$$

and the **upper sum** of f corresponding to P is defined to be

$$U(P;f) := \sum_{k=1}^{n} M_k(x_k - x_{k-1}).$$

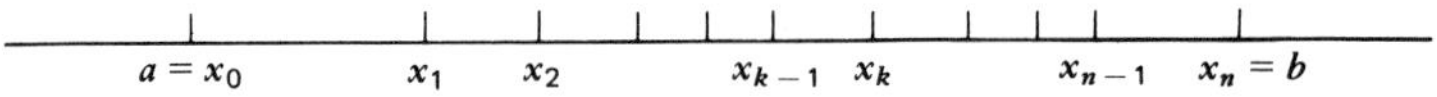

FIGURE 6.1.1 A partition of $I = [a,b]$.

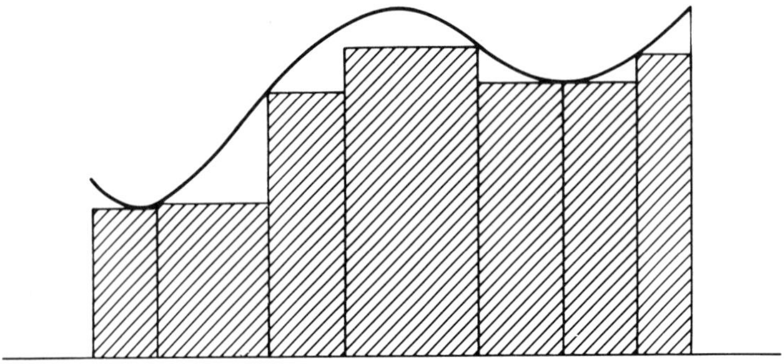

FIGURE 6.1.2 ***L(P;f)*, a lower sum.**

If f is a positive function, then the lower sum $L(P;f)$ can be interpreted as the area of the union of rectangles with base $[x_{k-1},x_k]$ and height m_k. (See Figure 6.1.2.) Similarly, the upper sum $U(P;f)$ can be interpreted as the area of the union of rectangles with bases $[x_{k-1},x_k]$ and heights M_k. (See Figure 6.1.3.) The geometric interpretation suggests that, for a given partition, the lower sum is less than or equal to the upper sum. We now show this to be the case.

6.1.1 Lemma. *If $f : I \to \mathbf{R}$ is bounded and P is any partition of I, then* $L(P;f) \leq U(P;f)$.

Proof. Let $P := (x_0,x_1, \ldots, x_n)$. Since $m_k \leq M_k$ for $k = 1, 2, \ldots, n$ and since $x_k - x_{k-1} > 0$ for $k = 1, 2, \ldots, n$, it follows that

$$L(P;f) = \sum_{k=1}^{n} m_k(x_k - x_{k-1}) \leq \sum_{k=1}^{n} M_k(x_k - x_{k-1}) = U(P;f).$$

Q.E.D.

If $P := (x_0,x_1, \ldots, x_n)$ and $Q := (y_0,y_1, \ldots, y_m)$ are partitions of I, we say that Q is a ***refinement*** of P if each partition point $x_k \in P$ also belongs to Q (that is, if $P \subseteq Q$). A refinement Q of a partition P can be obtained by adjoining a finite number of points to P. In this case, each one of the intervals $[x_{k-1},x_k]$ into which P divides I can be written as the union of intervals whose end points belong to Q; that is,

$$[x_{k-1},x_k] = [y_{j-1},y_j] \cup [y_j,y_{j+1}] \cup \cdots \cup [y_{h-1},y_h].$$

We now show that refining a partition increases lower sums and decreases upper sums.

6.1.2 Lemma. *If $f : I \to \mathbf{R}$ is bounded, if P is a partition of I, and if Q is a refinement of P, then*

$$L(P;f) \leq L(Q;f) \qquad \textit{and} \qquad U(Q;f) \leq U(P;f).$$

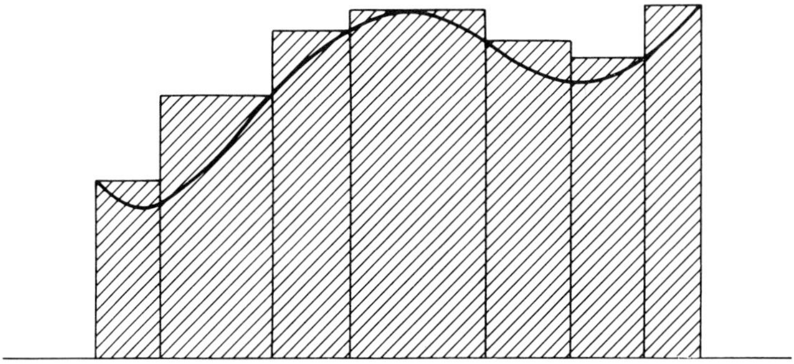

FIGURE 6.1.3 ***U(P;f)*, an upper sum.**

Proof. Let $P = (x_0, x_1, \ldots, x_n)$. We first examine the effect of adjoining one point to P. Let $z \in I$ satisfy $x_{k-1} < z < x_k$ and let P' be the partition

$$P' := (x_0, x_1, \ldots, x_{k-1}, z, x_k, \ldots, x_n)$$

obtained from P by adjoining z to P. Let m_k' and m_k'' be the numbers

$$m_k' := \inf\{f(x) : x \in [x_{k-1}, z]\}, \qquad m_k'' := \inf\{f(x) : x \in [z, x_k]\}.$$

Then $m_k \leq m_k'$ and $m_k \leq m_k''$ (why?) and therefore

$$m_k(x_k - x_{k-1}) = m_k(z - x_{k-1}) + m_k(x_k - z) \leq m_k'(z - x_{k-1}) + m_k''(x_k - z).$$

If we add the terms $m_j(x_j - x_{j-1})$ for $j \neq k$ to the above inequality, we obtain $L(P;f) \leq L(P';f)$.

Now if Q is any refinement of P (i.e., if $P \subseteq Q$), then Q can be obtained from P by adjoining a finite number of points to P one at a time. Hence, repeating the preceding argument, we infer that $L(P;f) \leq L(Q;f)$.

In the same way, if P' is as above, and if M_k' and M_k'' are the numbers

$$M_k' := \sup\{f(x) : x \in [x_{k-1}, z]\}, \qquad M_k'' := \sup\{f(x) : x \in [z, x_k]\},$$

then $M_k' \leq M_k$ and $M_k'' \leq M_k$, whence it follows that $U(P';f) \leq U(P;f)$. Similarly, if $P \subseteq Q$, then $U(Q;f) \leq U(P;f)$. Q.E.D.

We have shown in Lemma 6.1.1 that a lower sum is smaller than an upper sum if both sums correspond to the same partition, and in Lemma 6.1.2 that refining a partition increases lower sums and decreases upper sums. These two results are now combined to conclude that a lower sum is always smaller than an upper sum even if they correspond to different partitions.

6.1.3 Lemma. *Let $f : I \to \mathbf{R}$ be bounded. If P_1, P_2 are any two partitions of I, then $L(P_1;f) \leqslant U(P_2;f)$.*

Proof. Let $Q := P_1 \cup P_2$ be the partition obtained by combining the points of P_1 and P_2. Then Q is a refinement of both P_1 and P_2. Hence, by Lemmas 6.1.1 and 6.1.2, we conclude that

$$L(P_1;f) \leqslant L(Q;f) \leqslant U(Q;f) \leqslant U(P_2;f). \qquad \text{Q.E.D.}$$

Upper and Lower Integrals

We shall denote the collection of all partitions of the interval I by $\mathcal{P}(I)$. If $f : I \to \mathbf{R}$ is bounded, then each P in $\mathcal{P}(I)$ determines two numbers: $L(P;f)$ and $U(P;f)$. Thus, the collection $\mathcal{P}(I)$ determines two sets of numbers: the set of lower sums $L(P;f)$ for $P \in \mathcal{P}(I)$, and the set of upper sums $U(P;f)$ for $P \in \mathcal{P}(I)$. Moreover, by Lemma 6.1.3, any upper sum is an upper bound for the set of lower sums, and any lower sum is a lower bound for the set of upper sums. Hence, we are led to the following definition.

6.1.4 Definition. Let $I := [a,b]$ and let $f : I \to \mathbf{R}$ be a bounded function. The **lower integral of f on I** is the number

$$L(f) := \sup \{L(P;f) : P \in \mathcal{P}(I)\},$$

and the **upper integral of f on I** is the number

$$U(f) := \inf \{U(P;f) : P \in \mathcal{P}(I)\}.$$

Since f is a bounded function, we are assured of the existence of the numbers

$$m_I := \inf \{f(x) : x \in I\} \qquad \text{and} \qquad M_I := \sup \{f(x) : x \in I\}.$$

It is readily seen that for any $P \in \mathcal{P}(I)$, we have

$$m_I(b-a) \leqslant L(P;f) \leqslant U(P;f) \leqslant M_I(b-a).$$

Note that the left-hand and right-hand terms in the preceding inequality are simply the lower and upper sums corresponding to the trivial partition consisting of only the points a, b. Consequently, the lower and upper integrals $L(f)$ and $U(f)$ exist, and moreover

$$m_I(b-a) \leqslant L(f) \leqslant M_I(b-a), \qquad m_I(b-a) \leqslant U(f) \leqslant M_I(b-a).$$

The next inequality is also anticipated.

6.1.5 Theorem. *Let $I = [a,b]$ and let $f : I \to \mathbf{R}$ be a bounded function. Then the lower integral $L(f)$ and the upper integral $U(f)$ of f on I exist. Moreover,*

$$(\dagger) \qquad L(f) \leq U(f).$$

Proof. If P_1 and P_2 are any partitions of I, then it follows from Lemma 6.1.3 that $L(P_1;f) \leq U(P_2;f)$. Therefore the number $U(P_2;f)$ is an upper bound for the set $\{L(P;f) : P \in \mathcal{P}(I)\}$. Consequently $L(f)$, being the supremum of this set, satisfies

$$L(f) \leq U(P_2;f).$$

Since P_2 is an arbitrary partition of I, we conclude that $L(f)$ is a lower bound for the set $\{U(P;f) : P \in \mathcal{P}(I)\}$. Consequently, the infimum $U(f)$ of this set satisfies the inequality $(\dagger)$. Q.E.D.

The Riemann Integral

If I is a closed bounded interval and $f : I \to \mathbf{R}$ is a bounded function, we have proved in Theorem 6.1.5 that the lower integral $L(f)$ and the upper integral $U(f)$ always exist. Moreover, *we always have* $L(f) \leq U(f)$. However, it is possible that we might have $L(f) < U(f)$, as we will see in Example 6.1.7(d). On the other hand, there is a large class of functions for which $L(f) = U(f)$. Such functions are said to be "integrable", and the common value of $L(f)$ and $U(f)$ is called "the integral of f over I".

6.1.6 Definition. Let $I = [a,b]$ and let $f : I \to \mathbf{R}$ be a bounded function. Then f is said to be **Riemann integrable on** I if $L(f) = U(f)$. In this case the **Riemann integral of** f **over** I is defined to be the value $L(f) = U(f)$ and this number is ordinarily denoted by

$$\int_a^b f \quad \text{or} \quad \int_a^b f(x)\,dx.$$

In addition, we define

$$\int_b^a f = -\int_a^b f \quad \text{and} \quad \int_a^a f = 0.$$

Thus we see that if the Riemann integral of a function on an interval exists, then the integral is the *unique* real number that lies between the lower sums and the upper sums.

In the sequel we shall often drop the adjective "Riemann" and simply use the term "integral" to refer to the Riemann integral. There are several different theories of integration in real analysis, and the Riemann integral is one of these; but since we shall limit our attention to the Riemann integral in this book, any reference to "integral" or "integrability" will be unambiguous. For a treatment of the Riemann-Stieltjes integral, the reader should consult *The Elements of Real Analysis*; for a treatment of the Lebesgue integral, consult *The Elements of Integration* by R.G. Bartle.

6.1.7 Examples. (a) A constant function is integrable.

Let $f(x) := c$ for $x \in I := [a,b]$. If P is any partition of I, it is easy to see that $L(P;f) = c(b-a) = U(P;f)$. (See Exercise 6.1.1.) Therefore the lower and upper integrals are given by $L(f) = c(b-a) = U(f)$. Consequently, f is integrable on I and

$$\int_a^b f = \int_a^b c\,dx = c(b-a).$$

(b) The function $g(x) := x$ is integrable on $[0,1]$.

Let P_n be the partition of $I := [0,1]$ into n subintervals given by

$$P_n := \left(0, \frac{1}{n}, \frac{2}{n}, \ldots, \frac{n-1}{n}, \frac{n}{n} = 1\right).$$

Since g is an increasing function, its infimum and supremum on the subinterval $[(k-1)/n, k/n]$ are attained at the left and right end points, respectively, and are thus given by

$$m_k = \frac{k-1}{n}, \qquad M_k = \frac{k}{n}.$$

Moreover, since $x_k - x_{k-1} = 1/n$ for all $k = 1, 2, \ldots, n$, we have

$$L(P_n;g) = \big(0 + 1 + \cdots + (n-1)\big)/n^2, \qquad U(P_n;g) = \big(1 + 2 + \cdots + n\big)/n^2.$$

If we use the formula $1 + 2 + \cdots + m = m(m+1)/2$, for $m \in \mathbf{N}$ [see Example 1.3.3(a)], we obtain that

$$L(P_n;g) = \frac{(n-1)n}{2n^2} = \frac{1}{2}\left(1 - \frac{1}{n}\right), \qquad U(P_n;g) = \frac{n(n+1)}{2n^2} = \frac{1}{2}\left(1 + \frac{1}{n}\right).$$

Since the set of partitions $\{P_n : n \in \mathbf{N}\}$ is a subset of the set all partitions $\mathcal{P}(I)$ of I, it follows that

$$\tfrac{1}{2} = \sup\{L(P_n;g) : n \in \mathbf{N}\} \leq \sup\{L(P;g) : P \in \mathcal{P}(I)\} = L(g),$$

and also that

$$U(g) = \inf\{U(P;g) : P \in \mathcal{P}(I)\} \leq \inf\{U(P_n;g) : n \in \mathbf{N}\} = \tfrac{1}{2}.$$

Since $\frac{1}{2} \leqslant L(g) \leqslant U(g) \leqslant \frac{1}{2}$, we conclude that $L(g) = U(g) = \frac{1}{2}$. Therefore g is integrable on $I = [0,1]$ and

$$\int_0^1 g = \int_0^1 x\,dx = \frac{1}{2}.$$

(c) The function $h(x) := x^2$ is integrable on $I := [0,1]$.

Let P_n be as in (b). Since h is increasing on $[0,1]$, we have

$$m_k = ((k-1)/n)^2, \qquad M_k = (k/n)^2$$

for $k = 1, 2, \ldots, n$. Hence, we have

$$L(P_n;h) = \left(0^2 + 1^2 + \cdots + (n-1)^2\right)/n^3,$$
$$U(P_n;h) = \left(1^2 + 2^2 + \cdots + n^2\right)/n^3.$$

Using the formula $1^2 + 2^2 + \cdots + m^2 = \frac{1}{6}m(m+1)(2m+1)$ [see Example 1.3.3(b)], we obtain

$$L(P_n;h) = (n-1)n(2n-1)/6n^3 = \frac{1}{3}\left(1 - \frac{3}{2n} + \frac{1}{2n^2}\right),$$
$$U(P_n;h) = n(n+1)\,(2n+1)/6n^3 = \frac{1}{3}\left(1 + \frac{3}{2n} + \frac{1}{2n^2}\right),$$

for all $n \in \mathbf{N}$. Therefore, we see that

$$\tfrac{1}{3} = \sup\{L(P_n;h) : n \in \mathbf{N}\} \leqslant \sup\{L(P;h) : P \in \mathcal{P}(I)\} = L(h),$$

and also that

$$U(h) = \inf\{U(P;h) : P \in \mathcal{P}(I)\} \leqslant \inf\{U(P_n;h) : n \in \mathbf{N}\} = \tfrac{1}{3}.$$

Hence, we infer that $L(h) = U(h) = \frac{1}{3}$. Therefore h is integrable on $I = [0,1]$ and

$$\int_0^1 h = \int_0^1 x^2\,dx = \frac{1}{3}.$$

(d) A non-integrable function.

Let $I := [0,1]$ and let $f : I \to \mathbf{R}$ be the Dirichlet function [see Example 4.4.5(g)] defined by

$$f(x) := 1 \quad \text{for} \quad x \text{ rational,}$$
$$:= 0 \quad \text{for} \quad x \text{ irrational.}$$

If $P := (x_0, x_1, \ldots, x_n)$ is any partition of $[0,1]$, then since every nontrivial interval contains both rational numbers and irrational numbers (see the Density Theorem 2.4.10 and its corollary), we have $m_k = 0$ and $M_k = 1$. Therefore, we have

$$L(P;f) = 0, \qquad U(P;f) = 1,$$

and it follows that

$$L(f) = 0, \qquad U(f) = 1.$$

Since $L(f) \neq U(f)$, the function f is *not* integrable on $[0,1]$.

In dealing with the Riemann integral we are confronted with two types of questions. First, for a given bounded function on an interval, there is the question of the existence of the integral. Second, if the integral is known to exist, then the problem of evaluating the integral arises. We conclude this section by establishing a necessary and sufficient condition for the existence of the integral of a bounded function.

6.1.8 Riemann's Criterion for Integrability. *Let $I := [a,b]$ and let $f : I \to \mathbf{R}$ be a bounded function on I. Then f is integrable on I if and only if for each $\varepsilon > 0$ there is a partition P_ε of I such that*

$$(*) \qquad U(P_\varepsilon;f) - L(P_\varepsilon;f) < \varepsilon.$$

Proof. If f is integrable, then we have $L(f) = U(f)$. If $\varepsilon > 0$ is given, then from the definition of the lower integral as a supremum, there is a partition P_1 of I such that

$$L(f) - \varepsilon/2 < L(P_1;f).$$

Similarly, there is a partition P_2 of I such that

$$U(P_2;f) < U(f) + \varepsilon/2.$$

If we let $P_\varepsilon := P_1 \cup P_2$, then P_ε is a refinement of both P_1 and P_2. Consequently, by Lemmas 6.1.2 and 6.1.1, we have

$$\begin{aligned} L(f) - \varepsilon/2 < L(P_1;f) &\leq L(P_\varepsilon;f) \\ &\leq U(P_\varepsilon;f) \leq U(P_2;f) < U(f) + \varepsilon/2. \end{aligned}$$

Since $L(f) = U(f)$, we conclude that $(*)$ holds.

To establish the converse, we first observe that for any partition P we have $L(P;f) \leq L(f)$ so that $-L(f) \leq -L(P;f)$, and also $U(f) \leq U(P;f)$. Addition of the last two inequalities gives us

$$U(f) - L(f) \leq U(P;f) - L(P;f).$$

Now suppose that for each $\varepsilon > 0$ there exists a partition P_ε such that (*) holds. Then we have

$$U(f) - L(f) \leq U(P_\varepsilon;f) - L(P_\varepsilon;f) < \varepsilon.$$

Since $\varepsilon > 0$ is arbitrary, we conclude from Theorem 2.2.9 that $U(f) \leq L(f)$. Since (by Theorem 6.1.5) the inequality $L(f) \leq U(f)$ is valid, we have $L(f) = U(f)$. Hence f is integrable. Q.E.D.

6.1.9 Corollary. *Let $I := [a,b]$ and let $f : I \to \mathbf{R}$ be a bounded function. If $\{P_n : n \in \mathbf{N}\}$ is a sequence of partitions of I such that*

$$\lim_n \left(U(P_n;f) - L(P_n;f)\right) = 0,$$

then f is integrable and

$$\lim_n L(P_n;f) = \int_a^b f = \lim_n U(P_n;f).$$

A geometric visualization of the difference $U(P;f) - L(P;f)$ is provided by Figure 6.1.4.

The significance of the corollary is the fact that although the definition of the Riemann integral involves the set of all possible partitions of an interval, for a

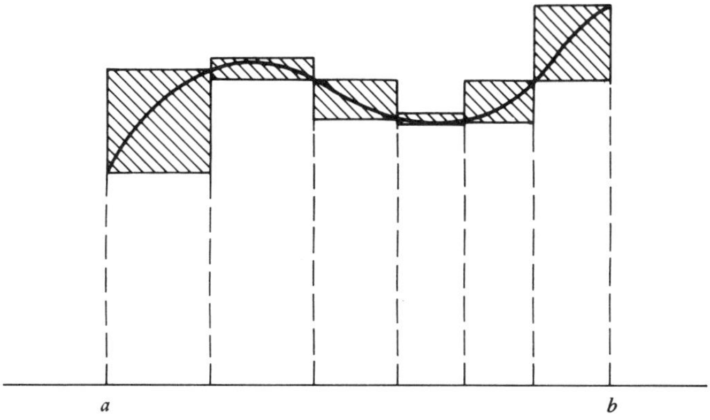

FIGURE 6.1.4 ***U(P;f) − L(P;f).***

given function, the existence of the integral and its value can often be determined by a special sequence of partitions. This was illustrated in Examples 6.1.7(b,c) which we now reconsider.

6.1.10 Examples. (a) Let $g(x) := x$ on $[0,1]$. If $P_n := (0, 1/n, \ldots, (n-1)/n, 1)$, then from the computations of Example 6.1.7(b) we have

$$\lim_n \left(U(P_n;f) - L(P_n;f)\right) = \lim_n \frac{1}{n} = 0$$

and hence that

$$\int_0^1 x\,dx = \lim_n U(P_n;f) = \lim_n \frac{1}{2}\left(1 + \frac{1}{n}\right) = \frac{1}{2}.$$

(b) If $h(x) := x^2$ on $[0,1]$ and if P_n is the partition in part (a), then it follows (why?) from Example 6.1.7(c) that

$$\int_0^1 x^2\,dx = \lim_n U(P_n;f) = \lim_n \frac{1}{3}\left(1 + \frac{3}{2n} + \frac{1}{2n^2}\right) = \frac{1}{3}.$$

Exercises for Section 6.1

1. Prove in detail that if $f(x) := c$ for $x \in [a,b]$, then $\int_a^b f = c(b-a)$.
2. Let $f : [0,2] \to \boldsymbol{R}$ be defined by $f(x) := 1$ if $x \neq 1$, and $f(1) := 0$. Show that f is integrable on $[0,2]$ and calculate its integral.
3. (a) Prove that if $g(x) := 0$ for $0 \leqslant x \leqslant \frac{1}{2}$ and $g(x) := 1$ for $\frac{1}{2} < x \leqslant 1$, then we have $\int_0^1 g = \frac{1}{2}$.

   (b) Does the conclusion hold if we change the value of g at the point $\frac{1}{2}$ to 7?
4. Let $h : [0,1] \to \boldsymbol{R}$ be defined by $h(x) := 0$ for x irrational, and $h(x) := x$ for x rational. Show that $L(h) = 0$ and $U(h) = \frac{1}{2}$. Hence h is *not* integrable on $[0,1]$.
5. Let $f(x) := x^3$ for $0 \leqslant x \leqslant 1$ and let P_n be the partition of Example 6.1.7(b). Calculate $L(P_n;f)$ and $U(P_n;f)$, and show that $\int_0^1 x^3\,dx = \frac{1}{4}$. (*Hint*: Use the formula $1^3 + 2^3 + \cdots + m^3 = [\frac{1}{2}m(m+1)]^2$.)
6. If $f : [a,b] \to \boldsymbol{R}$ is a bounded function such that $f(x) = 0$ except for x in $\{c_1, \ldots, c_n\}$ in $[a,b]$, show that f is integrable on $[a,b]$ and that $\int_a^b f = 0$.
7. Suppose that f is a bounded function on $[a,b]$ and that, for any number $c \in (a,b)$ the restriction of f to $[c,b]$ is integrable. Show that f is integrable on $[a,b]$ and that
$$\int_a^b f = \lim_{c \to a+} \int_c^b f.$$
8. Let $I := [a,b]$ and let $f : I \to \boldsymbol{R}$ be bounded and such that $f(x) \geqslant 0$ for all $x \in I$. Prove that $L(f) \geqslant 0$.

9. Let $I := [a,b]$, let $f : I \to \mathbf{R}$ be continuous, and let $f(x) \geqslant 0$ for all $x \in I$. Prove that if $L(f) = 0$, then $f(x) = 0$ for all $x \in I$.
10. Let $I := [a,b]$, let $f : I \to \mathbf{R}$ be continuous, and suppose that for every integrable function $g : I \to \mathbf{R}$, the product fg is integrable and $\int_a^b fg = 0$. Prove that $f(x) = 0$ for all $x \in I$.
11. Let $I := [0,1]$ and let h be the restriction to I of the function in Example 4.4.5(h). Prove that h is integrable and that $\int_0^1 h = 0$. Compare this result with that in Example 6.1.7(d).
12. Let $I := [a,b]$ and let $f_1, f_2 : I \to \mathbf{R}$ be bounded functions. Show that $L(f_1) + L(f_2) \leqslant L(f_1 + f_2)$. (*Hint*: Use Exercise 2.4.14.)
13. Give examples to show that strict inequality can hold in the preceding exercise.

SECTION 6.2 Properties of the Riemann Integral

In this section we shall establish some of the basic properties of the Riemann integral, including the important linearity and positivity properties. We shall also apply the Riemann Integrability Criterion 6.1.8 to deduce the existence of the integral for functions that are either monotone or continuous. Thus, the Riemann integral exists for a broad class of bounded functions.

The concept of integrability delineates a collection of functions—the class of integrable functions on an interval—and we shall conclude the section with an examination of the permanence properties of this class. The key result is that if an integrable function is composed with a continuous function, the resulting composite function is integrable. From this we infer that absolute values, powers, and products of integrable functions are integrable.

The following property is often referred to as the **linearity** property of the Riemann integral.

6.2.1 Theorem. *Let $I := [a,b]$ and let $f,g : I \to \mathbf{R}$ be integrable functions on I. If $k \in \mathbf{R}$, then the functions kf and $f + g$ are integrable on I, and*

$$\int_a^b kf = k\int_a^b f, \tag{1}$$

$$\int_a^b (f + g) = \int_a^b f + \int_a^b g. \tag{2}$$

Proof. If $k = 0$, the assertions about kf are trivial. We shall consider the case $k < 0$, leaving the slightly easier case $k > 0$ to the reader. Let $P := (x_0, x_1, \ldots, x_n)$ be a partition of I. Since $k < 0$, it is readily seen that

$$\inf \{kf(x) : x \in [x_{j-1},x_j]\} = k \sup \{f(x) : x \in [x_{j-1},x_j]\}$$

for $j = 1,2, \ldots, n$. Multiplying each of these terms by $x_j - x_{j-1}$ and summing, we obtain $L(P;kf) = kU(P;f)$. Therefore, since $k < 0$, we have

$$L(kf) = \sup\{L(P;kf) : P \in \mathcal{P}(I)\} = k \inf\{U(P;f) : P \in \mathcal{P}(I)\} = kU(f).$$

It is similarly shown that $U(P;kf) = kL(P;f)$ and hence that

$$U(kf) = \inf\{U(P;kf) : P \in \mathcal{P}(I)\} = k \sup\{L(P;f) : P \in \mathcal{P}(I)\} = kL(f).$$

Since f is integrable, then $U(f) = L(f)$, whence it follows that $L(kf) = kU(f) = kL(f) = U(kf)$. Thus kf is integrable on I and

$$\int_a^b kf = k \int_a^b f.$$

To establish the assertion about $f + g$, we utilize the following inequalities (see Exercise 2.4.14):

$$\inf\{f(x) : x \in [x_{j-1},x_j]\} + \inf\{g(x) : x \in [x_{j-1},x_j]\}$$
$$\leq \inf\{(f + g)(x) : x \in [x_{j-1},x_j]\}$$

and

$$\sup\{(f + g)(x) : x \in [x_{j-1},x_j]\}$$
$$\leq \sup\{f(x) : x \in [x_{j-1},x_j]\} + \sup\{g(x) : x \in [x_{j-1},x_j]\}.$$

It follows immediately that

$$L(P;f) + L(P;g) \leq L(P;f+g) \quad \text{and} \quad U(P;f+g) \leq U(P;f) + U(P;g)$$

for any partition $P \in \mathcal{P}(I)$. Now if $\varepsilon > 0$ is given, then since f is integrable there exists a partition $P_{1,\varepsilon}$ such that

$$U(P_{1,\varepsilon};f) \leq L(P_{1,\varepsilon};f) + \tfrac{1}{2}\varepsilon,$$

and since g is integrable there is a partition $P_{2,\varepsilon}$ such that

$$U(P_{2,\varepsilon};g) \leq L(P_{2,\varepsilon};g) + \tfrac{1}{2}\varepsilon.$$

If we let $P_\varepsilon := P_{1,\varepsilon} \cup P_{2,\varepsilon}$, then we obtain (why?)

$$U(P_\varepsilon;f+g) \leq U(P_\varepsilon;f) + U(P_\varepsilon;g)$$
$$\leq L(P_\varepsilon;f) + L(P_\varepsilon;g) + \varepsilon \leq L(P_\varepsilon;f+g) + \varepsilon.$$

Hence, by Riemann's Criterion 6.1.8, we see that $f + g$ is integrable, and furthermore that

$$\int_a^b (f+g) = \int_a^b f + \int_a^b g. \qquad \text{Q.E.D.}$$

6.2.2 Corollary. *If $f_i : I \to \mathbf{R}$ is an integrable function on $I := [a,b]$ and $k_i \in \mathbf{R}$ for $i = 1,2,\ldots, n$, then $\sum_{i=1}^{n} k_i f_i$ is integrable on I and*

$$\int_a^b \sum_{i=1}^{n} k_i f_i = \sum_{i=1}^{n} k_i \int_a^b f_i. \tag{3}$$

The next property is referred to as the **positivity** property of the integral.

6.2.3 Theorem. *Let $I := [a,b]$ and let $f : I \to \mathbf{R}$ be integrable on I. If $f(x) \geq 0$ for all $x \in I$, then*

$$\int_a^b f \geq 0. \tag{4}$$

Proof. If $P := (x_0, x_1, \ldots, x_n)$ is a partition of I, then

$$m_j := \inf\{f(x) : x \in [x_{j-1},x_j]\} \geq 0$$

for $j = 1,2, \ldots, n$, and therefore $L(P;f) \geq 0$. Consequently, we obtain

$$\int_a^b f = \sup\{L(P;f) : P \in \mathcal{P}(I)\} \geq 0. \qquad \text{Q.E.D.}$$

6.2.4 Corollary. *If $f, g : I \to \mathbf{R}$ are integrable on $I := [a,b]$, and if we have $f(x) \leq g(x)$ for all $x \in I$, then*

$$\int_a^b f \leq \int_a^b g. \tag{5}$$

Proof. By Theorem 6.2.1, the function $g - f$ is integrable on I and

$$\int_a^b (g-f) = \int_a^b g - \int_a^b f.$$

By hypothesis, we have $g - f \geqslant 0$, and therefore (from Theorem 6.2.3) we infer that

$$\int_a^b g - \int_a^b f \geqslant 0. \qquad \text{Q.E.D.}$$

6.2.5 Corollary. *If $f : I \to \mathbf{R}$ is integrable on $I := [a,b]$, and if we have $m \leqslant f(x) \leqslant M$ for all $x \in I$, then*

$$(6) \qquad m(b - a) \leqslant \int_a^b f \leqslant M(b - a).$$

Proof. It follows from Example 6.1.7(a) that

$$\int_a^b m\, dx = m(b - a), \qquad \int_a^b M\, dx = M(b - a).$$

Hence the assertion is an immediate consequence of Corollary 6.2.4. Q.E.D.

The following very useful property is concerned with the "additivity" of the integral with respect to the interval over which the integral is extended.

6.2.6 Theorem. *Let $I := [a,b]$ and let c satisfy $a < c < b$. Let $f : I \to \mathbf{R}$ be a bounded function. Then f is integrable on I if and only if it is integrable on both $I_1 := [a,c]$ and $I_2 := [c,b]$. In this case,*

$$(7) \qquad \int_a^b f = \int_a^c f + \int_c^b f.$$

We shall first establish the following preliminary result.

6.2.7 Lemma. *If $L_j(f)$ denotes the lower integral of f on I_j for $j = 1, 2$, then*

$$(8) \qquad L(f) = L_1(f) + L_2(f).$$

Proof. If P_j is a partition of I_j and $L_j(P_j;f)$ denotes the lower sum of f corresponding to the partition P_j of I_j, and if P is the partition of $I = I_1 \cup I_2$ formed by combining the points in P_1 and P_2, it is an exercise to show that

$$(*) \qquad L(P;f) = L_1(P_1;f) + L_2(P_2;f).$$

Now since $L_1(P_1;f) \leqslant L_1(f)$ and $L_2(P_2;f) \leqslant L_2(f)$ for all partitions P_j of I_j, it follows immediately that

$$L(P;f) \leqslant L_1(f) + L_2(f)$$

for every partition P of I that contains the point c. Consequently (see Exercise 6.2.6) it follows that

$$L(f) \leqslant L_1(f) + L_2(f).$$

To establish the opposite inequality, let $\varepsilon > 0$ be given and, for $j = 1,2$, let $P_{j,\varepsilon}$ be partitions of I_j such that

$$L_1(f) \leqslant L_1(P_{1,\varepsilon};f) + \tfrac{1}{2}\varepsilon, \qquad L_2(f) \leqslant L_2(P_{2,\varepsilon};f) + \tfrac{1}{2}\varepsilon.$$

If $P_\varepsilon := P_{1,\varepsilon} \cup P_{2,\varepsilon}$ is the corresponding partition of $I = I_1 \cup I_2$, then it is a consequence of $(*)$ that

$$\begin{aligned} L_1(f) + L_2(f) &\leqslant L_1(P_{1,\varepsilon};f) + L_2(P_{2,\varepsilon};f) + \varepsilon \\ &= L(P_\varepsilon;f) + \varepsilon \leqslant L(f) + \varepsilon. \end{aligned}$$

Since $\varepsilon > 0$ is arbitrary it follows that

$$L_1(f) + L_2(f) \leqslant L(f),$$

which, when combined with the opposite inequality, yields equation (8). Q.E.D.

We leave it as an exercise to the reader to show that if $U_j(f)$ denotes the upper integral of f on I_j, then

$$(9) \qquad U(f) = U_1(f) + U_2(f).$$

Proof of 6.2.6. If f is integrable on I_1 and I_2, then $L_1(f) = U_1(f)$ and $L_2(f) = U(f_2)$, whence it follows from (8) and (9) that $U(f) = U_1(f) + U_2(f) = L_1(f) + L_2(f) = L(f)$. Therefore f is integrable on I and (7) holds.

Conversely, if f is integrable on I, then it follows from (8) and (9) that

$$(10) \qquad L_1(f) + L_2(f) = L(f) = U(f) = U_1(f) + U_2(f).$$

Moreover, from Theorem 6.1.5 we know that $L_j(f) \leqslant U_j(f)$ for $j = 1, 2$. If either $L_1(f) < U_1(f)$ or $L_2(f) < U_2(f)$, we would have a contradiction to equation (10). Therefore $L_1(f) = U_1(f)$ and $L_2(f) = U_2(f)$, so that f is integrable on both I_1 and I_2 and equation (7) holds. Q.E.D.

6.2.8 Corollary. *If $P := (x_0, x_1, \ldots, x_n)$ is a partition of $I := [a,b]$ and f is integrable on I, then*

$$\int_a^b f = \sum_{k=1}^{n} \int_{x_{k-1}}^{x_k} f.$$

Existence of the Integral

We shall now return to the important question of the existence of the integral. To have a useful theory of integration there must be a broad and easily identifiable class of functions for which integrability is assured. As we shall see, a function that is either monotone or continuous has the desirable property of being integrable.

The main tool in establishing these existence results is the Riemann Integrability Criterion 6.1.8. For convenience we shall reformulate the condition slightly. If $f : I \to \mathbf{R}$ is a bounded function on $I := [a,b]$ and $P := (x_0, x_1, \ldots, x_n)$ is a partition of I, we shall use the familiar notation

$$m_k := \inf \{f(x) : x \in [x_{k-1}, x_k]\}, \qquad M_k := \sup \{f(x) : x \in [x_{k-1}, x_k]\}$$

for $k = 1, 2, \ldots, n$.

6.2.9 Integrability Theorem. *Let $I := [a,b]$ and let $f : I \to \mathbf{R}$ be a bounded function. Then the following assertions are equivalent.*

(a) *The function f is integrable on I.*

(b) *For each $\varepsilon > 0$ there is a partition $P_\varepsilon = (x_0, x_1, \ldots, x_n)$ of I such that*

$$\sum_{k=1}^{n} (M_k - m_k)(x_k - x_{k-1}) < \varepsilon. \tag{11}$$

(c) *For each $\varepsilon > 0$ there is a partition $P_\varepsilon = (x_0, x_1, \ldots, x_n)$ such that*

$$\sum_{k=1}^{n} w_k(x_k - x_{k-1}) < \varepsilon, \tag{12}$$

where $w_k := \sup \{f(x) - f(y) : x, y \in [x_{k-1}, x_k]\}$ for $k = 1, 2, \ldots, n$.

Proof. (a) $\Leftrightarrow$ (b). With m_k, M_k as defined above, we have

$$U(P_\varepsilon; f) - L(P_\varepsilon; f) = \sum_{k=1}^{n} (M_k - m_k)(x_k - x_{k-1})$$

so that condition (b) is equivalent to Riemann's Criterion 6.1.8.

(b) $\Leftrightarrow$ (c) This follows from the observation that

$$\sup \{f(x) : x \in [x_{k-1}, x_k]\} - \inf \{f(x) : x \in [x_{k-1}, x_k]\}$$
$$= \sup \{f(x) - f(y) : x, y \in [x_{k-1}, x_k]\}.$$

Hence $M_k - m_k = w_k$ for $k = 1, 2, \ldots, n$. Q.E.D.

Applying the integrability criterion to the class of monotone functions, we obtain the following existence theorem.

6.2.10 Integrability of Monotone Functions. *Let $I := [a,b]$ and let $f : I \to \mathbf{R}$ be monotone on I. Then f is integrable on I.*

Proof. Suppose that f is increasing on I. Let $P_n := (x_0, x_1, \ldots, x_n)$ be the partition of I into n equal parts, so that $x_k - x_{k-1} = (b-a)/n$ for $k = 1,2, \ldots, n$. Since f is increasing on $[x_{k-1}, x_k]$, it is clear that $m_k = f(x_{k-1})$ and $M_k = f(x_k)$. Therefore we have the "telescoping" sum:

$$\begin{aligned}\sum_{k=1}^{n} (M_k - m_k)(x_k - x_{k-1}) &= \frac{b-a}{n} \sum_{k=1}^{n} \big(f(x_k) - f(x_{k-1})\big) \\ &= \frac{b-a}{n} \big(f(x_1) - f(x_0) + f(x_2) - f(x_1) + \cdots + f(x_n) - f(x_{n-1})\big) \\ &= \frac{b-a}{n} \big(f(x_n) - f(x_0)\big) = \frac{b-a}{n} \big(f(b) - f(a)\big).\end{aligned}$$

Now if $\varepsilon > 0$ is given, we choose $n \in \mathbf{N}$ such that $n > (b-a)\big(f(b) - f(a)\big)/\varepsilon$. For the corresponding partition P_n we have

$$\sum_{k=1}^{n} (M_k - m_k)(x_k - x_{k-1}) < \varepsilon .$$

It follows from the Integrability Theorem 6.2.9(b) that f is integrable on I. Q.E.D.

We next show that continuity is also a sufficient condition for integrability.

6.2.11 Integrability of Continuous Functions. *Let $I := [a,b]$ and let $f : I \to \mathbf{R}$ be continuous on I. Then f is integrable on I.*

Proof. By the Uniform Continuity Theorem 4.7.3, f is uniformly continuous on I. Therefore, if $\varepsilon > 0$ is given, there exists $\delta(\varepsilon) > 0$ such that if $x,y \in I$ and $|x - y| < \delta(\varepsilon)$, then $|f(x) - f(y)| < \varepsilon/(b-a)$. Now let $n \in \mathbf{N}$ be such that $n > (b - a)/\delta(\varepsilon)$, and let $P_n := (x_0, x_1, \ldots, x_n)$ be the partition of I into n equal parts so that

$$x_k - x_{k-1} = (b-a)/n < \delta(\varepsilon).$$

Therefore we have

$$w_k := M_k - m_k = \sup \{f(x) - f(y) : x,y \in [x_{k-1}, x_k]\} \leqslant \varepsilon/(b-a)$$

for $k = 1, 2, \ldots, n$. Consequently, we have

$$\sum_{k=1}^{n} w_k(x_k - x_{k-1}) \leq n \cdot \frac{\varepsilon}{b-a} \cdot \frac{b-a}{n} = \varepsilon .$$

We conclude from the Integrability Theorem 6.2.9(c) that f is integrable on I.
Q.E.D.

It is an exercise (see Exercise 6.1.7) to show that if $f : [a,b] \to \mathbf{R}$ is a bounded function that is integrable on $[c,b]$ for every c in (a,b), then f is integrable on $[a,b]$ and

$$\int_a^b f = \lim_{c \to a+} \int_c^b f.$$

It follows from this that if $f : [a,b] \to \mathbf{R}$ is bounded and has at most a finite number of discontinuities on $[a,b]$, then f is integrable on $[a,b]$. (See Exercise 6.2.11.)

The Class of Integrable Functions

It was shown in Theorem 6.2.1 that a constant multiple of an integrable function is integrable. Similarly the sum of two integrable functions is integrable. We now show that certain other combinations of integrable functions are integrable. The most useful result in this direction will now be established.

6.2.12 The Composition Theorem. *Let $I := [a,b]$ and $J := [c,d]$ be intervals, and suppose that $f : I \to \mathbf{R}$ is integrable on I, that $\varphi : J \to \mathbf{R}$ is continuous, and that $f(I) \subseteq J$. Then the composition $\varphi \circ f : I \to \mathbf{R}$ is integrable on I.*

Proof. Let $\varepsilon > 0$ be given, let $K := \sup \{|\varphi(t)| : t \in J\}$ and let $\varepsilon' := \varepsilon/(b-a+2K)$. Since φ is uniformly continuous on J, there exists $\delta > 0$ such that $\delta < \varepsilon'$ and such that if $s,t \in J$ and $|s-t| < \delta$, then $|\varphi(s) - \varphi(t)| < \varepsilon'$.

Since f is integrable on $I = [a,b]$ and $\delta^2 > 0$, there exists a partition $P := (x_0, x_1, \ldots, x_n)$ of I such that

$$U(P;f) - L(P;f) < \delta^2.$$

We shall complete the proof by showing that for this partition P, we have

$$(\dagger) \qquad U(P;\varphi \circ f) - L(P;\varphi \circ f) \leq \varepsilon .$$

Since $\varepsilon > 0$ is arbitrary, this will show that $\varphi \circ f$ is integrable on I, by the Riemann Criterion 6.1.8.

To establish (†) we separate the set of indices of the partition P into two subsets. If m_k and M_k denote, as usual, the infimum and the supremum of f on $[x_{k-1},x_k]$, then we let

$$A := \{k : M_k - m_k < \delta\}, \qquad B := \{k : M_k - m_k \geq \delta\}.$$

If we let

$$\tilde{m}_k := \inf\{\varphi \circ f(x) : x \in [x_{k-1},x_k]\}, \qquad \tilde{M}_k := \sup\{\varphi \circ f(x) : x \in [x_{k-1},x_k]\},$$

then it is easily seen that

$$\tilde{M}_k - \tilde{m}_k = \sup\{\varphi \circ f(x) - \varphi \circ f(y) : x,y \in [x_{k-1},x_k]\}.$$

Now if $k \in A$ and $x,y \in [x_{k-1},x_k]$, then $|f(x) - f(y)| < \delta$, whence it follows that $|\varphi \circ f(x) - \varphi \circ f(y)| < \varepsilon'$. Therefore it follows that if $k \in A$, then $\tilde{M}_k - \tilde{m}_k \leq \varepsilon'$, from which we conclude that

$$\sum_{k \in A} (\tilde{M}_k - \tilde{m}_k)(x_k - x_{k-1}) \leq \varepsilon'(b - a).$$

On the other hand, if $k \in B$ then we can only say that $\tilde{M}_k - \tilde{m}_k \leq 2K$ so that

$$\sum_{k \in B} (\tilde{M}_k - \tilde{m}_k)(x_k - x_{k-1}) \leq 2K \sum_{k \in B} (x_k - x_{k-1}).$$

However, for $k \in B$ we have $\delta \leq M_k - m_k$ so that

$$\begin{aligned} \sum_{k \in B} (x_k - x_{k-1}) &\leq \frac{1}{\delta} \sum_{k \in B} (M_k - m_k)(x_k - x_{k-1}) \\ &\leq \frac{1}{\delta}\big(U(P;f) - L(P;f)\big) < \delta < \varepsilon'. \end{aligned}$$

Hence we have

$$\sum_{k \in B} (\tilde{M}_k - \tilde{m}_k)(x_k - x_{k-1}) \leq 2K\,\varepsilon'.$$

If we combine these two estimates, we obtain

$$\begin{aligned} U(P;\varphi \circ f) - L(P;\varphi \circ f) &= \sum_{k \in A} (\tilde{M}_k - \tilde{m}_k)(x_k - x_{k-1}) \\ &\quad + \sum_{k \in B} (\tilde{M}_k - \tilde{m}_k)(x_k - x_{k-1}) \\ &\leq \varepsilon'(b - a) + 2K\varepsilon' = \varepsilon. \end{aligned}$$

Thus we obtain (†), as desired. Q.E.D.

The effort expended in proving the Composition Theorem pays off with the following consequences.

6.2.13 Corollary. *Let $I := [a,b]$ and let $f : I \to \mathbf{R}$ be integrable on I. Then the function $|f|$ is integrable on I.*

Proof. Since f is integrable on I, there exists $K > 0$ such that $|f(x)| \leqslant K$ for all $x \in I$. Let $\varphi_1(t) := |t|$ for $t \in J := [-K,K]$; then φ_1 is continuous on J and $\varphi_1 \circ f = |f|$. It follows from the Composition Theorem that $|f|$ is integrable. Q.E.D.

6.2.14 Corollary. *Let $I := [a,b]$ and let $f : I \to \mathbf{R}$ be integrable on I. If $n \in \mathbf{N}$, then the function f^n is integrable on I.*

Proof. Let $|f(x)| \leqslant K$ for $x \in I$ and let $\varphi_2(t) := t^n$ for $t \in J := [-K,K]$. Then $\varphi_2 \circ f = f^n$ and the Composition Theorem applies. Q.E.D.

6.2.15 Corollary. *Let $I := [a,b]$ and let $f : I \to \mathbf{R}$ be integrable on I. If there exists $\delta > 0$ such that $f(x) \geqslant \delta$ for all $x \in I$, then $1/f$ is integrable on I.*

Proof. Let $\delta \leqslant f(x) \leqslant K$ for $t \in I$ and let $\varphi_3(t) = 1/t$ for $t \in J := [\delta,K]$. Then $\varphi_3 \circ f = 1/f$ and the Composition Theorem applies. Q.E.D.

Our next result is also a consequence of the Composition Theorem.

6.2.16 The Product Theorem. *Let $I := [a,b]$ and let $f, g : I \to \mathbf{R}$ be integrable on I. Then their product fg is integrable on I.*

Proof. It follows from Theorem 6.2.1 and from Corollary 6.2.14 (with $n = 2$) that $f + g$, $(f + g)^2$, f^2 and g^2 are integrable on I. Since we have

$$fg = \tfrac{1}{2}\,[(f + g)^2 - f^2 - g^2],$$

it follows from Theorem 6.2.1 that fg is integrable on I. Q.E.D.

The preceding results guarantee the existence of the integral for a very large class of functions. The next example shows that the hypothesis that φ is continuous cannot be entirely dispensed with.

6.2.17 Example. The composition of integrable functions need not be integrable.

Indeed, let $I := [0,1]$ and let $f : I \to \mathbf{R}$ be defined by $f(0) := 1$, $f(x) := 0$ if $x \in I$ is irrational, and $f(m/n) := 1/n$ if $m, n \in \mathbf{N}$ and m and n have no common integral factors. Then f is integrable on I (see Exercise 6.1.11). Let $g : I \to \mathbf{R}$ be defined by $g(0) := 0$ and $g(x) := 1$ for $x \in (0,1]$. Then g is integrable on I and is continuous at every point of I except 0. The composition $g \circ f(x) = 0$ if $x \in I$

is irrational, and $g \circ f(x) = 1$ if $x \in I$ is rational. Thus, $g \circ f$ is the Dirichlet function that was shown not to be integrable in Example 6.1.7(d).

Remark. We remarked after Theorem 6.2.11 that a bounded function that has at most a finite number of discontinuities is integrable, and the reader may suppose that this might characterize the collection of integrable functions. However, the function f in Example 6.2.17 illustrates that the collection of discontinuities of an integrable function need not be finite. In this connection, there is a theorem due to Henri Lebesgue that gives a necessary and sufficient condition for a bounded function to be integrable. In order to state this result we need to make a definition: A set $Z \subseteq \mathbf{R}$ is said to be a set of **measure zero** if for every $\varepsilon > 0$ there exists a countable collection of intervals $I_n := (a_n, b_n)$ with $a_n \leq b_n$ for $n \in \mathbf{N}$, such that $Z \subseteq \bigcup_{n=1}^{\infty} I_n$ and $\sum_{n=1}^{\infty} (b_n - a_n) < \varepsilon$. The theorem of Lebesgue is that a bounded function $f : [a,b] \to \mathbf{R}$ is Riemann integrable if and only if its set of discontinuities is a set of measure zero. A proof of Lebesgue's Theorem is sketched in *The Elements of Real Analysis* (Project 44.α).

Exercises for Section 6.2

1. Let $I := [a,b]$, let $f : I \to \mathbf{R}$ be bounded and let $k > 0$.
   (a) Show that $L(kf) = kL(f)$ and $U(kf) = kU(f)$.
   (b) Show that if f is integrable on I and $k > 0$, then kf is integrable on I and $\int_a^b kf = k \int_a^b f$.
2. Let $I := [a,b]$ and let f and g be bounded functions on I to $\mathbf{R}$. Suppose that $f(x) \leq g(x)$ for all $x \in I$. Show that $L(f) \leq L(g)$ and $U(f) \leq U(g)$.
3. If $I := [a,b]$, if $f : I \to \mathbf{R}$ is bounded on I, and if $m \leq f(x) \leq M$ for all $x \in I$, show that $m(b-a) \leq L(f) \leq U(f) \leq M(b-a)$.
4. Let $I := [a,b]$ and let f,g,h be bounded functions on I to $\mathbf{R}$. Suppose that $f(x) \leq g(x) \leq h(x)$ for $x \in \mathbf{R}$. Show that if f and h are integrable on I and if
$$A := \int_a^b f = \int_a^b h,$$
then g is integrable on I and $\int_a^b g = A$.
5. Establish the validity of $(*)$ in the proof of Lemma 6.2.7.
6. Let $I := [a,b]$ and let $c \in (a,b)$. Let $\mathcal{P}$ denote the set of all partitions of I and let $\mathcal{P}_c$ denote the set of all partitions of I that contain the point c. If $f := I \to \mathbf{R}$ is a bounded function on I, show that
$$L(f) = \sup \{L(P;f) : P \in \mathcal{P}_c\}.$$
7. Give a detailed proof of equation (9).
8. Let $a > 0$ and let $J := [-a,a]$. Let $f : I \to \mathbf{R}$ be bounded and let $\mathcal{P}^*$ be the set of all partitions P of J that contain 0 and are symmetric (that is, $x \in P$ if and only if $-x \in P$). Show that $L(f) = \sup \{L(P;f) : P \in \mathcal{P}^*\}$.
9. Suppose that $J := [-a,a]$, where $a > 0$, and that $f : J \to \mathbf{R}$ is integrable on J. Use the preceding exercise to establish the following results.
   (a) If f is even (that is, if $f(-x) = f(x)$ for all $x \in J$), then $\int_{-a}^a f = 2 \int_0^a f$.
   (b) If f is odd (that is, if $f(-x) = -f(x)$ for all $x \in J$), then $\int_{-a}^a f = 0$.

10. Let $I := [a,b]$ and let $f : I \to \boldsymbol{R}$ be increasing on I. If P_n is the partition of I into n equal parts, show that

$$0 \leqslant U(P_n;f) - \int_a^b f \leqslant \frac{b-a}{n}(f(b) - f(a)).$$

11. Show that if $f : [a,b] \to \boldsymbol{R}$ is bounded and has at most a finite number of discontinuities on $[a,b]$, then f is integrable on $[a,b]$. [*Hint*: Use Exercise 6.1.7 and Theorem 6.2.11.]
12. Give an example of a function $f : [0,1] \to \boldsymbol{R}$ that is not integrable on $[0,1]$, but is such that $|f|$ is integrable on $[0,1]$.
13. Let $I := [a,b]$ and let $f : I \to \boldsymbol{R}$ be integrable on I. Use the inequality

$$||f(x)| - |f(y)|| \leqslant |f(x) - f(y)|$$

for $x, y \in I$ to show that $|f|$ is also integrable on I, without using Theorem 6.2.12.
14. Let $I := [a,b]$, let $f : I \to \boldsymbol{R}$ be integrable on I, and let $|f(x)| \leqslant K$ for all $x \in I$. Use the inequality

$$|(f(x))^2 - (f(y))^2| \leqslant 2K|f(x) - f(y)|$$

for $x, y \in I$ to show that f^2 is integrable on I, without using Theorem 6.2.12.
15. If $I \subseteq \boldsymbol{R}$ is an interval, give an example of an integrable function f on I and a non-integrable function g such that fg is integrable on I.

SECTION 6.3 The Fundamental Theorem of Calculus

This section centers around the celebrated Fundamental Theorem of Calculus, which—in a sense that will be made precise—states that differentiation and integration are inverse operations. Part of this theorem's significance is that it provides a powerful tool for the evaluation of integrals, and we shall examine the basic substitution technique that is used extensively in introductory calculus courses.

In a broader sense, this section deals with the study of functions of the form $x \mapsto \int_a^x f$ where f is an integrable function on an interval $I := [a,b]$ and $x \in I$. To place the Fundamental Theorem in the general context of integration theory, we begin by showing how properties of this function obtained by integration are related to properties of the integrand f. The Fundamental Theorem then arises naturally in this context. We shall conclude the section by presenting several properties that the Riemann integral possesses, properties that are important in both theoretical and computational areas. In particular, we give a different form for the remainder in Taylor's Theorem.

We begin our discussion with a fundamental lemma.

6.3.1 Lemma. *If $I := [a,b]$, if $f : I \to \boldsymbol{R}$ is integrable, and if $|f(x)| \leqslant K$ for all $x \in I$, then*

$$\left|\int_a^b f\right| \leqslant \int_a^b |f| \leqslant K\,(b-a).$$

Proof. By Corollary 6.2.13, the integrability of f implies that of $|f|$. Since $-|f| \leqslant f \leqslant |f|$, it follows from Corollary 6.2.4 that

$$-\int_a^b |f| \leqslant \int_a^b f \leqslant \int_a^b |f|,$$

and hence we have $|\int_a^b f| \leqslant \int_a^b |f|$. The second inequality follows immediately from Corollary 6.2.5. Q.E.D.

Suppose that $f : I \to \mathbf{R}$ is a given integrable function on $I = [a,b]$. For each x such that $a \leqslant x \leqslant b$, the restriction of f to the interval $[a,x]$ is integrable. (See Theorem 6.2.6.) Therefore, we can define a function $F_a : I \to \mathbf{R}$ associated with f by

$$F_a(x) := \int_a^x f, \qquad x \in I.$$

The function F_a defined in this manner is called an **indefinite integral** of f on I.

An integrable function f need not be continuous. Nevertheless, an associated indefinite integral F_a of f is always continuous, as we now show.

6.3.2 Continuity Theorem. *If $I := [a,b]$ and $f : I \to \mathbf{R}$ is integrable on I, then the function $F_a : I \to \mathbf{R}$ defined by*

$$F_a(x) := \int_a^x f, \qquad x \in I,$$

is uniformly continuous on I.

Proof. If $x,y \in I$ satisfy $x < y$, then from Theorem 6.2.6 we infer that

$$\begin{aligned} F_a(y) - F_a(x) &= \int_a^y f - \int_a^x f \\ &= \int_a^x f + \int_x^y f - \int_a^x f = \int_x^y f. \end{aligned}$$

Since f is integrable on I, there exists a bound K for $|f|$, and hence from Lemma 6.3.1 we have

$$|F_a(x) - F_a(y)| \leqslant \int_x^y |f| \leqslant K(y - x).$$

Now if $\varepsilon > 0$ is given, and if we take $\delta(\varepsilon) := \varepsilon/K$, then for $x,y \in I$ such that $|x - y| < \delta$, it follows that $|F_a(x) - F_a(y)| < \varepsilon$. Thus, F_a is uniformly continuous on I. Q.E.D.

6.3.3 Example. Let $I := [-1,1]$ and define $f : I \to \boldsymbol{R}$ by $f(x) := 0$ for $-1 \leqslant x < 0$, and $f(x) := 1$ for $0 \leqslant x \leqslant 1$. Then for $x \in I$ we have $F(x) := \int_{-1}^{x} f = 0$ for $-1 \leqslant x \leqslant 0$, and $F(x) := x$ for $0 < x \leqslant 1$. (See Figure 6.3.1.) To see this, note that for $0 < x \leqslant 1$, the partition $P := (-1,0,x)$ of the interval $[-1,x]$ has associated sums $L(P;f) = x = U(P;f)$, and hence $F(x) = x$. If $-1 \leqslant x \leqslant 0$, then it is evident that $F(x) = 0$. We see that F is a continuous function on I, but that F fails to be differentiable at $x = 0$, the point of discontinuity of f.

The process of integration improves the status of a function from "integrable to continuous". We show next that the process of integration improves the status of a function from "continuous to differentiable".

6.3.4 Differentiation Theorem. *Let $f : I \to \boldsymbol{R}$ be integrable on the interval $I := [a,b]$ and let $F_a : I \to \boldsymbol{R}$ be defined by $F_a(x) := \int_a^x f$ for $x \in I$. Then F_a is differentiable at any point $c \in I$ at which f is continuous, and $F_a'(c) = f(c)$.*

Proof. Suppose that f is continuous at $c \in I$. Let $\varepsilon > 0$ be given and let $\delta = \delta(\varepsilon)$ be such that $|f(c + h) - f(c)| < \varepsilon$ whenever $c + h \in I$ and $|h| < \delta$. For any such h, we use the observation that $(1/h) \int_c^{c+h} 1 \, dx = 1$ and employ Lemma 6.3.1 to obtain

$$\begin{aligned}\left|\frac{F_a(c+h) - F_a(c)}{h} - f(c)\right| &= \left|\frac{1}{h}\int_c^{c+h} f(x)\,dx - f(c)\,\frac{1}{h}\int_c^{c+h} 1\,dx\right| \\ &= \frac{1}{|h|}\left|\int_c^{c+h} (f(x) - f(c))\,dx\right| \\ &\leqslant \frac{1}{|h|}\,\varepsilon\,|h| = \varepsilon.\end{aligned}$$

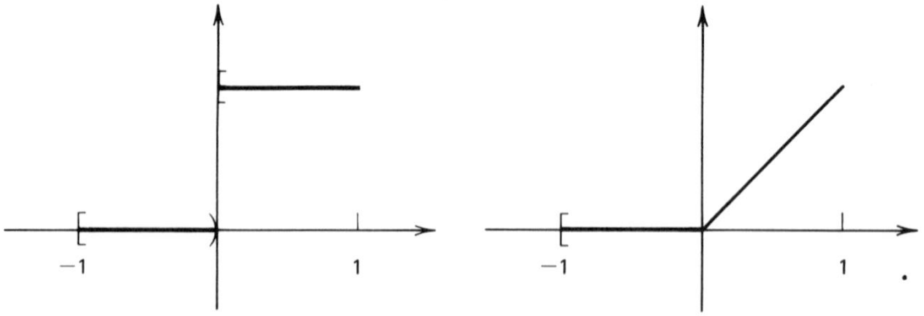

FIGURE 6.3.1 **(*a*) Graph of *f*. (*b*) Graph of *F*.**

Since $\varepsilon > 0$ is arbitrary, it follows that

$$F_a'(c) = \lim_{h \to 0} \frac{F_a(c+h) - F_a(c)}{h} = f(c). \qquad \text{Q.E.D.}$$

A function F is called an **antiderivative** (or a **primitive**) of a function f on an interval I if $F'(x) = f(x)$ for all $x \in I$. The preceding result implies that *if f is continuous on I, then the indefinite integral F_a is an antiderivative of f.* Thus a continuous function always possesses at least one antiderivative. This fact, in conjunction with the fact that any two antiderivatives of a given function differ by only a constant (see Corollary 5.2.6), results in the following relationship.

6.3.5 Fundamental Theorem of Integral Calculus. *Let f be continuous on an interval $I := [a,b]$. Then a function $F : I \to \mathbf{R}$ satisfies*

(1) $$F(x) - F(a) = \int_a^x f$$

if and only if $F'(x) = f(x)$ for all $x \in I$.

Proof. If (1) holds for all $x \in I$, then, using the notation of the Differentiation Theorem 6.3.4, we have $F(x) - F(a) = F_a(x)$ for all $x \in I$. Hence it follows that $F'(x) = F_a'(x) = f(x)$ for all $x \in I$.

Conversely, if $F : I \to \mathbf{R}$ is such that $F'(x) = f(x)$ for all $x \in I$, then $F'(x) = F_a'(x)$ for all $x \in I$. We deduce from Corollary 5.2.6 that a constant C exists such that $F(x) = F_a(x) + C$ for all $x \in I$. Since $F_a(a) = 0$, we see that $F(a) = C$, and hence

$$F(x) - F(a) = F_a(x) = \int_a^x f. \qquad \text{Q.E.D.}$$

6.3.6 Corollary. *If f is continuous on $I = [a,b]$ and if $F'(x) = f(x)$ for all $x \in I$, then*

(2) $$F(b) - F(a) = \int_a^b f.$$

Proof. Take $x = b$ in (1). Q.E.D.

Remarks. (a) The Fundamental Theorem can be given other formulations, but it always includes the evaluation of an integral by means of formulas such as (1) and (2).

(b) It is possible to evaluate the integral by a formula of this kind even when the integrand is not continuous. (See Theorem 6.4.7.)

(c) A function f may be integrable on an interval $I := [a,b]$, but not possess an antiderivative on I. For example, the function $f(x) := \text{sgn}\,(x)$ for $x \in I := [-1,1]$ has this property. (See Example 5.2.13.) Evidently the Fundamental Theorem cannot be applied in such a case.

(d) A function $F : I \to \mathbf{R}$ may be differentiable with derivative $F'(x) = f(x)$ for all $x \in I$, without f being integrable. For example, if F is defined on $I := [-1,1]$ by $F(x) := x^2 \sin(1/x^2)$ for $x \in I$, $x \neq 0$, and $F(0) := 0$, then F is differentiable at every point of I, but the derivative is unbounded and hence not integrable on I.

Evaluation of Integrals

We shall now briefly sample some of the standard "techniques of integration" that are based on the Fundamental Theorem of Integral Calculus. They should be familiar to the reader from an earlier calculus course.

6.3.7 Integration by Parts. *If f and g have continuous derivatives on $I := [a,b]$, then*

$$\int_a^b f'g = [f(b)g(b) - f(a)g(a)] - \int_a^b fg'. \tag{3}$$

Proof. Let $h := fg$. Then h and h' are continuous on I and $h' = f'g + fg'$. It follows from the Fundamental Theorem 6.3.5 that

$$\int_a^b (f'g + fg') = \int_a^b h' = h(b) - h(a)$$
$$= f(b)g(b) - f(a)g(a).$$

Equation (3) now follows immediately. Q.E.D.

The next two theorems provide the justification for the "change of variable" methods that are often used to evaluate integrals. These theorems are employed (usually implicitly) in the evaluation of integrals by means of the procedures involving the "manipulation of differentials" that is common in elementary calculus courses.

6.3.8 First Substitution Theorem. *Let $J := [\alpha,\beta]$ and let $\varphi : J \to \mathbf{R}$ have a continuous derivative on J. If f is continuous on $I := \varphi(J)$, then*

$$\int_\alpha^\beta f((\varphi(t))\varphi'(t)\; dt = \int_{\varphi(\alpha)}^{\varphi(\beta)} f(x)\; dx. \tag{4}$$

Proof. Let $c := \varphi(\alpha)$ and $d := \varphi(\beta)$; we know from Theorem 4.6.8 that $I := \varphi(J)$ is an interval containing the interval with end points c,d. Since f is continuous on I, we can define $F : I \to \boldsymbol{R}$ by

$$F(u) := \int_c^u f(x)\,dx \qquad \text{for} \quad u \in I.$$

Now consider the function $H : J \to \boldsymbol{R}$ defined by $H(t) := F((\varphi(t))$ for $t \in J$. It follows from the Chain Rule 5.1.6 that $H'(t) = F'((\varphi(t))\varphi'(t)$ and from the Differentiation Theorem 6.3.4 that $F'(u) = f(u)$, whence

$$H'(t) = f((\varphi(t))\varphi'(t) \qquad \text{for} \quad t \in J.$$

If we apply the Fundamental Theorem and use the fact that $H(\alpha) = F((\varphi(\alpha)) = F(c) = 0$, we conclude that

$$\int_\alpha^\beta f((\varphi(t))\varphi'(t)\,dt = H(\beta) - H(\alpha) = H(\beta).$$

On the other hand we also have

$$H(\beta) = F((\varphi(\beta)) = F(d) = \int_c^d f(x)\,dx.$$

Formula (4) follows from these two equations. Q.E.D.

6.3.9 Example. Consider the integral $\int_1^5 \frac{t}{1+t^2}\,dt$.

If we let $f(x) := \frac{1}{2}(1+x)^{-1}$ and $\varphi(t) := t^2$, then since $\varphi'(t) = 2t$, the integrand has the form $(f\circ\varphi(t))\varphi'(t)$. Since $\varphi([1,5]) = [1,25]$ and since f is continuous for $x > -1$, it is clear that f is continuous on $\varphi([1,5])$. Then the First Substitution Theorem 6.3.8 implies that the above integral is equal to

$$\begin{aligned} \frac{1}{2}\int_1^{25} \frac{1}{1+x}\,dx &= \frac{1}{2}\log(1+x)\Big|_1^{25} \\ &= \frac{1}{2}[\log 26 - \log 2] = \frac{1}{2}\log 13. \end{aligned}$$

6.3.10 Second Substitution Theorem. *Let $J := [\alpha,\beta]$ and let $\varphi : J \to \boldsymbol{R}$ have a continuous derivative such that $\varphi'(t) \neq 0$ for $t \in J$. Let $I := \varphi(J)$ and let $\psi : I \to \boldsymbol{R}$ be the function inverse to φ. If $f : I \to \boldsymbol{R}$ is continuous on I, then*

$$\text{(5)} \qquad \int_\alpha^\beta f(\varphi(t))\,dt = \int_{\varphi(\alpha)}^{\varphi(\beta)} f(x)\psi'(x)\,dx.$$

Proof. Since $\varphi'(t) \neq 0$ for $t \in J$, it follows that φ is strictly monotone. (See Theorem 5.2.7 and accompanying remarks.) Hence $\psi = \varphi^{-1}$ is defined on I. Moreover, by Theorem 5.1.9 it follows that ψ' exists and is continuous on I.

Now let $G : J \to \boldsymbol{R}$ be an antiderivative of the continuous function $f \circ \varphi$. Then $G \circ \psi$ is differentiable on I and

$$\begin{aligned}(G \circ \psi)'(x) &= G'(\psi(x))\psi'(x) = (f \circ \varphi)(\psi(x))\psi'(x) \\ &= f(\varphi \circ \psi(x))\psi'(x) = f(x)\psi'(x)\end{aligned}$$

for all $x \in I$. It follows from the Fundamental Theorem that

$$\begin{aligned}\int_{\varphi(\alpha)}^{\varphi(\beta)} f(x)\psi'(x)\,dx &= (G \circ \psi)(\varphi(\beta)) - (G \circ \psi)(\varphi(\alpha)) \\ &= G(\beta) - G(\alpha).\end{aligned}$$

On the other hand, since $G' = f \circ \varphi$, it also follows from the Fundamental Theorem that

$$\int_\alpha^\beta f(\varphi(t))\,dt = G(\beta) - G(\alpha).$$

If we combine the last two equations, we obtain (5). Q.E.D.

6.3.11 Example. Consider the integral $\int_1^4 1/(1 + \sqrt{t})\,dt$.

It is clear that the integrand has the form $f \circ \varphi(t)$ if we take $f(x) := 1/(1+x)$ and $\varphi(t) := \sqrt{t}$ for $t \in [1,4]$. Now $\varphi'(t) = \frac{1}{2}t^{-1/2} > 0$ for $t \in [1,4]$ and it is evident that $\psi(x) := x^2$ is the function inverse to φ on $[1,2] = \varphi([1,4])$. Since f is continuous on $[1,2]$, the Second Substitution Theorem can be applied to give

$$\begin{aligned}\int_1^4 \frac{1}{1+\sqrt{t}}\,dt &= \int_1^2 \frac{1}{1+x}\,2x\,dx \\ &= 2\int_1^2 \frac{x+1-1}{1+x}\,dx = 2\int_1^2 \{1 - (1+x)^{-1}\}\,dx \\ &= 2[x - \log(1+x)]\Big|_1^2 = 2[1 - \log \tfrac{3}{2}].\end{aligned}$$

6.3.12 Mean Value Theorem for Integrals. *Let f be continuous on $I := [a,b]$ and let p be integrable on I and such that $p(x) \geq 0$ for all $x \in I$. Then there exists a point $c \in I$ such that*

$$\int_a^b f(x)p(x)\,dx = f(c)\int_a^b p(x)\,dx. \tag{6}$$

Proof. It follows from Theorems 6.2.11 and 6.2.16 that the product fp is integrable. Let $m := \inf f(I)$ and $M := \sup f(I)$, so that $mp(x) \leqslant f(x)p(x) \leqslant Mp(x)$ for all $x \in I$. Therefore, by Corollary 6.2.4 we conclude that

$$m\int_a^b p \leqslant \int_a^b fp \leqslant M\int_a^b p.$$

If $\int_a^b p = 0$, we choose c arbitrarily; otherwise, we have

$$(7) \qquad m \leqslant \frac{\int_a^b fp}{\int_a^b p} \leqslant M.$$

Since f is continuous, it follows from Corollary 4.6.7 to the Bolzano Intermediate Value Theorem that there is a point $c \in I$ where $f(c)$ is equal to the quotient in inequality (7). Q.E.D.

6.3.13 Corollary. *If f is continuous on $I := [a,b]$, then there exists $c \in I$ such that*

$$(8) \qquad \int_a^b f = f(c)\,(b-a).$$

Proof. Take $p(x) := 1$ in (6). Q.E.D.

Integral Form for the Remainder

The reader will recall Taylor's Theorem 5.4.1, which enables one to calculate the value $f(b)$ in terms of the values $f(a), f'(a), \ldots, f^{(n)}(a)$ and a remainder term that involves $f^{(n+1)}$ evaluated at a point between a and b. For some applications it is more convenient to be able to express the remainder term as an integral involving $f^{(n+1)}$.

6.3.14 Taylor's Theorem. *Suppose that f and its derivatives $f', f'', \ldots, f^{(n)}, f^{(n+1)}$ are continuous on $[a,b]$ to $\mathbf{R}$. Then:*

$$(9) \qquad f(b) = f(a) + \frac{f'(a)}{1!}(b-a) + \cdots + \frac{f^{(n)}(a)}{n!}(b-a)^n + R_n,$$

where the remainder is given by

$$(10) \qquad R_n := \frac{1}{n!}\int_a^b (b-t)^n f^{(n+1)}(t)\,dt.$$

Proof. Integrate R_n by parts to get

$$R_n = \frac{1}{n!}\left\{(b-t)^n f^{(n)}(t)\Big|_{t=a}^{t=b} + n\int_a^b (b-t)^{n-1} f^{(n)}(t)\,dt\right\}$$

$$= -\frac{f^{(n)}(a)}{n!}(b-a)^n + \frac{1}{(n-1)!}\int_a^b (b-t)^{n-1} f^{(n)}(t)\,dt.$$

If we continue to integrate by parts in this way, we obtain formula (9). Q.E.D.

Instead of (10), it is often convenient to make the change of variable $t = (1-s)a + sb$ for $s \in [0,1]$ to obtain the formula

(11) $$R_n = \frac{(b-a)^{n+1}}{n!}\int_0^1 (1-s)^n f^{(n+1)}(a + (b-a)s)\,ds.$$

Exercises for Section 6.3

1. Let f be continuous on $I := [a,b]$ and let $H : I \to \boldsymbol{R}$ be defined by
$$H(x) := \int_x^b f \quad \text{for} \quad x \in I.$$
Find $H'(x)$ for $x \in I$.
2. Let $I := [a,b]$ and let $f : I \to \boldsymbol{R}$ be continuous on I. Let $J := [c,d]$ and let $v : J \to \boldsymbol{R}$ be differentiable on J and satisfy $v(J) \subseteq I$. Show that if $G : J \to \boldsymbol{R}$ is defined by
$$G(x) := \int_a^{v(x)} f \quad \text{for} \quad x \in J,$$
then $G'(x) = (f \circ v)(x)v'(x)$ for all $x \in J$.
3. Find F', when F is defined on $I := [0,1]$ as follows:

(a) $F(x) := \int_0^x \sin(t^2)\,dt$; (b) $F(x) := \int_0^{x^2} (1 + t^3)^{-1}\,dt$;

(c) $F(x) := \int_{x^2}^x \sqrt{1 + t^2}\,dt$; (d) $F(x) := \int_0^{\sin x} \cos t\,dt$.

4. Let $F : [0,3] \to \boldsymbol{R}$ be defined by $f(x) := x$ for $0 \leqslant x < 1$, $f(x) := 1$ for $1 \leqslant x < 2$, and $f(x) := x$ for $2 \leqslant x \leqslant 3$. Obtain an explicit expression for $F(x) := \int_0^x f$ as a function of x. Where is F differentiable? Evaluate $F'(x)$ at all points where F is differentiable.
5. One approach to the logarithm is to define $L : (0,\infty) \to \boldsymbol{R}$ by
$$L(x) := \int_1^x \frac{1}{t}\,dt \quad \text{for} \quad x > 0.$$

Verify the following properties of L.
(a) $L'(x) = 1/x$ for $x > 0$.
(b) $L(xy) = L(x) + L(y)$ for $x, y > 0$.
(c) $L(x^n) = nL(x)$ for $x > 0$, $n \in \boldsymbol{N}$.

6. Let $f : \boldsymbol{R} \to \boldsymbol{R}$ be continuous and let $\alpha > 0$. Define $g : \boldsymbol{R} \to \boldsymbol{R}$ by

$$g(x) := \int_{x-\alpha}^{x+\alpha} f \qquad \text{for} \quad x \in \boldsymbol{R}.$$

Show that g is differentiable and find g'.

7. Let $I := [0,1]$ and let $f : I \to \boldsymbol{R}$ be continuous. Suppose that

$$\int_0^x f = \int_x^1 f \qquad \text{for all} \quad x \in I.$$

Prove that $f(x) = 0$ for all $x \in I$.

8. Suppose that $f : [0,\infty) \to \boldsymbol{R}$ is continuous and that $f(x) \neq 0$ for all $x \geqslant 0$. If we have

$$(f(x))^2 = 2\int_0^x f \qquad \text{for all} \quad x > 0,$$

show that $f(x) = x$ for all $x \geqslant 0$.

9. Let $I := [a,b]$ and suppose that $f : I \to \boldsymbol{R}$ is continuous and $f(x) \geqslant 0$ for $x \in \boldsymbol{R}$. If $M := \sup\{f(x) : x \in I\}$, show that the sequence

$$\left(\left[\int_a^b f(x)^n \, dx\right]^{1/n}\right)$$

converges to M.

10. Evaluate the following integrals; justify each step.

(a) $\displaystyle\int_0^1 x\sqrt{1 + x^2}\, dx$; (b) $\displaystyle\int_0^1 \frac{\sqrt{x}}{1 + \sqrt{x}}\, dx$;

(c) $\displaystyle\int_1^2 \frac{\sqrt{x - 1}}{x}\, dx$; (d) $\displaystyle\int_1^4 \frac{\sqrt{1 + \sqrt{x}}}{\sqrt{x}}\, dx$.

11. Evaluate the following integrals; justify each step.

(a) $\displaystyle\int_0^2 \frac{1}{2 + \sqrt{x}}\, dx$; (b) $\displaystyle\int_1^5 x\sqrt{2x + 3}\, dx$;

(c) $\displaystyle\int_1^3 \frac{1}{x\sqrt{x + 1}}\, dx$; (d) $\displaystyle\int_1^4 \frac{\sqrt{x}}{x(x + 4)}\, dx$.

12. Let $f : [0,1] \to \boldsymbol{R}$ be continuous and let $g_n(x) := f(x^n)$ for $x \in [0,1]$, $n \in \boldsymbol{N}$. Prove that the sequence $(\int_0^1 g_n)$ converges to $f(0)$.

13. Let $I := [a,b]$ and let $g : I \to \boldsymbol{R}$ be continuous on I. Suppose that there exists $K > 0$ such that

$$|g(x)| \leq K \int_a^x |g| \qquad \text{for all} \quad x \in I.$$

Show that $g(x) = 0$ for all $x \in I$.

14. Let $I := [a,b]$ and let f,g be continuous on I and such that

$$\int_a^b f = \int_a^b g.$$

Prove that there exists $c \in I$ such that $f(c) = g(c)$.

15. Show that, under the hypotheses of 6.3.14, the remainder can be expressed in the form

$$R_n = \frac{(b-a)^{n+1}}{n!}(1-\theta)^n f^{(n+1)}(a + \theta(b-a))$$

for some number $\theta \in [0,1]$. [This form of the remainder is due to Cauchy.]

SECTION 6.4 The Integral as a Limit

In introductory calculus courses, the Riemann integral is often introduced in terms of limits of certain sums known as "Riemann sums", rather than in terms of upper and lower integrals as we have done. Fortunately, it is the case that these limiting approaches lead to the same concept of integration. One of the purposes of this section is to establish the equivalence of these theories of the Riemann integral. We shall also present a second version of the Fundamental Theorem of Integral Calculus. The section ends with a brief discussion of "improper" integrals.

6.4.1 Definition. Let $I := [a,b]$ and let $f : I \to \boldsymbol{R}$ be a bounded function. If $P := (x_0, x_1, \ldots, x_n)$ is a partition of I and if $(\xi_1, \xi_2, \ldots, \xi_n)$ are numbers such that $x_{k-1} \leq \xi_k \leq x_k$ for $k = 1, 2, \ldots, n$, then the sum

$$(1) \qquad S(P;f) := \sum_{k=1}^{n} f(\xi_k)\,(x_k - x_{k-1})$$

is called a **Riemann sum** for f, corresponding to the partition P and the **intermediate points** ξ_k.

Note. Although we shall not introduce a notation to indicate the dependence of the Riemann sums $S(P;f)$ on the intermediate points ξ_k, the reader should not forget this dependence. Indeed, there may be infinitely many values that one can obtain as a Riemann sum corresponding to P, by choosing different intermediate values.

For a positive function f on I, the sum (1) can be interpreted as the area of the union of rectangles with widths $x_k - x_{k-1}$ and heights $f(\xi_k)$. (See Figure 6.4.1.) If the partition P is very fine, it is reasonable to expect that the Riemann sum (1) will yield an approximation to the "area under the graph of f" and thus be close in value to the integral of f, provided this integral exists.

Indeed, it is clear that for any partition P of I, and for any choice of intermediate points ξ_k $(k = 1, \ldots, n)$, then

$$m_k \leq f(\xi_k) \leq M_k \qquad \text{for} \quad k = 1, \ldots, n,$$

where $m_k := \inf f(I_k)$ and $M_k := \sup f(I_k)$, $I_k := [x_{k-1}, x_k]$, so that it follows that

$$\sum_{k=1}^{n} m_k(x_k - x_{k-1}) \leq \sum_{k=1}^{n} f(\xi_k)(x_k - x_{k-1}) \leq \sum_{k=1}^{n} M_k(x_k - x_{k-1}).$$

Hence we have

$$L(P;f) \leq S(P;f) \leq U(P;f).$$

That is, any Riemann sum for f corresponding to P lies between the lower sum and the upper sum of f corresponding to P, no matter how we choose the intermediate points ξ_k.

We note that if the lower and upper bounds m_k and M_k are attained on $[x_{k-1}, x_k]$, for all $k = 1, \ldots, n$, then the lower and upper sums are equal to Riemann sums for particular choices of the intermediate points. However, in general, the upper and lower sums are not Riemann sums (since m_k and M_k may not be attained by f), even though they can be seen to be arbitrarily close to Riemann sums for carefully chosen intermediate points. (See Exercise 6.4.5.)

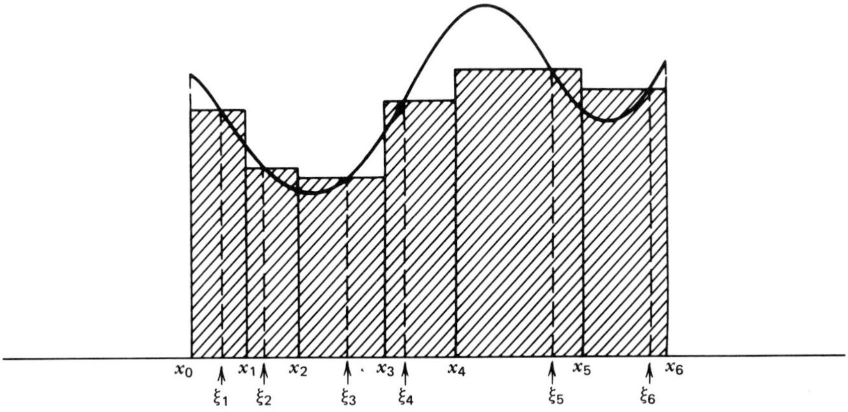

FIGURE 6.4.1 ***S(P;f)*, a Riemann sum.**

The question arises: Can we obtain the integral

$$\int_a^b f$$

as "the limit" of the Riemann sums $S(P;f)$ for f? In order to make this question more precise, we need to make more explicit what we mean by "the limit", since it should be clear to the reader that we have considered no limiting process thusfar that will accommodate the convergence of Riemann sums. However, we shall now show that there are at least two possible definitions of limit that can be given, and that the answer in both cases is affirmative.

6.4.2 Theorem. *Let $I := [a,b]$ and let $f : I \to \mathbf{R}$ be integrable on I in the sense of Definition* 6.1.6. *Then, if $\varepsilon > 0$ is given, there exists a partition P_ε of I such that if P is any partition that is a refinement of P_ε and if $S(P;f)$ is any Riemann sum for f, then*

$$\left| S(P;f) - \int_a^b f \right| < \varepsilon. \tag{2}$$

Proof. If f is integrable and $\varepsilon > 0$, then, by Riemann's Criterion 6.1.8, there is a partition P_ε of I such that $U(P_\varepsilon;f) - L(P_\varepsilon;f) < \varepsilon$. Moreover, if P is any partition such that $P \supseteq P_\varepsilon$ then it follows from Lemmas 6.1.2 and 6.1.3 that

$$L(P_\varepsilon;f) \leqslant L(P;f) \leqslant U(P;f) \leqslant U(P_\varepsilon;f)$$

so that $U(P;f) - L(P;f) < \varepsilon$. But if $S(P;f)$ is a Riemann sum for f corresponding to P and to any choice of intermediate points, then we have

$$L(P;f) \leqslant S(P;f) \leqslant U(P;f).$$

Moreover, we also have

$$L(P;f) \leqslant \int_a^b f \leqslant U(P;f).$$

It therefore follows that

$$\left| S(P;f) - \int_a^b f \right| \leqslant U(P;f) - L(P;f) < \varepsilon,$$

as asserted. Q.E.D.

The reader may ask if the converse to the above theorem is true. The answer is yes; in fact, one can prove the following theorem.

6.4.3 Theorem. *Let $I := [a,b]$ and let $f : I \to \mathbf{R}$ be a bounded function. Suppose there exists a number A such that for every $\varepsilon > 0$ there exists a partition P_ε such that if $P \supseteq P_\varepsilon$ and if $S(P;f)$ is any Riemann sum for f corresponding to P, then $|S(P;f) - A| < \varepsilon$. Then f is integrable on I in the sense of Definition* 6.1.6 *and* $A = \int_a^b f$.

A proof of this theorem can be based on the observation that given $\varepsilon > 0$ and a partition Q of I, there exists a Riemann sum for Q that is within ε of the upper sum $U(Q;f)$, and there exists another Riemann sum for Q that is within ε of the lower sum $L(Q;f)$. Since the proof is not particularly difficult, it will be left as an instructive exercise for the reader.

There is another sense in which the integral can be obtained as a limit of the Riemann sums.

6.4.4 Definition. If $I := [a,b]$ and if $P := (x_0, x_1, \ldots, x_n)$ is a partition of I, then the **norm** (or the **mesh**) **of** P, denoted by $||P||$, is defined by

$$||P|| := \sup \{x_1 - x_0,\ x_2 - x_1,\ \ldots,\ x_n - x_{n-1}\}.$$

In other words, $||P||$ is the maximum length of the subintervals into which I is divided by the partition P.

Note. It should be quite clear that two very different partitions of I can have the same norm. It is also clear that if $P \subseteq Q$, then $||Q|| \leqslant ||P||$; however, it does *not* follow that if $||Q|| \leqslant ||P||$, then $P \subseteq Q$. (Why?)

We now show that if f is integrable, then the integral of f is the limit of the Riemann sums as $||P|| \to 0$ in the sense of the following theorem.

6.4.5 Darboux's Theorem. *Let $I := [a,b]$ and let $f : I \to \mathbf{R}$ be integrable on I in the sense of Definition* 6.1.6. *Then, if $\varepsilon > 0$ is given, there is a $\delta(\varepsilon) > 0$ such that if P is any partition of I such that $||P|| < \delta(\varepsilon)$ and if $S(P;f)$ is any corresponding Riemann sum for f, then*

$$\left| S(P;f) - \int_a^b f \right| < \varepsilon. \tag{3}$$

Proof. Since f is integrable and $\varepsilon > 0$, it follows from Riemann's Criterion 6.1.8 that there is a partition $P_\varepsilon := (x_0, x_1, \ldots, x_n)$ such that $U(P_\varepsilon;f) - L(P_\varepsilon;f) < \varepsilon/3$. Moreover, if $P \supseteq P_\varepsilon$ then it will also be true that $U(P;f) - L(P;f) < \varepsilon/3$. Let $M := \sup \{|f(x)| : x \in I\}$ and let $\delta(\varepsilon) := \varepsilon/12nM$, where $n + 1$ is the number of points in P_ε. Now let $Q := (y_0, y_1, \ldots, y_m)$ be a partition of I with $||Q|| < \delta(\varepsilon)$ and let $Q^* := Q \cup P_\varepsilon$. It follows that $Q^* \supseteq P_\varepsilon$ and Q^* has at most $n - 1$ more points than Q; namely, those points among $x_1, \ldots, x_{n-1}$ that belong to P_ε but are not in Q.

We wish to compare $U(Q;f)$ and $U(Q^*;f)$. Since $Q^* \supseteq Q$ it follows that we have $U(Q;f) - U(Q^*;f) \geq 0$. If we write $Q^* = (z_0, z_1, \ldots, z_p)$, then it can be seen that $U(Q;f) - U(Q^*;f)$ can be written as the sum of at most $2(n-1)$ terms of the form

$$(M_j - M_k^*)(z_k - z_{k-1}),$$

where M_j is the supremum of f over the jth subinterval in Q and M_k^* is the supremum of f over the kth subinterval in Q^*. Since $|M_j - M_k^*| \leq 2M$ and $|z_k - z_{k-1}| \leq ||Q^*|| \leq ||Q|| \leq \delta(\varepsilon)$, we deduce that

$$0 \leq U(Q;f) - U(Q^*;f) \leq 2(n-1)2M\delta(\varepsilon) < \varepsilon/3.$$

Hence we have

$$U(Q;f) < U(Q^*;f) + \varepsilon/3.$$

An exactly similar argument implies that

$$L(Q^*;f) - \varepsilon/3 \leq L(Q;f).$$

If $S(Q;f)$ is any Riemann sum for f, then

$$L(Q;f) \leq S(Q;f) \leq U(Q;f);$$

moreover, we have

$$L(Q;f) \leq \int_a^b f \leq U(Q;f).$$

On the other hand, since $Q^* \supseteq P_\varepsilon$, it follows that $U(Q^*;f) < L(Q^*;f) + \varepsilon/3$. Consequently we infer that both $S(Q;f)$ and $\int_a^b f$ lie in the open interval

$$\big(L(Q^*;f) - \varepsilon/3,\ L(Q^*;f) + 2\varepsilon/3\big).$$

Therefore the inequality (3) is established. Q.E.D.

The converse to the preceding theorem is also true.

6.4.6 Theorem. *Let $I := [a,b]$ and let $f : I \to \mathbf{R}$ be a bounded function. Suppose there exists a number B such that for every $\varepsilon > 0$ there is a $\delta(\varepsilon) > 0$ such that if P is any partition of I with $||P|| < \delta(\varepsilon)$ and $S(P;f)$ is any corresponding Riemann sum, then $|S(P;f) - B| < \varepsilon$. Then f is integrable on I in the sense of Definition* 6.1.6 *and $B = \int_a^b f$.*

The proof of this result is considerably simpler than that of Darboux's Theorem 6.4.5; we leave it as an exercise for the reader.

Cauchy's Fundamental Theorem of Calculus

The Fundamental Theorem of Integral Calculus 6.3.5 provides a method of evaluating the integral of a continuous function f on an interval $I := [a,b]$ by means of an antiderivative F of f on I. However, as long as f is integrable and possesses an antiderivative, the continuity of the integrand f is not essential. The following formulation of the Fundamental Theorem of Calculus is due to Cauchy.

6.4.7 Theorem. *If f is integrable on $I := [a,b]$ and if $F'(x) = f(x)$ for all $x \in I$, then*

$$F(b) - F(a) = \int_a^b f.$$

Proof. Let $P = (x_0, x_1, \ldots, x_n)$ be any partition of I. It follows from the Mean Value Theorem 5.2.4 that there exist points $\xi_k \in [x_{k-1}, x_k]$ for $k = 1, 2, \ldots, n$ such that

$$F(x_k) - F(x_{k-1}) = F'(\xi_k)(x_k - x_{k-1}) = f(\xi_k)(x_k - x_{k-1}).$$

Therefore we have

$$F(b) - F(a) = \sum_{k=1}^{n} (F(x_k) - F(x_{k-1})) = \sum_{k=1}^{n} f(\xi_k)(x_k - x_{k-1}).$$

Hence, for any partition P, the Riemann sum $S(P;f)$ corresponding to the intermediate points $\xi_1, \xi_2, \ldots, \xi_n$ satisfies $S(P;f) = F(b) - F(a)$, and therefore

$$L(P;f) \leq F(b) - F(a) \leq U(P;f).$$

Since f is assumed to be integrable on I, it follows that

$$F(b) - F(a) = \int_a^b f. \qquad \text{Q.E.D.}$$

6.4.8 Example. Let F be defined on $I := [-2,2]$ by $F(x) := x^2 \sin(\pi/x)$ for $x \neq 0$ and $F(0) := 0$. Then F is differentiable at each point of I and the derivative is given by $F'(x) = 2x \sin(\pi/x) - \pi \cos(\pi/x)$ for $x \neq 0$ and $F'(0) = 0$. The function $f := F'$ is integrable on I, but it is discontinuous at $x = 0$ so that the earlier version of the Fundamental Theorem (see Theorem 6.3.5) does not apply. However, the hypotheses of Theorem 6.4.7 are satisfied and we may conclude that

$$\int_{-2}^{2} f(x)\,dx = F(2) - F(-2)$$

$$= 4\sin(\tfrac{1}{2}\pi) - 4\sin(-\tfrac{1}{2}\pi) = 8.$$

Improper Integrals

In the preceding discussion of the integral, there were two standing assumptions: the functions were required to be bounded and the domain of integration was required to be a bounded interval. If either of these conditions is not satisfied, then the foregoing integration theory does not apply without some change. Since there are important instances when it is desirable to relax one or both of these requirements, we shall briefly indicate the alterations that need to be made. A more thorough account can be found in *The Elements of Real Analysis*, p. 257–285.

We consider first the case of an unbounded function.

6.4.9 Definition. Let $f : (a,b] \to \boldsymbol{R}$ be such that f is integrable on the interval $[c,b]$ for every c in $(a,b]$. Suppose there is a real number A such that for each $\varepsilon > 0$ there exists $\delta > 0$ such that if $a < c < a + \delta$, then we have $|A - \int_c^b f| < \varepsilon$. In this case we say that A is the **improper integral** of f on $(a,b]$ and we denote the value A by

$$\int_a^b f \quad \text{or} \quad \int_a^b f(x)\,dx.$$

It is clear that the number A is the right-hand limit:

$$A = \lim_{c \to a+} \int_c^b f.$$

Thus it would be natural to denote the improper integral of f on $(a,b]$ by

$$A = \int_{a+}^b f.$$

However, it is customary *not* to write the plus sign on the lower limit of the integral.

Remark. If f is bounded on $[a,b]$ and if f is integrable on $[c,b]$ for every c satisfying $a < c \leq b$, then it follows that f is actually integrable on $[a,b]$. (See Exercise 6.1.7.) Moreover the value of the limit in the preceding definition coincides with the value of the integral of f over $[a,b]$. Thus the notion of improper integral is superfluous in this case. Note that f can be assigned an arbitrary value at a without affecting its integrability, or the value of its integral.

6.4.10 Examples. (a) The function $f : [0,1] \to \boldsymbol{R}$ defined by $f(x) := 1/\sqrt{x}$ for $0 < x \leq 1$, $f(0) := 0$, is unbounded and hence not integrable on $[0,1]$. However, for each c satisfying $0 < c < 1$, we have

$$\int_c^1 f = \int_c^1 \frac{1}{\sqrt{x}}\,dx = 2\sqrt{x}\,\bigg|_c^1 = 2(1 - \sqrt{c}).$$

Hence, if we let $c \to 0+$, we obtain that the improper integral of f on $(0,1]$ is

$$\int_0^1 f = \lim_{c\to 0} 2(1 - \sqrt{c}) = 2.$$

(b) If $g(x) := 1/x$ for $0 < x \leq 1$, then for each c with $0 < c < 1$, we have

$$\int_c^1 g = \int_c^1 \frac{1}{x}\,dx = \log 1 - \log\, c = -\log c.$$

Since $\log c$ is unbounded as $c \to 0+$, the improper integral of g on $(0,1]$ does *not* exist.

(c) Let $h(x) := x^{-\alpha}$ for x in $(0,1]$ where $\alpha > 0$, $\alpha \neq 1$. (If α is rational this function was discussed in Example 5.1.10(c); if α is irrational, it will be discussed in Section 7.3.) For $0 < c < 1$, we have

$$\int_c^1 h = \int_c^1 x^{-\alpha}\,dx = \frac{1}{1-\alpha}(1 - c^{1-\alpha}).$$

If $0 < \alpha < 1$, then $c^{1-\alpha} \to 0$ as $c \to 0+$ so that the improper integral of h exists. However, if $\alpha > 1$, then h does not have an improper integral.

The preceding definition involves an "impropriety" at the left end point of an interval. Analogous behavior at the right end point is treated in a similar manner. If a function f is unbounded in every neighborhood of an interior point p of $[a,b]$, then the improper integral of f over $[a,b]$ is said to exist if and only if the improper integrals of f over *both* intervals $[a,p)$ and $(p,b]$ exist, and in this case we define the improper integral of f over $[a,b]$ to be the sum

$$\int_a^p f + \int_p^b f.$$

For example, if $f(x) := 1/x^2$ for $x \in (0,1]$, $x \neq 0$, and $f(x) := 0$ for $x \in [-1,0]$, the improper integral of f over $[-1,1]$ does not exist because the improper integral of f over $(0,1]$ fails to exist.

We now turn to the case of improper integrals over unbounded intervals.

6.4.11 Definition. Let $a \in \mathbf{R}$ and let $f : [a,\infty) \to \mathbf{R}$ be such that for every $c > a$ the function f is integrable over the interval $[a,c]$. Suppose there is a number A such that for every $\varepsilon > 0$ there exists a real number M such that if $c > M$, then

$$\left| A - \int_a^c f \right| < \varepsilon .$$

In this case we say that A is the **improper integral** of f over (a,∞) and we denote the value A by

$$\int_a^\infty f \quad \text{or} \quad \int_a^\infty f(x)\, dx .$$

In other words, the improper integral of f is given by

$$\int_a^\infty f = \lim_{c\to\infty} \int_a^c f,$$

if the limit exists. It is sometimes said that the improper integral of f "converges" if the limit exists, and "diverges" otherwise.

6.4.12 Examples. (a) If $a > 0$ and if $f(x) := 1/x$ for x in $[a,\infty)$, then for each $c > a$ we have

$$\int_a^c \frac{1}{x}\, dx = \log c - \log a .$$

Since $\log c$ is unbounded as $c \to \infty$, the improper integral of f over (a,∞) does not exist.

(b) Let $\alpha > 0$, $\alpha \neq 1$, and let $g(x) := x^{-\alpha}$ for x in $[1,\infty)$. For any $c > 1$, we have

$$\int_1^c g = \int_1^c x^{-\alpha}\, dx = \frac{1}{\alpha - 1}\,(1 - c^{-\alpha+1}).$$

If $\alpha > 1$, then $c^{-\alpha+1} \to 0$ as $c \to \infty$ so that the improper integral of g over $(1,\infty)$ exists and has value $1/(\alpha - 1)$. If $0 < \alpha < 1$, then $c^{-\alpha+1}$ is unbounded as $c \to \infty$ and the improper integral of g over $[1,\infty)$ does not exist.

Improper integrals over intervals of the form $(-\infty,b]$ are treated in a similar manner. The case of a function f defined on $(-\infty,\infty)$ is treated by considering the improper integrals of f over both $(-\infty,b]$ and $[b,\infty)$ for any fixed $b \in \boldsymbol{R}$. If both of these improper integrals exist, then the improper integral of f over $(-\infty,\infty)$ exists and is defined to be

$$\int_{-\infty}^\infty f := \int_{-\infty}^b f + \int_b^\infty f.$$

For example,

$$\int_{-\infty}^{\infty} \frac{1}{1+x^2}\,dx = \int_{-\infty}^{0} \frac{1}{1+x^2}\,dx + \int_{0}^{\infty} \frac{1}{1+x^2}\,dx$$

$$= -\lim_{c\to-\infty} (\text{Arctan } c) + \lim_{c\to\infty} (\text{Arctan } c)$$

$$= -\left(-\frac{\pi}{2}\right) + \frac{\pi}{2} = \pi.$$

Exercises for Section 6.4

1. Let f be integrable on $[0,1]$. Show that

$$\lim_{n\to\infty}\left(\frac{1}{n}\sum_{k=1}^{n} f(k/n)\right) = \int_0^1 f.$$

2. Use Exercise 1 to express each of the following limits as an integral:

(a) $\displaystyle\lim_{n\to\infty}\left(\sum_{k=1}^{n}\frac{1}{n+k}\right)$; (b) $\displaystyle\lim_{n\to\infty}\left(\sum_{k=1}^{n}\frac{k}{n^2+k^2}\right)$.

3. Show that $\displaystyle\lim_{n\to\infty}\left(\sum_{k=1}^{n}\frac{n}{k^2+n^2}\right) = \pi/4.$

4. Use Exercise 1 to evaluate:

(a) $\displaystyle\lim\left(\frac{1}{n^8}\sum_{k=1}^{n} k^7\right)$; (b) $\displaystyle\lim\left(\frac{1}{n}\sum_{k=1}^{n}\sin\frac{k\pi}{n}\right)$.

5. Let f be a bounded function defined on $[a,b]$ and let $P := (x_0, x_1, \ldots, x_n)$ be a partition of $[a,b]$. Given $\varepsilon > 0$ show that there exist intermediate points $(\xi_1, \xi_2, \ldots, \xi_n)$ such that

$$U(P;f) - \sum_{k=1}^{n} f(\xi_k)(x_k - x_{k-1}) < \varepsilon.$$

Similarly, there exist intermediate points $(\eta_1, \eta_2, \ldots, \eta_n)$ such that

$$\sum_{k=1}^{n} f(\eta_k)(x_k - x_{k-1}) - L(P;f) < \varepsilon.$$

6. Let $f(x) := x$ for $x \in [0,b]$ and let $P := (x_0, x_1, \ldots, x_n)$ be a partition of $[0,b]$. Show that for intermediate points $\xi_k := \frac{1}{2}(x_k + x_{k-1})$, $k = 1, \ldots, n$, the corresponding Riemann sum satisfies $S(P;f) = \frac{1}{2}b^2$. Conclude that $\int_0^b x\,dx = \frac{1}{2}b^2$.

7. Let $g(x) := x^p$ for $x \in [0,b]$ where $p \in \mathbf{N}$, $p \geqslant 2$, let $P := (x_0, x_1, \ldots, x_n)$ be a partition of $[0,b]$, and let

$$\xi_k := [(x_k^p + x_k^{p-1}x_{k-1} + x_k^{p-2}x_{k-1}^2 + \cdots + x_{k-1}^p)/(p-1)]^{1/p}$$

for $k = 1, \ldots, n$. Show that $\xi_k \in [x_{k-1}, x_k]$ and find the value of the Riemann sum $S(P;g)$ for these intermediate points. Use this result to evaluate $\int_0^b x^p\,dx$. [*Hint*: $u^{p+1} - v^{p+1} = (u-v)(u^p + u^{p-1}v + u^{p-2}v^2 + \ldots + v^p)$.]

8. State why each integral is improper, and determine whether it is convergent or divergent. If it is convergent, compute its value.

(a) $\displaystyle\int_0^1 \log x\,dx$; (b) $\displaystyle\int_1^2 \frac{1}{x\log x}\,dx$;

(c) $\displaystyle\int_1^2 \frac{x}{\sqrt{x-1}}\,dx$; (d) $\displaystyle\int_0^1 x\log x\,dx$.

9. State why each integral is improper, and determine whether it is convergent or divergent. If it is convergent, compute its value.

(a) $\displaystyle\int_0^\infty e^{-x}\,dx$; (b) $\displaystyle\int_0^\infty \frac{1}{\sqrt{e^x}}\,dx$.

(c) $\displaystyle\int_2^\infty \frac{\log x}{x}\,dx$; (d) $\displaystyle\int_2^\infty \frac{1}{x(\log x)^2}\,dx$.

10. State why each integral is improper, and determine whether it is convergent or divergent. If it is convergent, compute its value.

(a) $\displaystyle\int_{-1}^1 \frac{1}{\sqrt{1-x^2}}\,dx$; (b) $\displaystyle\int_{-\infty}^\infty e^{-x}\,dx$;

(c) $\displaystyle\int_0^\infty \frac{1}{x^2}\,dx$; (d) $\displaystyle\int_0^\infty \frac{1}{\sqrt{x}\,(x+4)}\,dx$.

SECTION 6.5 Approximate Integration

The Fundamental Theorem of Calculus provides us with an easy method of evaluating an integral, but only if we can find an antiderivative for the integrand. This method is of no use at all when we cannot find one. There are a variety of techniques of approximating the value of an integral when an antiderivative cannot be found, and in this section we shall discuss several of the most elementary and most useful methods. For convenience, we shall limit our attention to continuous integrands.

One very elementary procedure to obtain quick estimates of the value of an integral $\int_a^b f(x)\,dx$ is based on the observation that if $g(x) \leqslant f(x) \leqslant h(x)$ for $x \in [a,b]$, then

$$\int_a^b g(x)\,dx \leqslant \int_a^b f(x)\,dx \leqslant \int_a^b h(x)\,dx.$$

If the integrals of g and h can be calculated, then we have an estimate for the value of the integral of f.

For example, suppose we wish to estimate the value of

$$\int_0^1 e^{-x^2}\,dx.$$

It is easy to show that $e^{-x} \leqslant e^{-x^2} \leqslant 1$ for $x \in [0,1]$, so that

$$\int_0^1 e^{-x}\,dx \leqslant \int_0^1 e^{-x^2}\,dx \leqslant \int_0^1 1\,dx.$$

Consequently, we have

$$1 - \frac{1}{e} \leqslant \int_0^1 e^{-x^2}\,dx \leqslant 1.$$

If we use the average of the bracketing values, we obtain the estimate $1 - 1/2e \approx 0.816$ for the integral with an error less than $1/2e < 0.184$. This estimate is crude, but it is obtained rapidly and may be quite satisfactory for our needs. If a better approximation is desired, we can attempt to find closer approximating functions g and h.

Taylor's Theorem 5.4.1 can be used to approximate e^{-x^2} by a polynomial. In using Taylor's Theorem, we must get bounds on the remainder term for our calculations to have significance. For example, if we apply Taylor's Theorem to e^{-y} for $0 \leqslant y \leqslant 1$, we get

$$e^{-y} = 1 - y + \tfrac{1}{2}y^2 - \tfrac{1}{6}y^3 + R_3$$

where $R_3 = y^4 e^{-c}/24$ where c is some number with $0 \leqslant c \leqslant 1$. Since we have no better information as to the location of c, we must be content with the estimate $0 \leqslant R_3 \leqslant y^4/24$. Hence we have

$$e^{-x^2} = 1 - x^2 + \tfrac{1}{2}x^4 - \tfrac{1}{6}x^6 + R_3$$

where $0 \leqslant R_3 \leqslant x^8/24$, for $x \in [0,1]$. Therefore, we get

$$\begin{aligned}\int_0^1 e^{-x^2}\,dx &= \int_0^1 (1 - x^2 + \tfrac{1}{2}x^4 - \tfrac{1}{6}x^6)\,dx + \int_0^1 R_3\,dx \\ &= 1 - \frac{1}{3} + \frac{1}{10} - \frac{1}{42} + \int_0^1 R_3\,dx.\end{aligned}$$

Since we have

$$0 \leqslant \int_0^1 R_3\,dx \leqslant \frac{1}{9\cdot 24} = \frac{1}{216} < 0.005$$

it follows that

$$\int_0^1 e^{-x^2}\,dx \approx \frac{26}{35} \ (\approx 0.7429),$$

with an error less than 0.005.

Upper and Lower Sums

It is natural to attempt to approximate integrals by considering the upper and lower sums (or the Riemann sums) that are used to define these integrals. However, a moment's reflection suggests that it will not be easy to evaluate upper and lower sums for general functions, since it is often not easy to determine the supremum and the infimum of a function on an interval. One case in which this is an easy matter, of course, is the case of a monotone function. In this case the supremum and the infimum will be obtained at the end points of the interval.

It is generally convenient to use partitions in which the points are equally spaced. Thus, if we are considering a continuous function f on the interval $[a,b]$, we consider partitions P_n of $[a,b]$ into n equal subintervals of length $h := (b-a)/n$ given by the partition points

$$a,\ a + h,\ a + 2h,\ \ldots,\ a + nh = b.$$

Suppose that f is continuous and increasing on $[a,b]$; then the lower sum of f corresponding to the partition P_n is

$$L(P_n;f) = h \sum_{k=0}^{n-1} f(a + kh),$$

while the upper sum of f corresponding to P_n is

$$U(P_n;f) = h \sum_{k=1}^{n} f(a + kh).$$

In this case the exact value of the integral $\int_a^b f(x)\,dx$ lies between $L(P_n;f)$ and $U(P_n;f)$. Unless we have reason to believe that one of these terms is closer to the integral than the other, we generally take their mean $\frac{1}{2}[L(P_n;f) + U(P_n;f)]$, which is readily seen to equal

(1) $$T_n(f) := h\left[\tfrac{1}{2}f(a) + \sum_{k=1}^{n-1} f(a + kh) + \tfrac{1}{2}f(b)\right],$$

as a reasonable approximation to the integral. In this case the error estimate is:

$$\left| \int_a^b f(x)\,dx - T_n(f) \right| \leq \tfrac{1}{2}[U(P_n;f) - L(P_n;f)]$$
$$= \tfrac{1}{2}h\,[f(b) - f(a)]$$
$$= \frac{[f(b) - f(a)](b-a)}{2n}.$$

An error estimate such as this is particularly useful since it gives an upper bound on the error in terms of quantities that are known at the outset. In particular, it can be used to determine how large we should choose n in order to have an approximation within a specified tolerance $\varepsilon > 0$.

6.5.1 Theorem. *If $f : [a,b] \to \boldsymbol{R}$ is monotone on $[a,b]$, and if $T_n(f)$ is defined by* (1), *then we have*

$$\text{(2)} \qquad \left| \int_a^b f(x)\,dx - T_n(f) \right| \leq \frac{|f(b)-f(a)|\,(b-a)}{2n}.$$

Proof. We have given the proof in the case where f is monotone increasing. We leave it to the reader to consider the case where f is decreasing. Q.E.D.

6.5.2 Example. If $f(x) = e^{-x^2}$ on $[0,1]$, then f is decreasing. It follows from (2) that if $n = 8$, then

$$\left| \int_0^1 e^{-x^2}\,dx - T_8(f) \right| \leq \frac{1-e^{-1}}{16} < 0.04,$$

and if $n = 16$, then

$$\left| \int_0^1 e^{-x^2}\,dx - T_{16}(f) \right| \leq \frac{1-e^{-1}}{32} < 0.02.$$

Actually, the approximation is considerably better than this, as we shall see in Example 6.5.5.

The Trapezoidal Rule

The method of numerical integration called the "Trapezoidal Rule" is based on approximating the continuous function $f:[a,b] \to \boldsymbol{R}$ by a piecewise linear function. Let $n \in \boldsymbol{N}$ and $h := (b-a)/n$; as before, we consider the partition $P_n := (a, a + h, a + 2h, \ldots, a + nh = b)$. We approximate f by the piecewise linear function g_n whose graph passes through the points $(a + kh, f(a + kh))$ where $k = 0,1, \ldots, n$. (See Figure 6.5.1.) It seems reasonable that the integral

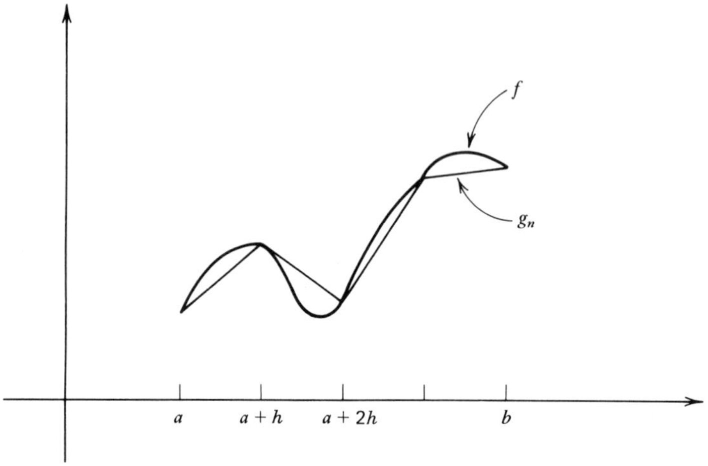

FIGURE 6.5.1 Approximation of *f* by g_n.

$\int_a^b f(x)\,dx$ will be "approximately equal to" the integral $\int_a^b g_n(x)\,dx$, provided that f is reasonably smooth and n is sufficiently large.

It is an exercise in elementary geometry to determine that the area of a trapezoid T with base of length h and sides of lengths l_1 and l_2 is $\frac{1}{2}h(l_1 + l_2)$. This can be interpreted as the length h of base times the average length $\frac{1}{2}(l_1 + l_2)$ of the sides of T. (See Figure 6.5.2.) Similarly one can show that since g_n is linear over the inverval $[a + kh, a + (k+1)h]$, the integral of g_n over that interval is given by

$$\int_{a+kh}^{a+(k+1)h} g_n(x)\,dx = \tfrac{1}{2}h[f(a+kh) + f(a+(k+1)h)]$$

where $k = 0,1, \ldots, n-1$. The integral of g_n over $[a,b]$ is then obtained by summing these values. Since each partition point in P_n except a and b belong to two adjacent subintervals, we obtain

$$\int_a^b g_n(x)\,dx = h[\tfrac{1}{2}f(a) + f(a+h) + \ldots + f(a+(n-1)h) + \tfrac{1}{2}f(b)].$$

Thus we see that the integral of the piecewise linear approximation g_n to f is equal to the sum

$$\text{(3)} \qquad T_n(f) := h\left[\tfrac{1}{2}f(a) + \sum_{k=1}^{n-1} f(a+kh) + \tfrac{1}{2}f(b)\right],$$

which is the same as we obtained previously as the mean of $L(P_n;f)$ and $U(P_n;f)$. We refer to $T_n(f)$ as the ***n*th trapezoidal approximation** to f on $[a,b]$. In the preceding theorem we obtained an error estimate when f is monotone. We now derive one without this restriction on f, but in terms of the second derivative f'' of f.

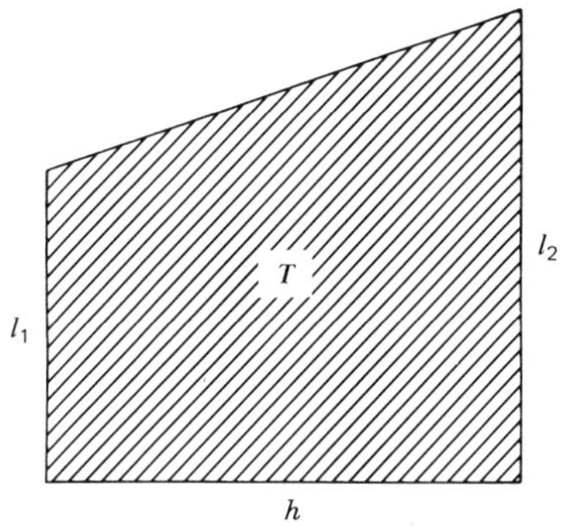

FIGURE 6.5.2 Area of trapezoid is $h(l_1 + l_2)/2$.

6.5.3 Theorem. *Let f, f', and f'' be continuous on $[a,b]$ and let $T_n(f)$ be the nth trapezoidal approximation (3). Then there exists a point $c \in [a,b]$ such that*

$$T_n(f) - \int_a^b f(x)\,dx = \frac{(b-a)h^2}{12} f''(c). \tag{4}$$

Proof. If $k = 1,2, \ldots, n$, let $a_k := a + (k-1)h$ and let $\phi_k : [0,h] \to \boldsymbol{R}$ be defined by

$$\phi_k(t) := \tfrac{1}{2}t[f(a_k) + f(a_k + t)] - \int_{a_k}^{a_k+t} f(x)\,dx$$

for $t \in [0,h]$. Note that $\phi_k(0) = 0$ and that (by the Differentiation Theorem 6.3.4)

$$\begin{aligned}\phi_k'(t) &= \tfrac{1}{2}[f(a_k) + f(a_k + t)] + \tfrac{1}{2}tf'(a_k + t) - f(a_k + t)\\ &= \tfrac{1}{2}[f(a_k) - f(a_k + t)] + \tfrac{1}{2}tf'(a_k + t).\end{aligned}$$

Consequently $\phi_k'(0) = 0$ and

$$\begin{aligned}\phi_k''(t) &= -\tfrac{1}{2}f'(a_k + t) + \tfrac{1}{2}f'(a_k + t) + \tfrac{1}{2}tf''(a_k + t)\\ &= \tfrac{1}{2}tf''(a_k + t).\end{aligned}$$

Now let A, B be defined by

$$A := \inf\{f''(x) : x \in [a,b]\}, \qquad B := \sup\{f''(x) : x \in [a,b]\}$$

so that we have

$$\tfrac{1}{2}At \leqslant \phi_k''(t) \leqslant \tfrac{1}{2}Bt$$

for $t \in [0,h]$, $k = 1, 2, \ldots, n$. Integrating, and applying the Fundamental Theorem of Calculus, we obtain, since $\phi_k'(0) = 0$, that

$$\tfrac{1}{4}At^2 \leq \phi_k'(t) \leq \tfrac{1}{4}Bt^2$$

for $t \in [0,h]$, $k = 1, 2, \ldots, n$. Integrating again and taking $t = h$, we obtain, since $\phi_k(0) = 0$, that

$$\tfrac{1}{12}Ah^3 \leq \phi_k(h) \leq \tfrac{1}{12}Bh^3$$

for $k = 1, 2, \ldots, n$. If we add these inequalities and note that

$$\sum_{k=1}^{n} \phi_k(h) = T_k(f) - \int_a^b f(x)\,dx,$$

we conclude that

$$\tfrac{1}{12}Ah^3 n \leq T_n(f) - \int_a^b f(x)\,dx \leq \tfrac{1}{12}Bh^3 n.$$

Since $h = (b-a)/n$, we have

$$\tfrac{1}{12}A(b-a)h^2 \leq T_n(f) - \int_a^b f(x)\,dx \leq \tfrac{1}{12}B(b-a)h^2.$$

Since f'' is continuous on $[a,b]$, it follows from the definitions of A and B and Bolzano's Intermediate Value Theorem 4.6.6 that there exists a point c in $[a,b]$ such that (4) holds. Q.E.D.

The equality (4) is interesting in that it gives both an upper and a lower bound for the difference $T_n(f) - \int_a^b f(x)\,dx$. For example, if $f''(x) \geq A > 0$ for all x in $[a,b]$, then (4) implies that this difference must always exceed $\tfrac{1}{12}A(b-a)h^2$. However, it is usually the upper bound that is of greater interest.

6.5.4 Corollary. *Let f, f', and f'' be continuous on $[a,b]$ and let $B_2 := \sup\{|f''(x)| : x \in [a,b]\}$. Then*

$$\left| T_n(f) - \int_a^b f(x)\,dx \right| \leq \frac{(b-a)h^2}{12} B_2. \tag{5}$$

We can also write (5) in the form

$$\left| T_n(f) - \int_a^b f(x)\,dt \right| \leq \frac{(b-a)^3}{12n^2} B_2. \tag{6}$$

This inequality can be used, when B_2 is known, to determine how large we must choose n in order to be certain of a desired degree of accuracy.

6.5.5 Example. If $f(x) := e^{-x^2}$ on $[0,1]$, then a calculation shows that $f''(x) = 2e^{-x^2}(2x^2 - 1)$. Therefore we have $B_2 \leq 2$. It follows from (6) that if $n = 8$, then

$$\left| T_8(f) - \int_0^1 e^{-x^2}\, dx \right| \leq \frac{2}{12 \cdot 64} = \frac{1}{384} \leq 0.003$$

and that if $n = 16$, then

$$\left| T_{16}(f) - \int_0^1 e^{-x^2}\, dx \right| \leq \frac{2}{12 \cdot 256} = \frac{1}{1536} < 0.00066.$$

This shows that the accuracy is considerably better in this case than was predicted in Example 6.5.2.

The Midpoint Rule

One obvious method of approximating the integral of f is to take the Riemann sums evaluated at the *midpoints* of the subintervals. Thus, if P_n is the partition

$$P_n := (a,\ a + h,\ a + 2h,\ \ldots,\ a + nh = b)$$

we have the **Midpoint Rule approximation** given by

$$\begin{aligned} M_n(f) &:= h[f(a + \tfrac{1}{2}h) + f(a + \tfrac{3}{2}h) + \cdots + f(a + (n - \tfrac{1}{2})h)] \\ &= h \sum_{k=1}^{n} f(a + (k - \tfrac{1}{2})h). \end{aligned} \tag{7}$$

Another method is to use piecewise linear functions that are *tangent* to the graph of f at the midpoints of these subintervals. At first glance, this latter method appears to have the disadvantage that we will need to know the slope of the tangent line to the graph of f at each of the midpoints $a + (k - \tfrac{1}{2})h$ $(k = 1, 2, \ldots, n)$. However, it is an exercise in geometry to show that the area of the trapezoid whose top is this tangent line at the midpoint $a + (k - \tfrac{1}{2})h$ equals the area of the rectangle whose height is $f(a + (k - \tfrac{1}{2})h)$. (See Figure 6.5.3.) Thus, this area is given by (7), and we see that the "Tangent Trapezoid Rule" turns out to be the same as the "Midpoint Rule". We now show that the Midpoint Rule gives slightly better accuracy than the Trapezoidal Rule.

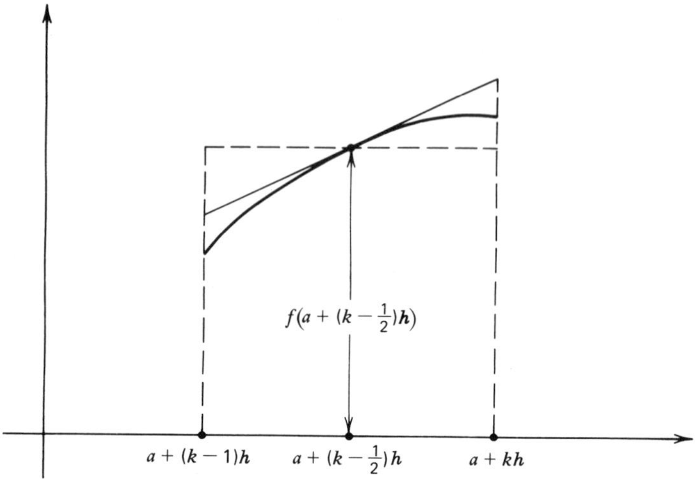

FIGURE 6.5.3 The tangent trapezoid.

6.5.6 Theorem. *Let f, f', and f'' be continuous on $[a,b]$ and let $M_n(f)$ be the nth midpoint approximation (7). Then there exists a point $\gamma \in [a,b]$ such that*

$$\int_a^b f(x)\,dx - M_n(f) = \frac{(b-a)h^2}{24} f''(\gamma). \tag{8}$$

Proof. If $k = 1, 2, \ldots, n$, let $c_k := a + (k - \frac{1}{2})h$, and let $\psi_k : [0,\frac{1}{2}h] \to \boldsymbol{R}$ be defined by

$$\psi_k(t) := \int_{c_k-t}^{c_k+t} f(x)\,dx - f(c_k)2t$$

for $t \in [0,\frac{1}{2}h]$. Note that $\psi_k(0) = 0$ and that since

$$\psi_k(t) = \int_{c_k}^{c_k+t} f(x)\,dx - \int_{c_k}^{c_k-t} f(x)\,dx - f(c_k)\,2t,$$

we have

$$\begin{aligned}\psi_k'(t) &= f(c_k + t) - f(c_k - t)\,(-1) - 2f(c_k)\\ &= [f(c_k + t) + f(c_k - t)] - 2f(c_k).\end{aligned}$$

Consequently $\psi_k'(0) = 0$ and

$$\begin{aligned}\psi_k''(t) &= f'(c_k + t) + f'(c_k - t)\,(-1)\\ &= f'(c_k + t) - f'(c_k - t).\end{aligned}$$

By the Mean Value Theorem 5.2.4, there exists a point $c_{k,t}$ with $|c_k - c_{k,t}| < t$ such that $\psi_k''(t) = 2tf''(c_{k,t})$. If we let A and B be as in the proof of Theorem 6.5.3, we have $2tA \leq \psi_k''(t) \leq 2tB$ for $t \in [0,h/2]$, $k = 1,2,\ldots,n$. It follows as before that

$$\tfrac{1}{3}At^3 \leq \psi_k(t) \leq \tfrac{1}{3}Bt^3$$

for all $t \in [0,\frac{1}{2}h]$, $k = 1,2,\ldots,n$. If we put $t = \frac{1}{2}h$, we get

$$\frac{1}{24}Ah^3 \leq \psi_k(\tfrac{1}{2}h) \leq \frac{1}{24}Bh^3.$$

If we add these inequalities and note that

$$\sum_{k=1}^{n} \psi_k(\tfrac{1}{2}h) = \int_a^b f(x)\,dx - M_n(f),$$

we conclude that

$$\frac{1}{24}Ah^3n \leq \int_a^b f(x)\,dx - M_n(f) \leq \frac{1}{24}Bh^3n.$$

If we use the fact that $h = (b-a)/n$ and apply Bolzano's Intermediate Value Theorem 4.6.6 to f'' on $[a,b]$ we conclude that there exists a point $\gamma \in [a,b]$ such that (8) holds. Q.E.D.

6.5.7 Corollary. *Let f, f', and f'' be continuous on $[a,b]$ and let $B_2 := \sup\{|f''(x)| : x \in [a,b]\}$. Then*

$$\left| M_n(f) - \int_a^b f(x)\,dx \right| \leq \frac{(b-a)h^2}{24} B_2. \tag{9}$$

We can also write (9) in the form

$$\left| M_n(f) - \int_a^b f(x)\,dx \right| \leq \frac{(b-a)^3}{24n^2} B_2. \tag{10}$$

Simpson's Rule

The approximation method we shall now introduce usually gives a better approximation than the Trapezoidal or Midpoint Rule and requires very little extra calculation. Whereas the Trapezoidal and Midpoint Rules approximate the function f by piecewise linear functions, Simpson's Rule approximates f by a function whose graph is the union of parts of parabolas.

To help motivate the formula, the reader may show that if three points $(-h,y_0)$, $(0,y_1)$, (h,y_2) are given, then the quadratic function $q(x) = Ax^2 + Bx + C$ that passes through these three points has the property that

$$\int_{-h}^{h} q(x)\,dx = \tfrac{1}{3}h(y_0 + 4y_1 + y_2).$$

Now let f be a continuous function on $[a,b]$ and let $n \in \mathbf{N}$ be *even*. We let $h = (b-a)/n$. On each "double subinterval"

$$[a,a + 2h], [a + 2h,a + 4h], \ldots, [b - 2h,b],$$

we approximate f by a quadratic function that agrees with f at the points

$$y_0 = f(a), \qquad y_1 = f(a + h), \qquad y_2 = f(a + 2h), \ldots .$$

In view of the above relation this leads to the nth **Simpson approximation** defined by

$$(11) \quad S_n(f) := \tfrac{1}{3}h[f(a) + 4f(a + h) + 2f(a + 2h) + 4f(a + 3h) + 2f(a + 4h) + \cdots + 2f(b - 2h) + 4f(b - h) + f(b)].$$

Note that the coefficients of the values of f at the $n + 1$ partition points are 1, 4, 2, 4, 2, . . . , 2, 4, 1.

We now establish a theorem that gives information about the accuracy of the Simpson approximation.

6.5.8 Theorem. *Let f, f', f'', f''' and $f^{(4)}$ be continuous on $[a,b]$ and let $n \in \mathbf{N}$ be even. If $S_n(f)$ is the nth Simpson approximation* (11), *then there exists a point $c \in [a,b]$ such that*

$$(12) \qquad S_n(f) - \int_a^b f(x)\,dx = \frac{(b-a)h^4}{180} f^{(4)}(c).$$

Proof. If $k = 0, 1, \ldots, \tfrac{1}{2}n - 1$, let $c_k := a + (2k+1)h$ and let $\varphi_k : [0,h] \to \boldsymbol{R}$ be defined by

$$\varphi_k(t) := \tfrac{1}{3}t[f(c_k - t) + 4f(c_k) + f(c_k + t)] - \int_{c_k - t}^{c_k + t} f(x)\,dx.$$

Evidently $\varphi_k(0) = 0$ and

$$\varphi_k'(t) = \tfrac{1}{3}t[-f'(c_k - t) + f'(c_k + t)] - \tfrac{2}{3}[f(c_k - t) - 2f(c_k) + f(c_k + t)],$$

so that $\varphi_k'(0) = 0$ and

$$\varphi_k''(t) = \tfrac{1}{3}t[f''(c_k - t) + f''(c_k + t)] - \tfrac{1}{3}[-f'(c_k - t) + f'(c_k + t)],$$

so that $\varphi_k''(0) = 0$ and

$$\varphi_k'''(t) = \tfrac{1}{3}t[f'''(c_k + t) - f'''(c_k - t)].$$

Hence it follows from the Mean Value Theorem 5.2.4 that there exists $\gamma_{k,t}$ with $|c_k - \gamma_{k,t}| \leqslant t$ such that $\varphi_k'''(t) = \frac{2}{3}t^2 f^{(4)}(\gamma_{k,t})$. If we let A and B be defined by

$$A := \inf\{f^{(4)}(x) : x \in [a,b]\} \quad \text{and} \quad B := \sup\{f^{(4)}(x) : x \in [a,b]\},$$

then we have

$$\tfrac{2}{3}At^2 \leqslant \varphi_k'''(t) \leqslant \tfrac{2}{3}Bt^2$$

for $t \in [0,h]$, $k = 0,1, \ldots, \frac{1}{2}n-1$. After three integrations, this inequality becomes

$$\frac{1}{90}At^5 \leqslant \varphi_k(t) \leqslant \frac{1}{90}Bt^5$$

for all $t \in [0,h]$, $k = 0,1, \ldots, \frac{1}{2}n-1$. If we put $t = h$, we get

$$\frac{1}{90}Ah^5 \leqslant \varphi_k(h) \leqslant \frac{1}{90}Bh^5$$

for $k = 0,1, \ldots, \frac{1}{2}n-1$.

If we add these $\frac{1}{2}n$ inequalities and note that

$$\sum_{k=0}^{\frac{1}{2}n-1} \varphi_k(h) = S_n(f) - \int_a^b f(x)\,dx,$$

we conclude that

$$\frac{1}{90}Ah^5\,\frac{n}{2} \leqslant S_n(f) - \int_a^b f(x)\,dx \leqslant \frac{1}{90}Bh^5\,\frac{n}{2}.$$

Since $h = (b-a)/n$, it follows from Bolzano's Intermediate Value Theorem 4.6.6 (applied to $f^{(4)}$) that there exists a point $c \in [a,b]$ such that the relation (12) holds. Q.E.D.

6.5.9 Corollary. *Let f, f', f'', f''', and $f^{(4)}$ be continuous on $[a,b]$ and let $B_4 := \sup\{|f^{(4)}(x)| : x \in [a,b]\}$. Then*

$$(13) \qquad \left| S_n(f) - \int_a^b f(x)\,dx \right| \leqslant \frac{(b-a)h^4}{180} B_4.$$

We can also write (13) in the form

$$(14) \qquad \left| S_n(f) - \int_a^b f(x)\,dx \right| \leqslant \frac{(b-a)^5}{180n^4} B_4.$$

6.5.10 Example. If $f(x) := e^{-x^2}$ on $[0,1]$ then a calculation shows that

$$f^{(4)}(x) = 4e^{-x^2}[4x^4 - 12x^2 + 3],$$

whence it follows that $|f^{(4)}(x)| \leqslant 20$ for $x \in [0,1]$. Thus $B_4 \leqslant 20$. It follows from (14) that if $n = 8$ then

$$\left| S_8(f) - \int_0^1 e^{-x^2}\,dx \right| \leqslant \frac{1}{180 \cdot 8^4} \cdot 20 = \frac{1}{36864} < 0.00003,$$

and that if $n = 16$ then

$$\left| S_{16}(f) - \int_0^1 e^{-x^2}\,dx \right| \leqslant \frac{1}{589824} < 0.0000017.$$

6.5.11 Remarks. (a) The nth midpoint approximation $M_n = M_n(f)$ can easily be used to step up to the $(2n)$th trapezoidal and Simpson approximations by using the formulas established in Exercises 6.5.9 and 6.5.10. In fact, once the initial trapezoidal approximation T_1 is calculated, only the M_n need be found. A rapid and effective procedure for the approximation of an integral can be based on the following sequence of calculations: $T_1 = \frac{1}{2}(b - a)[f(a) + f(b)]$; $M_1 = (b - a)\, f(\frac{1}{2}(a + b))$, $T_2 = \frac{1}{2}M_1 + \frac{1}{2}T_1$, $S_2 = \frac{2}{3}M_1 + \frac{1}{3}T_1$; M_2, $T_4 = \frac{1}{2}M_2 + \frac{1}{2}T_2$, $S_4 = \frac{2}{3}M_2 + \frac{1}{3}T_2$; M_4, T_8, S_8; . . .

(b) If $f''(x) \geqslant 0$ [or $f''(x) \leqslant 0$] for all $x \in [a,b]$, then the value of the integral lies between M_n and T_{2n} (see Exercise 6.5.8), so that an estimate of the error is immediate.

Exercises for Section 6.5

1. Use the trapezoidal approximation with $n = 4$ to evaluate $\log 2 = \int_1^2 (1/x)\,dx$. Show that $0.686 \leqslant \log 2 \leqslant 0.699$ and that

$$0.0013 < \frac{1}{768} \leqslant T_4 - \log 2 \leqslant \frac{1}{96} < 0.0105.$$

2. Use the Simpson approximation with $n = 2$ to evaluate $\log 2 = \int_1^2 (1/x)\,dx$. Show that $0.6930 \leqslant \log 2 \leqslant 0.6936$ and that

$$0.000016 < \frac{1}{2^5} \cdot \frac{1}{1920} \leqslant S_4 - \log 2 \leqslant \frac{1}{1920} < 0.000521.$$

3. Let $f(x) := (1 + x^2)^{-1}$ for $x \in [0,1]$. Show that $f''(x) = 2(3x^2 - 1)(1 + x^2)^{-3}$ and that $|f''(x)| \leqslant 2$ for $x \in [0,1]$. Use the trapezoidal approximation with $n = 4$ to evaluate $\pi/4 = \int_0^1 f(x)\,dx$. Show that

$$\left|T_4(f) - \frac{\pi}{4}\right| \leqslant \frac{1}{96} < 0.011.$$

4. If the trapezoidal approximation $T_n(f)$ is used to approximate $\pi/4$ as in Exercise 3, show that we must take $n \geqslant 409$ in order to be sure that the error is less than 10^{-6}.
5. Let f be as in Exercise 3. Show that $f^{(4)}(x) = 24(5x^4 - 10x^2 + 1)(1 + x^2)^{-5}$ and that $|f^{(4)}(x)| \leqslant 96$ for $x \in [0,1]$. Use Simpson's approximation with $n = 4$ to evaluate $\pi/4$. Show that

$$\left|S_4(f) - \frac{\pi}{4}\right| \leqslant \frac{1}{480} < 0.0021.$$

6. If the Simpson approximation $S_n(f)$ is used to approximate $\pi/4$ as in Exercise 5, show that we must take $n \geqslant 28$ in order to be sure that the error is less than 10^{-6}.
7. If p is polynomial of degree at most 3, show that the Simpson approximations are exact.
8. Show that if $f''(x) \geqslant 0$ on $[a,b]$ (that is, if f is convex on $[a,b]$), then for any natural numbers m, n we have

$$M_n(f) \leqslant \int_a^b f(x)\,dx \leqslant T_m(f).$$

If $f''(x) \leqslant 0$ on $[a,b]$, this inequality is reversed.

9. Show that $T_{2n}(f) = \frac{1}{2}[M_n(f) + T_n(f)]$.
10. Show that $S_{2n}(f) = \frac{2}{3} M_n(f) + \frac{1}{3} T_n(f)$.
11. Show that one has the estimate

$$\left|S_n(f) - \int_a^b f(x)\,dx\right| \leqslant \frac{(b - a)^2}{18n^2} B_2,$$

where $B_2 \geqslant |f''(x)|$ for all $x \in [a,b]$.

12. Note that $\int_0^1 (1 - x^2)^{1/2}\, dx = \pi/4$. Explain why the error estimates given by formulas (4), (8), and (12) cannot be used. Show that if $h(x) := (1 - x^2)^{1/2}$ for x in $[0,1]$, then

$$M_n(h) \leq \frac{\pi}{4} \leq T_n(h).$$

Calculate $M_8(h)$ and $T_8(h)$.

13. If h is as in Exercise 12, explain why

$$K := \int_0^{1/\sqrt{2}} h(x)\, dx = \frac{\pi}{8} + \frac{1}{4}.$$

Show that $|h''(x)| \leq 2^{3/2}$ and that $|h^{(4)}(x)| \leq 9 \cdot 2^{7/2}$ for $x \in [0, 1/\sqrt{2}]$. Show that $|K - T_n(h)| \leq 1/12n^2$ and that $|K - S_n(h)| \leq 1/10n^4$. Use these results to calculate π.

In Exercises 14–20, approximate the indicated integrals, giving estimates for the error. Use a calculator (or a computer) to obtain a high degree of precision.

14. $\displaystyle\int_0^2 (1 + x^4)^{1/2}\, dx$

15. $\displaystyle\int_0^2 (4 + x^3)^{1/2}\, dx$

16. $\displaystyle\int_0^1 \frac{dx}{1 + x^3}$

17. $\displaystyle\int_0^\pi \frac{\sin x}{x}\, dx$

18. $\displaystyle\int_0^{\pi/2} \frac{dx}{1 + \sin x}$

19. $\displaystyle\int_0^{\pi/2} \sqrt{\sin x}\, dx$

20. $\displaystyle\int_0^1 \cos(x^2)\, dx$

CHAPTER SEVEN

SEQUENCES OF FUNCTIONS

In previous chapters we have often made use of sequences of real numbers. In this chapter we shall consider sequences whose terms are *functions* rather than real numbers. Sequences of functions arise naturally in real analysis and are especially useful in obtaining approximations to a given function and defining new functions from known ones.

In Section 7.1 we shall introduce two different notions of convergence for a sequence of functions: pointwise convergence and uniform convergence. The latter type of convergence is very important and will be the main focus of our attention. The reason for this focus is the fact that, as is shown in Section 7.2, uniform convergence "preserves" certain properties in the sense that if each term of a uniformly convergent sequence of functions possesses these properties, then the limit function also possesses the properties.

In Section 7.3 we shall apply the concept of uniform convergence to define and derive the basic properties of the exponential and logarithmic functions. Section 7.4 is devoted to a similar treatment of the trigonometric functions.

SECTION 7.1 Pointwise and Uniform Convergence

Let $A \subseteq \boldsymbol{R}$ be given and suppose that for each $n \in \boldsymbol{N}$ there is a function $f_n : A \to \boldsymbol{R}$; we shall say that (f_n) is a **sequence of functions** on A to $\boldsymbol{R}$. Clearly, for each $x \in A$, such a sequence gives rise to a sequence of real numbers, namely the sequence

$$(1) \qquad (f_n(x))$$

that is obtained by evaluating each of the functions at the point x. For certain values of $x \in A$ the sequence (1) may converge, and for other values of $x \in A$ this sequence may diverge. For each number $x \in A$ for which the sequence (1) converges, there is, by Theorem 3.1.5, a uniquely determined real number, namely, $\lim (f_n(x))$. In general, the value of this limit, when it exists, will depend on the choice of the point $x \in A$. Thus, there arises in this way a function whose domain consists of all numbers $x \in A$ for which the sequence (1) converges.

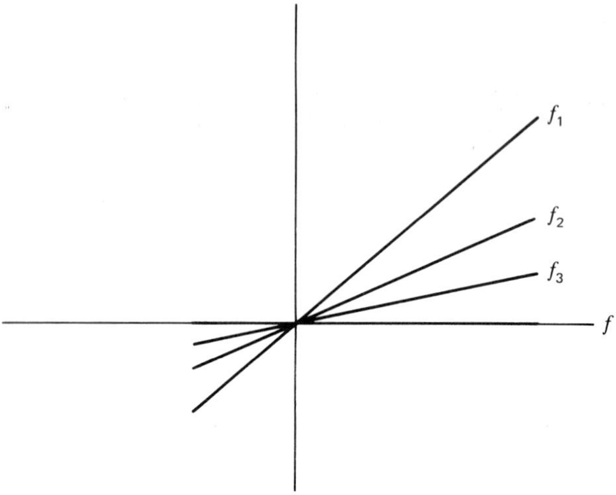

FIGURE 7.1.1 $f_n(x) = \dfrac{x}{n}$.

7.1.1 Definition. Let (f_n) be a sequence of functions on $A \subseteq \boldsymbol{R}$ to $\boldsymbol{R}$, let $A_0 \subseteq A$ and let $f : A_0 \to \boldsymbol{R}$. We say that the **sequence** (f_n) **converges on** A_0 **to** f if, for each $x \in A_0$, the sequence $(f_n(x))$ converges to $f(x)$ in $\boldsymbol{R}$. In this case we call f the **limit on** A_0 **of the sequence** (f_n). When such a function f exists, we say that the sequence (f_n) **is convergent on** A_0, or that (f_n) **converges pointwise on** A_0.

It follows from Theorem 3.1.5 that, except for a possible restriction of the domain A_0, the limit function is uniquely determined. Ordinarily we choose A_0 to be the largest set possible; that is, we take A_0 to be the set of all $x \in A$ for which the sequence (1) is convergent in $\boldsymbol{R}$.

In order to symbolize that the sequence (f_n) converges on A_0 to f, we sometimes write

$$f = \lim (f_n) \quad \text{on} \quad A_0, \qquad \text{or} \qquad f_n \to f \quad \text{on} \quad A_0.$$

Sometimes, when f_n and f are given by formulas we write

$$f(x) = \lim f_n(x) \quad \text{for} \quad x \in A_0, \qquad \text{or} \qquad f_n(x) \to f(x) \quad \text{for} \quad x \in A_0.$$

7.1.2 Examples. (a) $\lim (x/n) = 0$ for $x \in \boldsymbol{R}$.

For $n \in \boldsymbol{N}$, let $f_n(x) := x/n$ and let $f(x) := 0$ for $x \in \boldsymbol{R}$. By Example 3.1.7(a), we have $\lim (1/n) = 0$. Hence it follows from Theorem 3.2.3 that $\lim (f_n(x)) = 0$ for all $x \in \boldsymbol{R}$. (See Figure 7.1.1.)

(b) $\lim (x^n)$.

Let $g_n(x) := x^n$ for $x \in \boldsymbol{R}$, $n \in \boldsymbol{N}$. (See Figure 7.1.2.) Clearly, if $x = 1$, then the sequence $(g_n(1)) = (1)$ converges to 1. It follows from Example 3.1.11(c) that

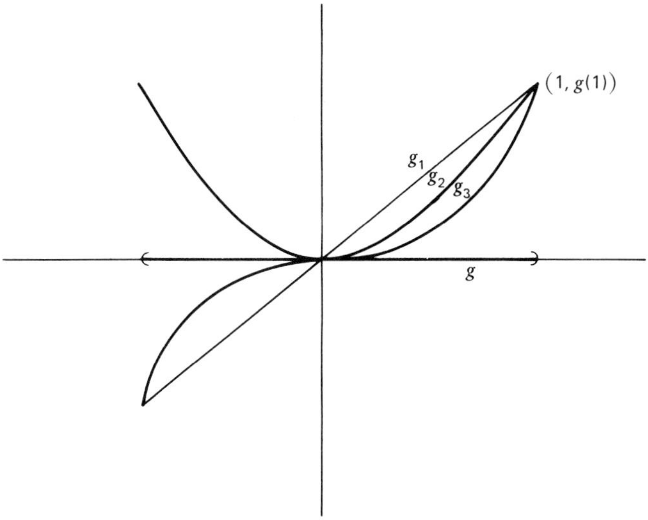

FIGURE 7.1.2 $g_n(x) = x^n$.

$\lim (x^n) = 0$ for $0 \leq x < 1$ and it is readily seen that this is also true for $-1 < x < 0$. If $x = -1$, then $g_n(-1) = (-1)^n$ and it was seen in Example 3.2.8(b) that the sequence is divergent. Similarly, if $|x| > 1$ then the sequence (x^n) is not bounded and so it is not convergent in $\boldsymbol{R}$. We conclude that if

$$\begin{aligned} g(x) &:= 0 \quad \text{for} \quad -1 < x < 1, \\ &:= 1 \quad \text{for} \quad x = 1, \end{aligned}$$

then the sequence (g_n) converges to g on the set $(-1,1]$.

(c) $\lim \left((x^2 + nx)/n\right) = x$ for $x \in \boldsymbol{R}$.

Let $h_n(x) := (x^2 + nx)/n$ for $x \in \boldsymbol{R}$, $n \in \boldsymbol{N}$, and let $h(x) := x$ for $x \in \boldsymbol{R}$. (See Figure 7.1.3.) Since we have $h_n(x) = (x^2/n) + x$, it follows from Example 3.1.7(a) and Theorem 3.2.3 that $h_n(x) \to x = h(x)$ for all $x \in \boldsymbol{R}$.

(d) $\lim \left((1/n) \sin (nx + n)\right) = 0$ for $x \in \boldsymbol{R}$.

Let $F_n(x) := (1/n) \sin (nx + n)$ for $x \in \boldsymbol{R}$, $n \in \boldsymbol{N}$, and let $F(x) := 0$ for $x \in \boldsymbol{R}$. (See Figure 7.1.4.) Since $|\sin y| \leq 1$ for all $y \in \boldsymbol{R}$ we have

$$(2) \qquad |F_n(x) - F(x)| = \left|\frac{1}{n} \sin (nx + n)\right| \leq \frac{1}{n}$$

for all $x \in \boldsymbol{R}$. Therefore it follows that $\lim \left(F_n(x)\right) = 0 = F(x)$ for all $x \in \boldsymbol{R}$. The reader should note that, given any $\varepsilon > 0$, by choosing n sufficiently large we can make $|F_n(x) - F(x)| < \varepsilon$ for all values of x simultaneously!

Partly to reinforce Definition 7.1.1 and partly to prepare the way for the important notion of uniform convergence, we reformulate Definition 7.1.1 as follows.

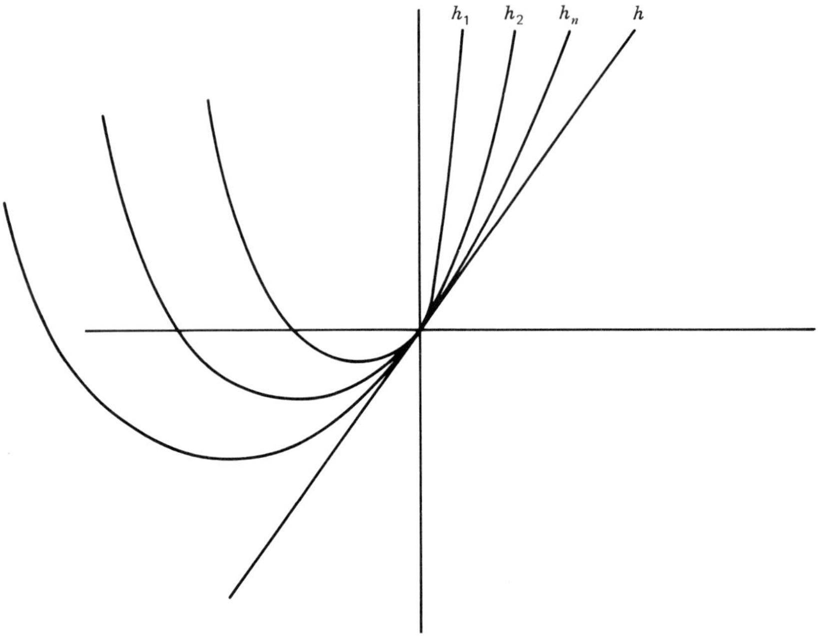

FIGURE 7.1.3 $h_n(x) = \dfrac{x^2 + nx}{n}$.

7.1.3 Lemma. *A sequence (f_n) of functions on $A \subseteq \boldsymbol{R}$ to $\boldsymbol{R}$ converges to a function $f : A_0 \to \boldsymbol{R}$ on A_0 if and only if for each $\varepsilon > 0$ and each $x \in A_0$ there is a natural number $K(\varepsilon,x)$ such that if $n \geqslant K(\varepsilon,x)$, then*

$$(3) \qquad |f_n(x) - f(x)| < \varepsilon .$$

We leave it to the reader to show that this is equivalent to Definition 7.1.1. We wish to emphasize that the value of $K(\varepsilon,x)$ will depend, in general, on *both* $\varepsilon > 0$ and $x \in A_0$. The reader should confirm the fact that in Examples 1.7.2(a–c), the value of $K(\varepsilon,x)$ required to obtain an inequality such as (3) does depend on both $\varepsilon > 0$ and $x \in A_0$. The intuitive reason for this is that the convergence of the sequence is "significantly faster" at some points than it is at others. However, in Example 7.1.2(d), as we have seen in inequality (2), if we choose n sufficiently large, we can make $|F_n(x) - F(x)| < \varepsilon$ for all values of $x \in \boldsymbol{R}$. It is precisely this rather subtle difference that distinguishes between the notion of the "ordinary convergence" of a sequence of functions (as defined in Definition 7.1.1) and the notion of the "uniform convergence".

Uniform Convergence

7.1.4 Definition. A sequence (f_n) of functions on $A \subseteq \boldsymbol{R}$ to $\boldsymbol{R}$ **converges uniformly on** $A_0 \subseteq A$ to a function $f : A_0 \to \boldsymbol{R}$ if for each $\varepsilon > 0$ there is a natural

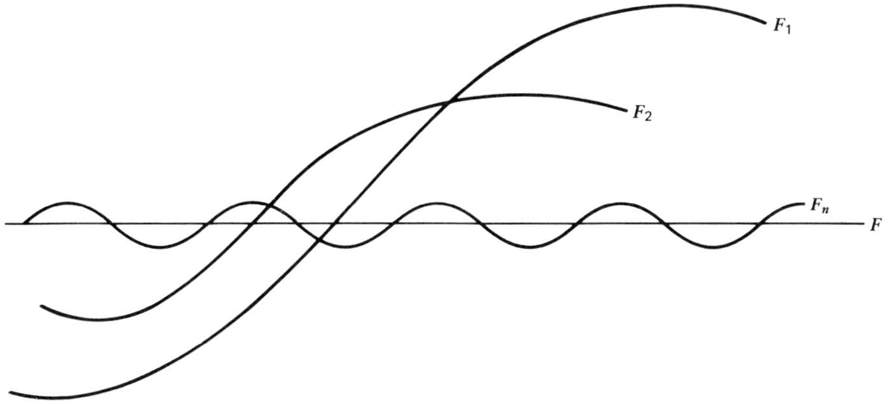

FIGURE 7.1.4 $F_n(x) = \dfrac{\sin(nx + n)}{n}$.

number $K(\varepsilon)$ (depending on ε but **not** on $x \in A_0$) such that if $n \geq K(\varepsilon)$ and $x \in A_0$, then

$$(4) \qquad |f_n(x) - f(x)| < \varepsilon .$$

In this case we say that the sequence (f_n) is **uniformly convergent on** A_0. Sometimes we write

$$f_n \rightrightarrows f \quad \text{on} \quad A_0, \qquad \text{or} \qquad f_n(x) \rightrightarrows f(x) \qquad \text{for} \quad x \in A_0.$$

It is an immediate consequence of the definitions that if the sequence (f_n) is uniformly convergent on A_0 to f, then this sequence also converges pointwise on A_0 to f in the sense of Definition 7.1.1. That the converse is not always true is seen by a careful examination of Examples 7.1.2(a–c); other examples will be given below. It is sometimes useful to have the following necessary and sufficient condition for a sequence (f_n) to *fail* to converge uniformly on A_0 to f.

7.1.5 Lemma. *A sequence (f_n) of functions on $A \subseteq \mathbf{R}$ to $\mathbf{R}$ does not converge uniformly on $A_0 \subseteq A$ to a function $f : A_0 \to \mathbf{R}$ if and only if for some $\varepsilon_0 > 0$ there is a subsequence (f_{n_k}) of (f_n) and a sequence (x_k) in A_0 such that*

$$(5) \qquad |f_{n_k}(x_k) - f(x_k)| \geq \varepsilon_0 \qquad \textit{for all} \quad k \in \mathbf{N}.$$

The proof of this result requires only that the reader negate Definition 7.1.4; we leave this to the reader as an important exercise. We now show how this result can be used.

7.1.6 Examples. (a) Consider Example 7.1.2(a). If we let $n_k = k$ and $x_k = k$, then $f_{n_k}(k) = 1$ so that $|f_{n_k}(x_k) - f(x_k)| = |1 - 0| = 1$. Therefore the sequence (f_n) does not converge uniformly on $\mathbf{R}$ to f.

(b) Consider Example 7.1.2(b). If $n_k = k$ and $x_k = (\frac{1}{2})^{1/k}$, then $|g_{n_k}(x_k) - g(x_k)| = |\frac{1}{2} - 0| = \frac{1}{2}$. Therefore the sequence (g_n) does not converge uniformly on $(-1,1]$ to g.

(c) Consider Example 7.1.2(c). If $n_k = k$ and $x_k = -k$, then $h_{n_k}(x_k) = 0$ and $h(x_k) = -k$ so that $|h_{n_k}(x_k) - h(x_k)| = k$. Therefore the sequence (h_n) does not converge uniformly on $\boldsymbol{R}$ to h.

The Uniform Norm

In discussing uniform convergence, it is often convenient to use the notion of the uniform norm on a set of bounded functions.

7.1.7 Definition. If $A \subseteq \boldsymbol{R}$ and $\varphi : A \to \boldsymbol{R}$ is a function, we say that φ is **bounded on** A if the set $\varphi(A)$ is a bounded subset of $\boldsymbol{R}$. If φ is bounded we define the **uniform norm of** φ **on** A by

(6) $$\|\varphi\|_A := \sup\{|\varphi(x)| : x \in A\}.$$

Note that it follows that if $\varepsilon > 0$, then

(7) $$\|\varphi\|_A \leq \varepsilon \quad \Leftrightarrow \quad |\varphi(x)| \leq \varepsilon \qquad \text{for all} \quad x \in A.$$

7.1.8 Lemma. *A sequence (f_n) of bounded functions on $A \subseteq \boldsymbol{R}$ converges uniformly on A to f if and only if $\|f_n - f\|_A \to 0$.*

Proof. ($\Rightarrow$) If (f_n) converges uniformly on A to f, then by Definition 7.1.4, given any $\varepsilon > 0$ there exists $K(\varepsilon)$ such that if $n \geq K(\varepsilon)$ and $x \in A$ then

$$|f_n(x) - f(x)| \leq \varepsilon.$$

Therefore the function f is bounded, and it follows that $\|f_n - f\|_A \leq \varepsilon$ whenever $n \geq K(\varepsilon)$. Since $\varepsilon > 0$ is arbitrary this implies that $\|f_n - f\|_A \to 0$.

($\Leftarrow$) If $\|f_n - f\|_A \to 0$, then given $\varepsilon > 0$ there is a natural number $H(\varepsilon)$ such that if $n \geq H(\varepsilon)$ then $\|f_n - f\|_A \leq \varepsilon$. It follows from (7) that $|f_n(x) - f(x)| \leq \varepsilon$ for all $n \geq H(\varepsilon)$ and $x \in A$. Therefore (f_n) converges uniformly on A to f. Q.E.D.

We now illustrate the use of Lemma 7.1.8 as a tool in examining a sequence of bounded functions for uniform convergence.

7.1.9 Examples. (a) We cannot apply Lemma 7.1.8 to the sequence in Example 7.1.2(a) since the function $f_n(x) - f(x) = x/n$ is not bounded on $\boldsymbol{R}$. For the sake of illustration, let $A := [0,1]$. Although the sequence (x/n) did not converge uniformly on $\boldsymbol{R}$ to the zero function, we shall show that the convergence is uniform on A. To see this, we observe that

$$\|f_n - f\|_A = \sup\{|x/n - 0| : 0 \leqslant x \leqslant 1\} = \frac{1}{n}$$

so that $\|f_n - f\|_A \to 0$. Therefore (f_n) is uniformly convergent on A to f.

(b) Let $g_n(x) := x^n$ for $x \in A := [0,1]$ and $n \in \mathbf{N}$, and let $g(x) := 0$ for $0 \leqslant x < 1$ and $g(1) := 1$. The functions $g_n(x) - g(x)$ are bounded on A and

$$\|g_n - g\|_A = \sup\left\{\begin{array}{ll} x^n, & 0 \leqslant x < 1 \\ 0\,, & x = 1 \end{array}\right\} = 1$$

for any $n \in \mathbf{N}$. Since $\|g_n - g\|_A$ does *not* converge to 0, we infer that the sequence (g_n) does *not* converge uniformly on A to g.

(c) We cannot apply Lemma 7.1.8 to the sequence in Example 7.1.2(c) since the function $h_n(x) - h(x) = x^2/n$ is not bounded on $\mathbf{R}$. Instead, let $A := [0,8]$ and consider

$$\|h_n - h\|_A = \sup\{x^2/n : 0 \leqslant x \leqslant 8\} = 64/n.$$

Therefore, the sequence (h_n) converges uniformly on A to h.

(d) Referring to Example 7.1.2(d), we see from relation (2) that $\|F_n - F\|_{\mathbf{R}} \leqslant 1/n$. Hence (F_n) converges uniformly on $\mathbf{R}$ to F.

By making use of the uniform norm, we can obtain a necessary and sufficient condition for uniform convergence that is often useful.

7.1.10 Cauchy Criterion for Uniform Convergence. *Let (f_n) be a sequence of bounded functions on $A \subseteq \mathbf{R}$. Then this sequence converges uniformly on A to a bounded function f if and only if for each $\varepsilon > 0$ there is a number $H(\varepsilon)$ in $\mathbf{N}$ such that for all $m,n \geqslant H(\varepsilon)$, then $\|f_m - f_n\|_A \leqslant \varepsilon$.*

Proof. ($\Rightarrow$) If $f_n \rightrightarrows f$ on A, then given $\varepsilon > 0$ there exists a natural number $K(\tfrac{1}{2}\varepsilon)$ such that if $n \geqslant K(\tfrac{1}{2}\varepsilon)$ then $\|f_n - f\|_A \leqslant \tfrac{1}{2}\varepsilon$. Hence, if both $m,n \geqslant K(\tfrac{1}{2}\varepsilon)$, then we conclude that

$$|f_m(x) - f_n(x)| \leqslant |f_m(x) - f(x)| + |f_n(x) - f(x)| \leqslant \tfrac{1}{2}\varepsilon + \tfrac{1}{2}\varepsilon = \varepsilon$$

for all $x \in A$. Therefore $\|f_m - f_n\|_A \leqslant \varepsilon$, for $m,n \geqslant K(\tfrac{1}{2}\varepsilon) =: H(\varepsilon)$.

($\Leftarrow$) Conversely, suppose that for $\varepsilon > 0$ there is $H(\varepsilon)$ such that if $m,n \geqslant H(\varepsilon)$, then $\|f_m - f_n\|_A \leqslant \varepsilon$. Therefore, for each $x \in A$ we have

$$(8) \qquad |f_m(x) - f_n(x)| \leqslant \|f_m - f_n\|_A \leqslant \varepsilon \quad \text{for} \quad m,n \geqslant H(\varepsilon).$$

It follows that $(f_n(x))$ is a Cauchy sequence in $\mathbf{R}$; therefore, by Theorem 3.5.4, it is a convergent sequence. We define $f : A \to \mathbf{R}$ by

$$f(x) := \lim (f_n(x)) \qquad \text{for} \quad x \in A.$$

If we let $n \to \infty$ in (8), it follows from Theorem 3.2.6 that for each $x \in A$ we have

$$|f_m(x) - f(x)| \leq \varepsilon \qquad \text{for} \quad m \geq H(\varepsilon).$$

Therefore the sequence (f_n) converges uniformly on A to f. Q.E.D.

Exercises for Section 7.1

1. Show that $\lim (x/(x + n)) = 0$ for all $x \in \mathbf{R}$, $x \geq 0$.
2. Show that $\lim (nx/(1 + n^2x^2)) = 0$ for all $x \in \mathbf{R}$.
3. Evaluate $\lim (nx/(1 + nx))$ for $x \in \mathbf{R}$, $x \geq 0$.
4. Evaluate $\lim (x^n/(1 + x^n))$ for $x \in \mathbf{R}$, $x \geq 0$.
5. Evaluate $\lim ((\sin nx)/(1 + nx))$ for $x \in \mathbf{R}$, $x \geq 0$.
6. Show that $\lim (\text{Arctan } nx) = (\pi/2) \operatorname{sgn} x$ for $x \in \mathbf{R}$.
7. Evaluate $\lim (e^{-nx})$ for $x \in \mathbf{R}$, $x \geq 0$.
8. Show that $\lim (xe^{-nx}) = 0$ for $x \in \mathbf{R}$, $x \geq 0$.
9. Show that $\lim (x^2e^{-nx}) = 0$ and that $\lim (n^2x^2e^{-nx}) = 0$ for $x \in \mathbf{R}$, $x \geq 0$.
10. Show that $\lim ((\cos \pi x)^{2n})$ exists for all $x \in \mathbf{R}$. What is its limit?
11. Show that if $a > 0$, then the convergence of the sequence is Exercise 1 in uniform on the interval $[0,a]$, but is not uniform on the interval $[0,\infty)$.
12. Show that if $a > 0$, then the convergence of the sequence in Exercise 2 is uniform on the interval $[a,\infty)$, but is not uniform on the interval $[0,\infty)$.
13. Show that if $a > 0$, then the convergence of the sequence in Exercise 3 is uniform on the interval $[a,\infty)$, but is not uniform on the interval $[0,\infty)$.
14. Show that if $0 < b < 1$, then the convergence of the sequence in Exercise 4 is uniform on the interval $[0,b]$, but is not uniform on the interval $[0,1]$.
15. Show that if $a > 0$, then the convergence of the sequence in Exercise 5 is uniform on the interval $[a,\infty)$, but is not uniform on the interval $[0,\infty)$.
16. Show that if $a > 0$, then the convergence of the sequence in Exercise 6 is uniform on the interval $[a,\infty)$, but is not uniform on the interval $(0,\infty)$.
17. Show that if $a > 0$, then the convergence of the sequence in Exercise 7 is uniform on the interval $[a,\infty)$, but is not uniform on the interval $[0,\infty)$.
18. Show that the convergence of the sequence in Exercise 8 is uniform on $[0,\infty)$.
19. Show that the sequence (x^2e^{-nx}) converges uniformly on $[0,\infty)$.
20. Show that if $a > 0$, then the sequence $(n^2x^2e^{-nx})$ converges uniformly on the interval $[a,\infty)$, but that it does not converge uniformly on the interval $[0,\infty)$.

SECTION 7.2 Interchange of Limits

It is often useful to know whether the limit of a sequence of functions is a continuous function, or a differentiable function, or an integrable function. Unfortunately, it is not always the case that the limit of a sequence of functions possesses these useful properties.

7.2.1 Examples. (a) Let $g_n(x) := x^n$ for $x \in [0,1]$ and $n \in \mathbf{N}$. Then, as we have noted in Example 7.1.2(b), the sequence (g_n) converges pointwise to the function

$$g(x) := 0 \quad \text{for} \quad 0 \leqslant x < 1,$$
$$:= 1 \quad \text{for} \quad x = 1.$$

Although all of the functions g_n are continuous at $x = 1$, the limit function g is not continuous at $x = 1$. Recall that it was shown in Example 7.1.6(b) that this sequence does not converge uniformly to g on $[0,1]$.

(b) Each of the functions $g_n(x) = x^n$ in part (a) has a continuous derivative on $[0,1]$. However, the limit function g does not have a derivative at $x = 1$, since it is not continuous at that point.

(c) Let $f_n : [0,1] \to \mathbf{R}$ be defined for $n \geqslant 2$ by

$$f_n(x) := n^2 x \quad \text{for} \quad 0 \leqslant x \leqslant 1/n,$$
$$:= -n^2(x - 2/n) \quad \text{for} \quad 1/n \leqslant x \leqslant 2/n,$$
$$:= 0 \quad \text{for} \quad 2/n \leqslant x \leqslant 1.$$

(See Figure 7.2.1.) It is clear that each of the functions f_n is continuous on $[0,1]$; hence it is integrable. Either by means of a direct calculation, or by referring to the significance of the integral as an area, we obtain

$$\int_0^1 f_n(x)\, dx = 1 \quad \text{for} \quad n \geqslant 2.$$

The reader may show that $f_n(x) \to 0$ for all $x \in [0,1]$; hence the limit function f vanishes identically and is continuous (and hence integrable), and $\int_0^1 f(x)\, dx = 0$. Therefore we have the uncomfortable situation:

$$\int_0^1 f(x)\, dx = 0 \neq 1 = \lim \int_0^1 f_n(x)\, dx.$$

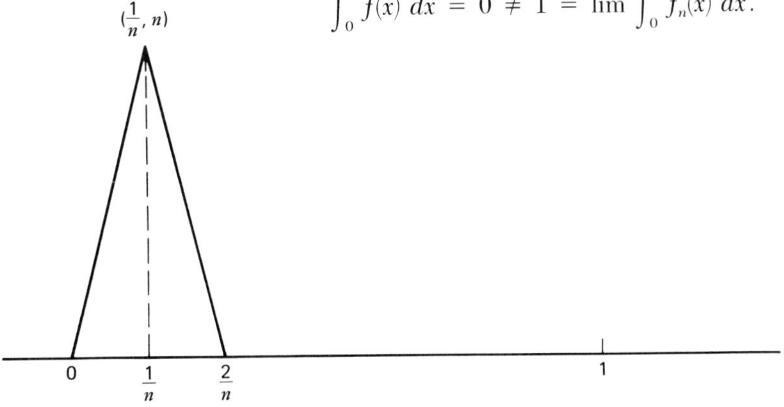

FIGURE 7.2.1 Example 7.2.1(c).

(d) If the reader considers the functions f_n in part (c) to be "artificial", he may prefer to consider the sequence (h_n) defined by $h_n(x) := 2nxe^{-nx^2}$ for $x \in [0,1]$, $n \in \mathbf{N}$. Since $h_n = H_n'$, where $H_n(x) := -e^{-nx^2}$, we have

$$\int_0^1 h_n(x)\,dx = H_n(1) - H_n(0) = 1 - e^{-n}.$$

Moreover, it is an exercise to show that $h(x) := \lim\,(h_n(x)) = 0$ for all $x \in [0,1]$. Hence we have

$$\int_0^1 h(x)\,dx \neq \lim \int_0^1 h_n(x)\,dx.$$

Although the extent of the discontinuity of the limit function in Example 7.2.1(a) is not very great, it is evident that more complicated examples can be constructed that will produce more extensive discontinuity. In any case, we must abandon the hope that the limit of a convergent sequence of continuous [respectively, differentiable, integrable] functions will be continuous [respectively, differentiable, integrable].

It will now be seen that the additional hypothesis of uniform convergence is sufficient to guarantee that the limit of a sequence of continuous functions be continuous. Similar results will also be established for sequences of differentiable and integrable functions.

Interchange of Limit and Continuity

7.2.2 Theorem. *Let (f_n) be a sequence of continuous functions on a set $A \subseteq \mathbf{R}$ and suppose that (f_n) converges uniformly on A to a function $f : A \to \mathbf{R}$. Then f is continuous on A.*

Proof. By hypothesis, given $\varepsilon > 0$ there exists a natural number $H := H(\frac{1}{3}\varepsilon)$ such that if $n \geq H$ then $|f_n(x) - f(x)| < \frac{1}{3}\varepsilon$ for all $x \in A$. Now let $c \in A$ be arbitrary; we shall show that f is continuous at c. By the Triangle Inequality we have

$$\begin{aligned} |f(x) - f(c)| &\leq |f(x) - f_H(x)| + |f_H(x) - f_H(c)| + |f_H(c) - f(c)| \\ &\leq \tfrac{1}{3}\varepsilon + |f_H(x) - f_H(c)| + \tfrac{1}{3}\varepsilon. \end{aligned}$$

Since f_H is continuous at c, there exists a number $\delta := \delta(\frac{1}{3}\varepsilon, c, f_H) > 0$ such that if $|x - c| < \delta$ and $x \in A$, then $|f_H(x) - f_H(c)| < \frac{1}{3}\varepsilon$. (See Figure 7.2.2.) Therefore, if $|x - c| < \delta$ and $x \in A$, then we have $|f(x) - f(c)| < \varepsilon$. Since $\varepsilon > 0$ is arbitrary, this establishes the continuity of f at the arbitrary point $c \in A$. Q.E.D.

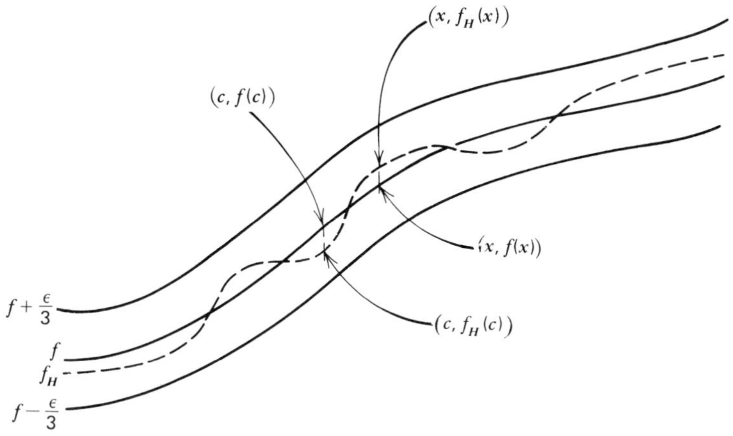

FIGURE 7.2.2 $|f_H(x) - f_H(c)| < \varepsilon/3$.

Remark. Although the uniform convergence of the sequence of continuous functions is sufficient to guarantee the continuity of the limit function, it is *not* necessary. (See Exercise 7.2.2.)

Interchange of Limit and Derivative

We mentioned in Section 5.1 that Weierstrass showed that the function defined by the series

$$f(x) := \sum_{k=0}^{\infty} 2^{-k} \cos (3^k x)$$

is continuous at every point but does not have a derivative at any point in $\boldsymbol{R}$. By considering the partial sums of this series, we obtain a sequence of functions (f_n) that possess a derivative at every point and are uniformly convergent to f. Thus, even though the sequence of differentiable functions (f_n) is uniformly convergent, it does not follow that the limit function is differentiable.

We shall now show that if *the sequence of derivatives* (f'_n) is uniformly convergent, then all is well. If one adds the hypothesis that the derivatives are continuous, then it is possible to give a short proof, based on the integral. (See Exercise 7.2.8.) However, if the derivatives are not assumed to be continuous, a somewhat more delicate argument is required.

7.2.3 Theorem. *Let $J \subseteq \boldsymbol{R}$ be a finite interval and let (f_n) be a sequence of functions on J to $\boldsymbol{R}$. Suppose that there exists $x_0 \in J$ such that $(f_n(x_0))$ converges, and that the sequence (f'_n) of derivatives exists on J and converges uniformly on*

J to a function g. Then the sequence (f_n) converges uniformly on J to a function f that has a derivative at every point of J and $f' = g$.

Proof. Let $a < b$ be the end points of J and let $x \in J$ be arbitrary. If m, $n \in \mathbf{N}$, we apply the Mean Value Theorem 5.2.4 to the difference $f_m - f_n$ on the interval with end points x_0, x. We conclude that there exists a point y (depending on m, n) such that

$$f_m(x) - f_n(x) = f_m(x_0) - f_n(x_0) + (x - x_0)\{f'_m(y) - f'_n(y)\}.$$

Hence we have

$$\|f_m - f_n\|_J \leqslant |f_m(x_0) - f_n(x_0)| + (b - a)\|f'_m - f'_n\|_J.$$

From Theorem 7.1.10, it follows from this inequality, and the hypothesis that $(f_n(x_0))$ is convergent and that (f'_n) is uniformly convergent on J, that (f_n) is uniformly convergent on J. We denote the limit of the sequence (f_n) by f. Since the f_n are all continuous and the convergence is uniform, it follows from Theorem 7.2.2 that f is continuous on J.

To establish the existence of the derivative of f at a point $c \in J$, we apply the Mean Value Theorem 5.2.4 to $f_m - f_n$ on an interval with end points c, x. We conclude that there exists a point z (depending on m, n) such that

$$\{f_m(x) - f_n(x)\} - \{f_m(c) - f_n(c)\} = (x - c)\{f'_m(z) - f'_n(z)\}.$$

Hence, if $x \neq c$, we have

$$\left|\frac{f_m(x) - f_m(c)}{x - c} - \frac{f_n(x) - f_n(c)}{x - c}\right| \leqslant \|f'_m - f'_n\|_J.$$

Since (f'_n) converges uniformly on J, if $\varepsilon > 0$ is given there exists $H(\varepsilon)$ such that if m, $n \geqslant H(\varepsilon)$ and $x \neq c$, then

$$\left|\frac{f_m(x) - f_m(c)}{x - c} - \frac{f_n(x) - f_n(c)}{x - c}\right| \leqslant \varepsilon.$$

If we take the limit in this inequality with respect to m and use Theorem 3.2.6, we have

$$\left|\frac{f(x) - f(c)}{x - c} - \frac{f_n(x) - f_n(c)}{x - c}\right| \leqslant \varepsilon,$$

provided that $x \neq c$, $n \geqslant H(\varepsilon)$. Since $g(c) = \lim (f'_n(c))$, there exists $N(\varepsilon)$ such that if $n \geqslant N(\varepsilon)$, then $|f'_n(c) - g(c)| < \varepsilon$. Now let $K := \sup\{H(\varepsilon), N(\varepsilon)\}$. Since $f'_K(c)$ exists, there exists $\delta_K(\varepsilon) > 0$ such that if $0 < |x - c| < \delta_K(\varepsilon)$, then

$$\left| \frac{f_K(x) - f_K(c)}{x - c} - f'_K(c) \right| < \varepsilon .$$

Combining these inequalities, we conclude that if $0 < |x - c| < \delta_K(\varepsilon)$, then

$$\left| \frac{f(x) - f(c)}{x - c} - g(c) \right| < 3\varepsilon.$$

Since $\varepsilon > 0$ is arbitrary, this shows that $f'(c)$ exists and equals $g(c)$. Since $c \in J$ is arbitrary, we conclude that $f' = g$ on J. Q.E.D.

Interchange of Limit and Integral

We have seen in Example 7.2.1(c) that if (f_n) is a sequence of integrable functions that converges on $[a,b]$ to an integrable function f, then it need not happen that

$$(*) \qquad \int_a^b f(x)\, dx = \lim \int_a^b f_n(x)\, dx.$$

We now show that the uniformity of the convergence is sufficient to guarantee that this equality holds.

7.2.4 Theorem. *Let (f_n) be a sequence of functions that are integrable on $[a,b]$ and suppose that (f_n) converges uniformly on $[a,b]$ to f. Then f is integrable on $[a,b]$ and $(*)$ holds.*

Proof. Let $J := [a,b]$ and let $\varepsilon > 0$ be given. Then there exists $K(\varepsilon)$ such that if $k \geqslant K(\varepsilon)$ then $\| f - f_k \|_J < \varepsilon/4(b-a)$.

Let $K := K(\varepsilon)$; since f_K is integrable, by the Riemann Criterion 6.1.8, there exists a partition $P_\varepsilon = (x_0, x_1, \ldots, x_n)$ of J such that

$$(\dagger) \qquad U(P_\varepsilon; f_K) - L(P_\varepsilon; f_K) < \tfrac{1}{2}\varepsilon.$$

Since $|f(x) - f_K(x)| \leqslant \varepsilon/4(b-a)$ for all $x \in J$ it follows that

$$U(P_\varepsilon; f) \leqslant U(P_\varepsilon; f_K) + \tfrac{1}{4}\varepsilon.$$

Moreover we have

$$L(P_\varepsilon; f_K) - \tfrac{1}{4}\varepsilon \leqslant L(P_\varepsilon; f).$$

Therefore we conclude that

$$U(P_\varepsilon;f) - L(P_\varepsilon;f) \leq U(P_\varepsilon;f_K) - L(P_\varepsilon;f_K) + \tfrac{1}{2}\varepsilon$$
$$< \tfrac{1}{2}\varepsilon + \tfrac{1}{2}\varepsilon = \varepsilon,$$

where we have used (†). Since $\varepsilon > 0$ is arbitrary, it follows from the Riemann Criterion that f is integrable on J.

To establish (*) we employ Corollary 6.2.13:

$$\left|\int_a^b f(x)\,dx - \int_a^b f_n(x)\,dx\right| = \left|\int_a^b \{f(x) - f_n(x)\}\,dx\right|$$
$$\leq \int_a^b |f(x) - f_n(x)|\,dx \leq \|f - f_n\|_J(b-a).$$

Since $\lim \|f - f_n\|_J = 0$, the conclusion follows. Q.E.D.

The hypothesis that the convergence of the sequence (f_n) is uniform is rather severe, and restricts the utility of this result. We shall now state a result that does not restrict the convergence so heavily but requires the integrability of the limit function. The proof of this result is quite delicate and will be omitted.

7.2.5 Bounded Convergence Theorem. *Let (f_n) be a sequence of functions that are integrable on $[a,b]$ and suppose that (f_n) converges on $[a,b]$ to an integrable function f. Suppose also that there exists $B > 0$ such that $|f_n(x)| \leq B$ for all $x \in [a,b]$, $n \in \mathbf{N}$. Then (*) holds.*

Exercises for Section 7.2

1. Show that the sequence $((x^n/(1 + x^n))$ does not converge uniformly on $[0,2]$ by showing that the limit function is not continuous on $[0,2]$.
2. Prove that the sequence in Example 7.2.1(c) is an example of a sequence of continuous functions that converges nonuniformly to a continuous limit.
3. Construct a sequence of functions on $[0,1]$ each of which is discontinuous at every point of $[0,1]$ and which converges uniformly to a function that is continuous at every point.
4. Suppose (f_n) is a sequence of continuous functions on an interval I that converges uniformly on I to a function f. If $(x_n) \subseteq I$ converges to $x_0 \in I$, show that $\lim (f_n(x_n)) = f(x_0)$.
5. Let $f : \mathbf{R} \to \mathbf{R}$ be uniformly continuous on $\mathbf{R}$ and if $f_n(x) := f(x + 1/n)$ for $x \in \mathbf{R}$, show that (f_n) converges uniformly on $\mathbf{R}$ to f.
6. Let $f_n(x) := x^n/n$ for $x \in [0,1]$. Show that the sequence (f_n) of differentiable functions converges uniformly to a differentiable function f on $[0,1]$, and that the sequence (f_n') converges on I to a function g, but that $g(1) \neq f'(1)$.

7. Let $g_n(x) := e^{-nx}/n$ for $x \geq 0$, $n \in \mathbf{N}$. Consider the relation between $\lim (g_n)$ and $\lim (g_n')$.
8. Let $I := [a,b]$ and let (f_n) be a sequence of functions on $I \to \mathbf{R}$ that converges on I to f. Suppose that each derivative f_n' is continuous on I and that the sequence (f_n') is uniformly convergent to g on I. Prove that

$$f(x) - f(a) = \int_a^x g(t)\,dt$$

and that $f'(x) = g(x)$ for all $x \in I$.
9. Show that $\lim \int_1^2 e^{-nx^2}\,dx = 0$.
10. If $a > 0$, show that

$$\lim \int_a^{\pi} \frac{\sin nx}{nx}\,dx = 0.$$

What happens if $a = 0$?
11. Let $f_n(x) := nx/(1 + nx)$ for $x \in [0,1]$. Show that (f_n) converges nonuniformly to an integrable function f and that

$$\int_0^1 f(x)\,dx = \lim \int_0^1 f_n(x)\,dx.$$

12. Let $g_n(x) := nx(1-x)^n$ for $x \in [0,1]$, $n \in \mathbf{N}$. Discuss the convergence of (g_n) and $(\int_0^1 g_n\,dx)$.
13. Let $\{r_1, r_2, \ldots, r_n, \ldots\}$ be an enumeration of the rational numbers in $I := [0,1]$, and let $f_n : I \to \mathbf{R}$ be defined to be 1 if $x = r_1, \ldots, r_n$ and equal to 0 otherwise. Show that f_n is Riemann integrable for each $n \in \mathbf{N}$, that $f_1(x) \leq f_2(x) \leq \cdots \leq f_n(x) \leq \cdots$ and that $f(x) := \lim (f_n(x))$ is the Dirichlet function, which is not Riemann integrable on $[0,1]$.

SECTION 7.3 The Exponential and Logarithmic Functions

In this section we shall introduce the exponential and logarthmic functions and shall derive some of their most important properties. In earlier sections of this book, we assumed some familiarity with these functions for the purpose of discussing examples. However, it is necessary at some point to place these important functions on a firm foundation, to establish their existence and determine their basic properties. We shall do that here. There are several alternative approaches one can take to accomplish this goal. We shall proceed by first proving the existence of a function that has itself as derivative. From this basic result, we obtain some of the main properties of the exponential function. The logarithm function is then introduced as the inverse of the exponential function, and this inverse relation is used to derive some of the properties of the logarithm function.

The Exponential Function

We begin by establishing the key existence result for the exponential function.

7.3.1 Theorem. *There exists a function* $E : \mathbf{R} \to \mathbf{R}$ *such that:*
(i) $E'(x) = E(x)$ *for all* $x \in \mathbf{R}$,
(ii) $E(0) = 1$.

Proof. We inductively define a sequence (E_n) of continuous functions as follows:

$$E_1(x) := 1 + x, \tag{1}$$

$$E_{n+1}(x) := 1 + \int_0^x E_n(t)\,dt, \tag{2}$$

for all $n \in \mathbf{N}$, $x \in \mathbf{R}$. Clearly E_1 is continuous on $\mathbf{R}$ and hence is integrable over any bounded interval. If E_n has been defined and is continuous on $\mathbf{R}$, then it is integrable over any bounded interval, so that E_{n+1} is well-defined by the above formula. Moreover, it follows from the Differentiation Theorem 6.3.4 that E_{n+1} is differentiable at any point $x \in \mathbf{R}$ and that

$$E'_{n+1}(x) = E_n(x) \qquad \text{for} \quad n \in \mathbf{N}. \tag{3}$$

An easy induction argument (which we leave to the reader) shows that

$$E_n(x) = 1 + \frac{x}{1!} + \frac{x^2}{2!} + \cdots + \frac{x^n}{n!} \qquad \text{for} \quad x \in \mathbf{R}. \tag{4}$$

Let $A > 0$ be given; then if $|x| \leqslant A$ and $m \geqslant n > 2A$, we have

$$\begin{aligned} |E_m(x) - E_n(x)| &= \left| \frac{x^{n+1}}{(n+1)!} + \cdots + \frac{x^m}{m!} \right| \\ &\leqslant \frac{A^{n+1}}{(n+1)!} \left[1 + \frac{A}{n} + \cdots + \left(\frac{A}{n}\right)^{m-n} \right] \\ &< \frac{A^{n+1}}{(n+1)!}\, 2. \end{aligned} \tag{5}$$

Since $\lim (A^n/n!) = 0$, it follows that the sequence (E_n) converges uniformly on the interval $[-A, A]$ where $A > 0$ is arbitrary. In particular this means that $(E_n(x))$ converges for each $x \in \mathbf{R}$. We define $E : \mathbf{R} \to \mathbf{R}$ by

$$E(x) := \lim_n E_n(x) \qquad \text{for} \quad x \in \mathbf{R}.$$

Since each $x \in \boldsymbol{R}$ is contained in some interval $[-A, A]$, it follows from Theorem 7.2.2 that E is continuous at x. Moreover, it is clear from (1) and (2) that $E_n(0) = 1$ for all $n \in \boldsymbol{N}$. Therefore $E(0) = 1$, which proves (ii).

On any interval $[-A, A]$ we have the uniform convergence of the sequence (E_n). In view of (3), we also have the uniform convergence of the sequence (E_n') of derivatives. It therefore follows from Theorem 7.2.3 that the limit function E is differentiable on $[-A, A]$ and that

$$E'(x) = \lim \left(E_n'(x)\right) = \lim \left(E_{n-1}(x)\right) = E(x)$$

for all $x \in [-A, A]$. Since $A > 0$ is arbitrary, statement (i) is established.

Q.E.D.

7.3.2 Corollary. *The function E has a derivative of every order and $E^{(n)}(x) = E(x)$ for all $n \in \boldsymbol{N}$, $x \in \boldsymbol{R}$.*

Proof. If $n = 1$, the statement is merely property (i). It follows for arbitrary $n \in \boldsymbol{N}$ by induction. Q.E.D.

7.3.3 Corollary. *If $x > 0$, then $1 + x < E(x)$.*

Proof. It is clear from (4) that if $x > 0$, then the sequence $\left(E_n(x)\right)$ is strictly increasing. Hence $E_1(x) < E(x)$ for all $x > 0$. Q.E.D.

It is next shown that the function E, whose existence was established by Theorem 7.3.1, is in fact unique.

7.3.4 Theorem. *The function $E : \boldsymbol{R} \to \boldsymbol{R}$ satisfying properties* (i) *and* (ii) *of Theorem* 7.3.1 *is unique.*

Proof. Let E_1 and E_2 be two functions on $\boldsymbol{R}$ to $\boldsymbol{R}$ that satisfy properties (i) and (ii) of Theorem 7.3.1 and let $F := E_1 - E_2$. Then

$$F'(x) = E_1'(x) - E_2'(x) = E_1(x) - E_2(x) = F(x)$$

for all $x \in \boldsymbol{R}$ and

$$F(0) = E_1(0) - E_2(0) = 1 - 1 = 0.$$

It is clear (by induction) that F has derivatives of all orders and indeed that $F^{(n)}(x) = F(x)$ for $n \in \boldsymbol{N}$, $x \in \boldsymbol{R}$.

Let $x \in \boldsymbol{R}$ be arbitrary, and let I_x be the closed interval with end points 0, x. Since F is continuous on I_x there exists $K > 0$ such that $|F(t)| \leqslant K$ for all $t \in I_x$. If we apply Taylor's Theorem 5.4.1 to F on the interval I_x and use the fact that $F^{(k)}(0) = F(0) = 0$ for all $k \in \boldsymbol{N}$, it follows that for each $n \in \boldsymbol{N}$ there is a point $c_n \in I_x$ such that

$$F(x) = F(0) + \frac{F'(0)}{1!}x + \cdots + \frac{F^{(n-1)}(0)}{(n-1)!}x^{n-1} + \frac{F^{(n)}(c_n)}{n!}x^n$$
$$= \frac{F(c_n)}{n!}x^n.$$

Therefore we have

$$|F(x)| \leqslant \frac{K|x|^n}{n!} \qquad \text{for all} \quad n \in \boldsymbol{N}.$$

But since $\lim\limits_n (|x|^n/n!) = 0$, we conclude that $F(x) = 0$. Since $x \in \boldsymbol{R}$ is arbitrary, we infer that $E_1(x) - E_2(x) = F(x) = 0$ for all $x \in \boldsymbol{R}$. Q.E.D.

The standard terminology and notation for the function E (which we now know exists and is unique) is given in the following definition.

7.3.5 Definition. The unique function $E : \boldsymbol{R} \to \boldsymbol{R}$ such that $E'(x) = E(x)$ for all $x \in \boldsymbol{R}$ and $E(0) = 1$, is called the **exponential function.** The number $e := E(1)$ is called **Euler's number.** We shall frequently write

$$\exp(x) := E(x) \qquad \text{or} \qquad e^x := E(x) \qquad \text{for} \quad x \in \boldsymbol{R}.$$

The number e can be obtained as a limit, and thereby approximated, in several different ways. [See Exercises 7.3.1 and 7.3.10, and Example 3.3.3(f).]

The use of the notation e^x for $E(x)$ is justified by property (v) in the next theorem, where it is noted that if r is a rational number, then $E(r)$ and e^r coincide. (Rational exponents were discussed in Section 4.8.) Thus, the function E can be viewed as extending the idea of exponentiation from rational numbers to arbitrary real numbers. For a definition of a^x for $a > 0$ and arbitrary $x \in \boldsymbol{R}$, see Definition 7.3.10.

7.3.6 Theorem. *The exponential function satisfies the following properties:*

(iii) $E(x) \neq 0$ *for all* $x \in \boldsymbol{R}$;
(iv) $E(x+y) = E(x)E(y)$ *for all* $x, y \in \boldsymbol{R}$;
(v) $E(r) = e^r$ *for all* $r \in \boldsymbol{Q}$.

Proof. (iii) Let $\alpha \in \boldsymbol{R}$ be such that $E(\alpha) = 0$, and let J_α be the closed interval with end points $0, \alpha$. Let $K \geqslant |E(t)|$ for all $t \in J_\alpha$. If we apply Taylor's Theorem 5.4.1, we conclude that for each $n \in \boldsymbol{N}$ there exists a point $c_n \in J_\alpha$ such that

$$1 = E(0) = E(\alpha) + \frac{E'(\alpha)}{1!}(-\alpha) + \cdots + \frac{E^{(n-1)}(\alpha)}{(n-1)!}(-\alpha)^{n-1} + \frac{E^{(n)}(c_n)}{n!}(-\alpha)^n$$

$$= \frac{E(c_n)}{n!}(-\alpha)^n.$$

Thus we have

$$0 < 1 < \frac{K}{n!}|\alpha|^n \qquad \text{for} \quad n \in \mathbf{N}.$$

But since $\lim (|\alpha|^n/n!) = 0$, this is a contradiction.

(iv) Let y be fixed; by (iii) we have $E(y) \neq 0$. Let $G : \mathbf{R} \to \mathbf{R}$ be defined by

$$G(x) := \frac{E(x+y)}{E(y)} \qquad \text{for} \quad x \in \mathbf{R}.$$

Evidently we have

$$G'(x) = \frac{E'(x+y)}{E(y)} = \frac{E(x+y)}{E(y)} = G(x)$$

for all $x \in \mathbf{R}$, and

$$G(0) = \frac{E(0+y)}{E(y)} = 1.$$

It follows from the uniqueness of E, proved in Theorem 7.3.4, that $G(x) = E(x)$ for all $x \in \mathbf{R}$. Hence $E(x+y) = E(x)E(y)$ for all $x \in \mathbf{R}$. Since $y \in \mathbf{R}$ is arbitrary, we obtain (iv).

It follows from (iv) and induction that if $n \in \mathbf{N}$, $x \in \mathbf{R}$, then

$$E(nx) = E(x)^n.$$

If we let $x = 1/n$, this relation implies that

$$e = E(1) = E\left(n \cdot \frac{1}{n}\right) = E\left(\frac{1}{n}\right)^n,$$

whence it follows that $E(1/n) = e^{1/n}$. Also we have $E(-m) = 1/E(m) = 1/e^m = e^{-m}$ for $m \in \mathbf{N}$. Therefore, if $m \in \mathbf{Z}$, $n \in \mathbf{N}$, we have

$$E(m/n) = \left(E(1/n)\right)^m = \left(e^{1/n}\right)^m = e^{m/n}.$$

This establishes (v). Q.E.D

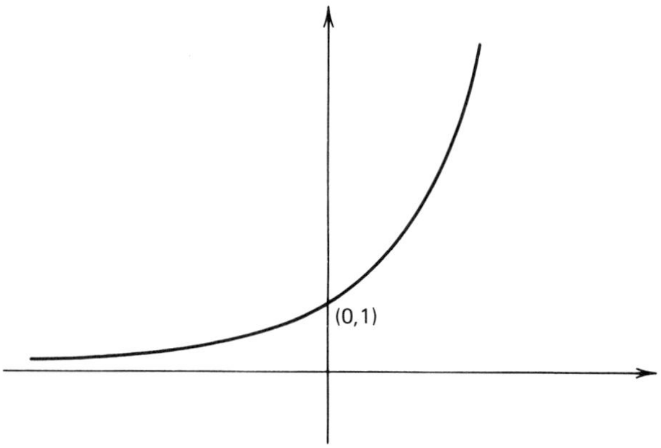

FIGURE 7.3.1 Graph of *E*.

7.3.7 Theorem. *The exponential function E is strictly increasing on* $\boldsymbol{R}$ *and has range equal to* $\{y \in \boldsymbol{R} : y > 0\}$. *Further, we have*

(vi) $\lim_{x \to -\infty} E(x) = 0$ *and* $\lim_{x \to \infty} E(x) = \infty$.

Proof. We know that $E(0) = 1 > 0$ and $E(x) \neq 0$ for all $x \in \boldsymbol{R}$. Since E is continuous on $\boldsymbol{R}$, it follows from Bolzano's Intermediate Value Theorem 4.6.6 that $E(x) > 0$ for all $x \in \boldsymbol{R}$. Therefore $E'(x) = E(x) > 0$ for $x \in \boldsymbol{R}$, so that E is strictly increasing on $\boldsymbol{R}$.

It follows from Corollary 7.3.3 that $2 < e$; consequently $2^n < e^n = E(n)$, whence it follows that $\lim E(n) = \infty$. Since E is strictly increasing, we conclude that

$$\lim_{x \to \infty} E(x) = \infty.$$

Also since $0 < E(-n) = e^{-n} < 2^{-n}$, it follows that $\lim E(-n) = 0$, so that

$$\lim_{x \to -\infty} E(x) = 0.$$

Therefore, by the Intermediate Value Theorem 4.4.6, every $y \in \boldsymbol{R}$ with $y > 0$ belongs to the range of E. Q.E.D.

The Logarithm Function

We have seen that the exponential function E is a strictly increasing, differentiable function with domain $\boldsymbol{R}$ and range $\{y \in \boldsymbol{R} : y > 0\}$. (See Figure 7.3.1.) It follows that E has an inverse function.

7.3.8 Definition. The function inverse to $E : \boldsymbol{R} \to \boldsymbol{R}$ is called the **logarithm** (or the **natural logarithm**). It will be denoted by L, or by log. (See Figure 7.3.2.)

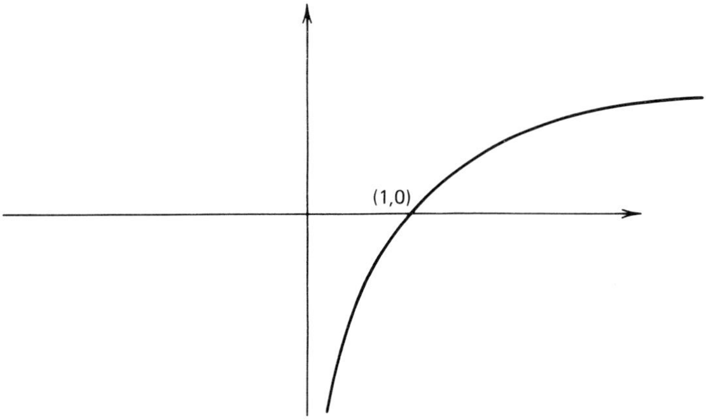

FIGURE 7.3.2 Graph of *L*.

Since E and L are inverse functions, we have

$$L \circ E(x) = x \qquad \text{for all} \quad x \in \boldsymbol{R}$$

and

$$E \circ L(y) = y \qquad \text{for all} \quad y \in \boldsymbol{R},\ y > 0.$$

These formulas may also be written in the form

$$\log e^x = x, \qquad e^{\log y} = y.$$

7.3.9 Theorem. *The logarithm is a strictly increasing function L with domain* $\{x \in \boldsymbol{R} : x > 0\}$ *and range* $\boldsymbol{R}$. *The derivative of L is given by*

(vii) $L'(x) = 1/x \quad$ *for* $\quad x > 0$.

The logarithm satisfies the functional equation

(viii) $L(xy) = L(x) + L(y) \quad$ *for* $\quad x > 0,\ y > 0$.

Moreover, we have

(ix) $L(1) = 0, \quad L(e) = 1,$

(x) $\lim_{x \to 0+} L(x) = -\infty, \quad \lim_{x \to \infty} L(x) = \infty.$

Proof. That L is strictly increasing with domain $\{x \in \boldsymbol{R} : x > 0\}$ and range $\boldsymbol{R}$ follows from the fact that E is strictly increasing with domain $\boldsymbol{R}$ and range $\{y \in \boldsymbol{R} : y > 0\}$.

(vii) Since $E'(x) = E(x) > 0$, it follows from Theorem 5.1.9 that L is differentiable on $(0,\infty)$ and that

$$L'(x) = \frac{1}{E' \circ L(x)} = \frac{1}{E \circ L(x)} = \frac{1}{x} \qquad \text{for} \quad x \in (0,\infty).$$

(viii) If $x > 0$, $y > 0$, let $u := L(x)$ and $v := L(y)$. Then we have $x = E(u)$ and $y = E(v)$. It follows from property (iv) of Theorem 7.3.6 that

$$xy = E(u)\,E(v) = E(u + v),$$

so that $L(xy) = L \circ E(u + v) = u + v = L(x) + L(y)$. This establishes (viii).

The properties in (ix) follow from the relations $E(0) = 1$ and $E(1) = e$.

To establish property (x), we first note that since $2 < e$, then $\lim (e^n) = \infty$ and $\lim (e^{-n}) = 0$. Since $L(e^n) = n$ and $L(e^{-n}) = -n$ it follows from the fact that L is strictly increasing that

$$\lim_{x \to \infty} L(x) = \lim L(e^n) = \infty \qquad \text{and} \qquad \lim_{x \to 0+} L(x) = \lim L(e^{-n}) = -\infty.$$

Q.E.D.

Power Functions

In Definition 4.8.7, we discussed the power function $x \mapsto x^r$, $x > 0$, where r is a rational number. By using the exponential and logarithm functions, we can extend the notion of power functions from rational to arbitrary real powers.

7.3.10 Definition. If $\alpha \in \boldsymbol{R}$ and $x > 0$, the number x^α is defined to be

$$x^\alpha := e^{\alpha \log x}.$$

The function $x \mapsto x^\alpha$ for $x > 0$ is called the **power function** with exponent α.

Note. If $x > 0$ and $\alpha = m/n$ where $m \in \boldsymbol{Z}$, $n \in \boldsymbol{N}$, then we defined $x^\alpha := (x^m)^{1/n}$ in Section 4.8. Hence we have

$$\log x^\alpha = \log (x^m)^{1/n} = \frac{1}{n} \log (x^m) = \frac{m}{n} \log x = \alpha \log x,$$

whence it follows that

$$x^\alpha = e^{\log x^\alpha} = e^{\alpha \log x}$$

Hence Definition 7.3.10 is consistent with the definition given in Section 4.8.

We now state some properties of the power functions. Their proofs are immediate consequences of the properties of the exponential and logarithm functions and will be left to the reader.

7.3.11 Theorem. *If $\alpha \in \boldsymbol{R}$ and x, y belong to $(0,\infty)$, then:*

(a) $1^\alpha = 1$; (b) $x^\alpha > 0$;

(c) $(xy)^\alpha = x^\alpha y^\alpha$; (d) $(x/y)^\alpha = x^\alpha/y^\alpha$.

7.3.12 Theorem. *If $\alpha, \beta \in \mathbf{R}$ and $x \in (0,\infty)$, then:*
(a) $x^{\alpha+\beta} = x^{\alpha}x^{\beta}$; (b) $(x^{\alpha})^{\beta} = x^{\alpha\beta} = (x^{\beta})^{\alpha}$;
(c) $x^{-\alpha} = 1/x^{\alpha}$; (d) *if* $\alpha < \beta$, *then* $x^{\alpha} < x^{\beta}$.

The next result concerns the differentiability of the power functions.

7.3.13 Theorem. *Let $\alpha \in \mathbf{R}$. Then the function $x \mapsto x^{\alpha}$ on $(0,\infty)$ to $\mathbf{R}$ is continuous and differentiable, and*

$$Dx^{\alpha} = \alpha x^{\alpha-1} \quad \text{for} \quad x \in (0,\infty).$$

Proof. By the Chain Rule we have

$$\begin{aligned} Dx^{\alpha} &= De^{\alpha \log x} = e^{\alpha \log x} \cdot D(\alpha \log x) \\ &= x^{\alpha} \cdot \frac{\alpha}{x} = \alpha x^{\alpha-1} \end{aligned}$$

for all $x \in (0,\infty)$. Q.E.D.

It will be seen in an exercise that if $\alpha > 0$ the power function $x \mapsto x^{\alpha}$ is strictly increasing on $(0,\infty)$ to $\mathbf{R}$, and that if $\alpha < 0$ the function $x \mapsto x^{\alpha}$ is strictly decreasing. (What happens if $\alpha = 0$?)

The graphs of the functions $x \mapsto x^{\alpha}$ on $(0,\infty)$ to $\mathbf{R}$ are similar to those in Figure 4.8.9.

The function $\log_a$

If $a > 0$, $a \neq 1$, it is sometimes useful to define the function $\log_a$.

7.3.14 Definition. Let $a > 0$, $a \neq 1$. We define

$$\log_a(x) := \frac{\log x}{\log a} \quad \text{for} \quad x \in (0,\infty).$$

For $x \in (0,\infty)$, the number $\log_a(x)$ is called the **logarithm of x to the base a**. The case $a = e$ yields the logarithm (or natural logarithm) function of Definition 7.3.8. The case $a = 10$ gives the base 10 logarithm (or common logarithm) function $\log_{10}$ often used in computations. Properties of the functions $\log_a$ will be given in the exercises.

Exercises for Section 7.3

1. Show that if $x > 0$ and if $n > 2x$, then

$$\left| e^x - \left(1 + \frac{x}{1!} + \cdots + \frac{x^n}{n!}\right) \right| < \frac{2x^{n+1}}{(n+1)!}.$$

Use this formula to show that $2\frac{2}{3} < e < 2\frac{3}{4}$; hence e is not an integer.

2. Calculate e correct to 5 decimal places.
3. Show that if $0 \leqslant x \leqslant a$ and $n \in \mathbf{N}$, then

$$1 + \frac{x}{1!} + \cdots + \frac{x^n}{n!} \leqslant e^x \leqslant 1 + \frac{x}{1!} + \cdots + \frac{x^{n-1}}{(n-1)!} + \frac{e^a x^n}{n!}.$$

4. Show that if $n \geqslant 2$, then

$$0 < en! - \left(1 + 1 + \frac{1}{2!} + \cdots + \frac{1}{n!}\right)n! < \frac{e}{n+1} < 1.$$

   Use this inequality to prove that e is not a rational number.
5. If $x \geqslant 0$ and $n \in \mathbf{N}$, show that

$$\frac{1}{x+1} = 1 - x + x^2 - x^3 + \cdots + (-x)^{n-1} + \frac{(-x)^n}{1+x}.$$

   Use this to show that

$$\log(x+1) = x - \frac{x^2}{2} + \frac{x^3}{3} - \cdots + (-1)^{n-1}\frac{x^n}{n} + \int_0^x \frac{(-t)^n}{1+t}\,dt$$

   and that

$$\left|\log(x+1) - \left(x - \frac{x^2}{2} + \frac{x^3}{3} - \cdots + (-1)^{n-1}\frac{x^n}{n}\right)\right| \leqslant \frac{x^{n+1}}{n+1}.$$

6. Use the formula in the preceding exercise to calculate log 1.1 and log 1.4 accurate to four decimal places. How large must one choose n in this inequality to calculate log 2 accurate to four decimal places?
7. Show that $\log(e/2) = 1 - \log 2$. Use this result to calculate log 2 accurate to four decimal places.
8. Let $f : \mathbf{R} \to \mathbf{R}$ be such that $f'(x) = f(x)$ for all $x \in \mathbf{R}$. Show that there exists $K \in \mathbf{R}$ such that $f(x) = Ke^x$ for all $x \in \mathbf{R}$.
9. Let $a_k > 0$ for $k = 1, \ldots, n$ and let $A := (a_1 + \cdots + a_n)/n$ be the arithmetic mean of these numbers. For each k, put $x_k := a_k/A - 1$ in the inequality $1 + x \leqslant e^x$ (valid for $x \geqslant 0$). Multiply the resulting terms to prove the Arithmetic-Geometric Mean Inequality

$$(*) \qquad (a_1 \cdots a_n)^{1/n} \leqslant \frac{1}{n}(a_1 + \cdots + a_n).$$

   Moreover, show that equality holds in $(*)$ if and only if $a_1 = a_2 = \cdots = a_n$.
10. Evaluate $L'(1)$ by using the sequence $(1 + 1/n)$ to prove that $e = \lim((1 + 1/n)^n)$.
11. Establish the assertions in Theorem 7.3.12.
12. Establish the assertions in Theorem 7.3.13.
13. (a) Show that if $\alpha > 0$, then the function $x \mapsto x^\alpha$ is strictly increasing on $(0,\infty)$ to $\mathbf{R}$ and that

$$\lim_{x\to 0} x^\alpha = 0 \qquad \text{and} \qquad \lim_{x\to\infty} x^\alpha = \infty.$$

(b) Show that if $\alpha < 0$, then the function $x \mapsto x^\alpha$ is strictly decreasing on $(0,\infty)$ to $\boldsymbol{R}$ and that

$$\lim_{x\to 0} x^\alpha = \infty \qquad \text{and} \qquad \lim_{x\to\infty} x^\alpha = 0.$$

14. Prove that if $a > 0$, $a \neq 1$, then $a^{\log_a x} = x$ for all $x \in (0,\infty)$ and $\log_a (a^y) = y$ for all $y \in \boldsymbol{R}$. Show that the function $x \mapsto \log_a x$ on $(0,\infty)$ to $\boldsymbol{R}$ is inverse to the function $y \mapsto a^y$ on $\boldsymbol{R}$.
15. If $a > 0$, $a \neq 1$, show that the function $x \mapsto \log_a x$ is differentiable on $(0,\infty)$ and that

$$D \log_a x = 1/(x \log a) \qquad \text{for} \quad x \in (0,\infty).$$

16. If $a > 0$, $a \neq 1$, and x and y belong to $(0,\infty)$, show that $\log_a (xy) = \log_a x + \log_a y$.
17. If $a > 0$, $a \neq 1$, and $b > 0$, $b \neq 1$, show that

$$\log_a x = \frac{\log b}{\log a} \log_b x \qquad \text{for} \quad x \in (0,\infty).$$

In particular, show that $\log_{10} x = (\log e/\log 10) \log x = (\log_{10} e) \log x$ for $x \in (0,\infty)$.

SECTION 7.4 The Trigonometric Functions

Along with the exponential and logarithmic functions, there is another very important collection of transcendental functions known as the trigonometric functions. These are the sine, cosine, tangent, cotangent, secant, and cosecant functions. In elementary courses, they are usually introduced on a geometric basis in terms of either triangles or the unit circle. In this section, we introduce the trigonometric functions in an analytical manner and then establish some of their basic properties. In particular, the various properties of the trigonometric functions that were used in examples in earlier parts of this book will be derived rigorously in this section.

It suffices to deal with the sine and cosine since the other four trigonometric functions are defined in terms of these two. Our approach to the sine and cosine is similar in spirit to our approach to the exponential function in that we first establish the existence of functions that satisfy certain differentiation properties.

7.4.1 Theorem. *There exist functions $C : \boldsymbol{R} \to \boldsymbol{R}$ and $S : \boldsymbol{R} \to \boldsymbol{R}$ such that*
(i) $C''(x) = -C(x)$ and $S''(x) = -S(x)$ for all $x \in \boldsymbol{R}$;
(ii) $C(0) = 1$, $C'(0) = 0$, and $S(0) = 0$, $S'(0) = 1$.

Proof. We define the sequences (C_n) and (S_n) of continuous functions inductively as follows:

$$(1) \qquad C_1(x) := 1, \qquad S_1(x) := x,$$

$$(2) \qquad S_n(x) := \int_0^x C_n(t)\,dt,$$

$$(3) \qquad C_{n+1}(x) := 1 - \int_0^x S_n(t)\,dt,$$

for all $n \in \boldsymbol{N}$, $x \in \boldsymbol{R}$.

One sees by induction that the functions C_n and S_n are continuous on $\boldsymbol{R}$ and hence they are integrable over any bounded interval; thus these functions are well-defined by the above formulas. Moreover, it follows from the Differentiation Theorem 6.3.4 that S_n and C_{n+1} are differentiable at every point and that

$$(4) \qquad S_n'(x) = C_n(x) \qquad \text{and} \qquad C_{n+1}'(x) = -S_n(x) \qquad \text{for} \quad n \in \boldsymbol{N},\ x \in \boldsymbol{R}.$$

Easy induction arguments (which we leave to the reader) show that

$$C_{n+1}(x) = 1 - \frac{x^2}{2!} + \frac{x^4}{4!} - \cdots + (-1)^n \frac{x^{2n}}{(2n)!},$$

$$S_{n+1}(x) = x - \frac{x^3}{3!} + \frac{x^5}{5!} - \cdots + (-1)^n \frac{x^{2n+1}}{(2n+1)!}.$$

Let $A > 0$ be given. Then if $|x| \leqslant A$ and $m \geqslant n > 2A$, we have that (since $A/2n < 1/4$):

$$(5) \qquad |C_m(x) - C_n(x)| = \left| \frac{x^{2n}}{(2n)!} - \frac{x^{2n+2}}{(2n+2)!} + \cdots \pm \frac{x^{2m-2}}{(2m-2)!} \right|$$
$$\leqslant \frac{A^{2n}}{(2n)!}\left[1 + \left(\frac{A}{2n}\right)^2 + \cdots + \left(\frac{A}{2n}\right)^{2m-2n-2}\right]$$
$$< \frac{A^{2n}}{(2n)!}\left(\frac{16}{15}\right).$$

Since $\lim \left(A^{2n}/(2n)!\right) = 0$, the sequence (C_n) converges uniformly on the interval $[-A, A]$, where $A > 0$ is arbitrary. In particular, this means that $(C_n(x))$ converges for each $x \in \boldsymbol{R}$. We define $C : \boldsymbol{R} \to \boldsymbol{R}$ by

$$C(x) := \lim C_n(x) \qquad \text{for} \quad x \in \boldsymbol{R}.$$

It follows from Theorem 7.2.2 that C is continuous on $\boldsymbol{R}$ and, since $C_n(0) = 1$ for all $n \in \boldsymbol{N}$, that $C(0) = 1$.

If $|x| \leqslant A$ and $m \geqslant n > 2A$, it follows from (2) that

$$S_m(x) - S_n(x) = \int_0^x \{C_m(t) - C_n(t)\}\,dt.$$

If we use (5) and Corollary 6.2.13, we conclude that

$$|S_m(x) - S_n(x)| \leqslant \frac{A^{2n}}{(2n)!}\left(\frac{16}{15}A\right),$$

whence it follows that the sequence (S_n) converges uniformly on $[-A,A]$. We define $S : \boldsymbol{R} \to \boldsymbol{R}$ by

$$S(x) := \lim S_n(x) \qquad \text{for} \quad x \in \boldsymbol{R}.$$

It follows from Theorem 7.2.2 that S is continuous on $\boldsymbol{R}$ and, since $S_n(0) = 0$ for all $n \in \boldsymbol{N}$, that $S(0) = 0$.

Since $C_n'(x) = -S_{n-1}(x)$ for $n > 1$, it follows from the above that the sequence (C_n') converges uniformly on $[-A,A]$. Hence by Theorem 7.2.3, the limit function C is differentiable on $[-A,A]$ and

$$C'(x) = \lim C_n'(x) = \lim\left(-S_{n-1}(x)\right) = -S(x) \qquad \text{for} \quad x \in [-A,A].$$

Since $A > 0$ is arbitrary, we have

(6) $$C'(x) = -S(x) \qquad \text{for} \quad x \in \boldsymbol{R}.$$

A similar argument, based on the fact that $S_n'(x) = C_n(x)$, shows that S is differentiable on $\boldsymbol{R}$ and that

(7) $$S'(x) = C(x) \qquad \text{for all} \quad x \in \boldsymbol{R}.$$

It follows from (6) and (7) that

$$C''(x) = -\left(S(x)\right)' = -C(x) \qquad \text{and} \qquad S''(x) = \left(C(x)\right)' = -S(x)$$

for all $x \in \boldsymbol{R}$. Moreover, we have

$$C'(0) = -S(0) = 0, \qquad S'(0) = C(0) = 1.$$

Thus statements (i) and (ii) are proved. Q.E.D.

7.4.2 Corollary. *If C, S are the functions in Theorem 7.4.1, then*
(iii) $C'(x) = -S(x)$ *and* $S'(x) = C(x)$ *for* $x \in \boldsymbol{R}$.
Moreover, these functions have derivatives of all orders.

Proof. The formulas (iii) were established in (6) and (7). The existence of the higher order derivatives follows by induction. Q.E.D.

7.4.3 Corollary. *The functions C and S satisfy the Pythagorean Identity*

(iv) $\left(C(x)\right)^2 + \left(S(x)\right)^2 = 1$ *for* $x \in \boldsymbol{R}$

Proof. Let $f(x) := \left(C(x)\right)^2 + \left(S(x)\right)^2$ for $x \in \boldsymbol{R}$, so that

$$f'(x) = 2C(x)\left(-S(x)\right) + 2S(x)\left(C(x)\right) = 0 \qquad \text{for} \quad x \in \boldsymbol{R}.$$

Thus it follows that $f(x)$ is a constant for all $x \in \boldsymbol{R}$. But since $f(0) = 1 + 0 = 1$, we conclude that $f(x) = 1$ for all $x \in \boldsymbol{R}$. Q.E.D.

We next establish the uniqueness of the functions C and S.

7.4.4 Theorem. *The functions C and S satisfying properties* (i) *and* (ii) *of Theorem* 7.4.1 *are unique.*

Proof. Let C_1 and C_2 be two functions on $\boldsymbol{R}$ to $\boldsymbol{R}$ that satisfy $C_j''(x) = -C_j(x)$ for all $x \in \boldsymbol{R}$ and $C_j(0) = 1$, $C_j'(0) = 0$ for $j = 1, 2$. If we let $D := C_1 - C_2$, then $D''(x) = -D(x)$ for $x \in \boldsymbol{R}$ and $D(0) = 0$ and $D^{(k)}(0) = 0$ for all $k \in \boldsymbol{N}$.

Now let $x \in \boldsymbol{R}$ be arbitrary, and let I_x be the interval with end points $0, x$. Since $D = C_1 - C_2$ and $T := S_1 - S_2 = C_2' - C_1'$ are continuous on I_x, there exists $K > 0$ such that $|D(t)| \leqslant K$ and $|T(t)| \leqslant K$ for all $t \in I_x$. If we apply Taylor's Theorem 5.4.1 to D on I_x and use the fact that $D(0) = 0$, $D^{(k)}(0) = 0$ for $k \in \boldsymbol{N}$, it follows that for each $n \in \boldsymbol{N}$ there is a point $c_n \in I_x$ such that

$$\begin{aligned} D(x) &= D(0) + \frac{D'(0)}{1!}x + \cdots + \frac{D^{(n-1)}(0)}{(n-1)!}x^{n-1} + \frac{D^{(n)}(c_n)}{n!}x^n \\ &= \frac{D^{(n)}(c_n)}{n!}x^n. \end{aligned}$$

Now either $D^{(n)}(c_n) = \pm D(c_n)$ or $D^{(n)}(c_n) = \pm T(c_n)$. In either case we have

$$|D(x)| \leqslant \frac{K|x|^n}{n!}.$$

But since $\lim \left(|x|^n/n!\right) = 0$, we conclude that $D(x) = 0$. Since $x \in \boldsymbol{R}$ is arbitrary, we infer that $C_1(x) - C_2(x) = 0$ for all $x \in \boldsymbol{R}$.

A similar argument shows that if S_1 and S_2 are two functions on $\boldsymbol{R} \to \boldsymbol{R}$ such that $S_j''(x) = -S_j(x)$ for all $x \in \boldsymbol{R}$ and $S_j(0) = 0$, $S_j'(0) = 1$ for $j = 1, 2$, then we have $S_1(x) = S_2(x)$ for all $x \in \boldsymbol{R}$. Q.E.D.

Now that existence and uniqueness of the functions C and S have been established, we shall give these functions their familiar names.

7.4.5 Definition. The unique functions $C : \boldsymbol{R} \to \boldsymbol{R}$ and $S : \boldsymbol{R} \to \boldsymbol{R}$ such that $C''(x) = -C(x)$ and $S''(x) = -S(x)$ for all $x \in \boldsymbol{R}$ and $C(0) = 1$, $C'(0) = 0$ and $S(0) = 0$, $S'(0) = 1$, are called the **cosine function** and the **sine function**, respectively. We ordinarily write

$$\cos x := C(x) \qquad \text{and} \qquad \sin x := S(x) \qquad \text{for} \quad x \in \boldsymbol{R}.$$

The differentiation properties in (i) of Theorem 7.4.1 do not by themselves lead to uniquely determined functions. We have the following relationship.

7.4.6 Theorem. *If $f : \boldsymbol{R} \to \boldsymbol{R}$ is such that*

$$f''(x) = -f(x) \qquad \textit{for} \quad x \in \boldsymbol{R},$$

then there exist real numbers α, β such that

$$f(x) = \alpha C(x) + \beta S(x) \qquad \textit{for} \quad x \in \boldsymbol{R}.$$

Proof. Let $g(x) := f(0)C(x) + f'(0)S(x)$ for $x \in \boldsymbol{R}$. It is readily seen that $g''(x) = -g(x)$ and that $g(0) = f(0)$, and since

$$g'(x) = -f(0)S(x) + f'(0)C(x),$$

that we have $g'(0) = f'(0)$. Therefore the function $h := f - g$ has the property that $h''(x) = -h(x)$ for all $x \in \boldsymbol{R}$ and $h(0) = 0$, $h'(0) = 0$. Thus it follows from the proof of the preceding theorem that $h(x) = 0$ for all $x \in \boldsymbol{R}$. Therefore $f(x) = g(x)$ for all $x \in \boldsymbol{R}$. Q.E.D.

We shall now derive a few of the basic properties of the cosine and sine functions.

7.4.7 Theorem. *The function C is even and S is odd in the sense that*

(v) $C(-x) = C(x)$ *and* $S(-x) = -S(x)$ *for* $x \in \boldsymbol{R}$.

If $x, y \in \boldsymbol{R}$, then we have the "addition formulas"

(vi) $C(x + y) = C(x)C(y) - S(x)S(y)$, $\quad S(x + y) = S(x)C(y) + C(x)S(y)$.

Proof. (v) If $\varphi(x) := C(-x)$ for $x \in \boldsymbol{R}$, then a calculation shows that $\varphi''(x) = -\varphi(x)$ for $x \in \boldsymbol{R}$. Moreover, $\varphi(0) = 1$ and $\varphi'(0) = 0$ so that $\varphi = C$. Hence, $C(-x) = C(x)$ for all $x \in \boldsymbol{R}$. In a similar way one shows that $S(-x) = -S(x)$ for all $x \in \boldsymbol{R}$.

(vi) Let $y \in \boldsymbol{R}$ be given and let $f(x) := C(x + y)$ for $x \in \boldsymbol{R}$. A calculation shows that $f''(x) = -f(x)$ for $x \in \boldsymbol{R}$. Hence, by Theorem 7.4.6, there exists real numbers α, β such that

$$f(x) = C(x + y) = \alpha C(x) + \beta S(x) \qquad \text{and}$$
$$f'(x) = -S(x + y) = -\alpha S(x) + \beta C(x)$$

for $x \in \boldsymbol{R}$. If we let $x = 0$, we obtain

$$C(y) = \alpha, \qquad -S(y) = \beta,$$

whence the first formula in (vi) follows. The second formula is proved similarly. Q.E.D.

The following inequalities were used earlier (for example, in 4.2.8).

7.4.8 Theorem. *If $x \in \boldsymbol{R}$, $x \geqslant 0$, then we have*
(vii) $-x \leqslant S(x) \leqslant x$;
(viii) $1 - \dfrac{x^2}{2} \leqslant C(x) \leqslant 1$;
(ix) $x - \dfrac{x^3}{6} \leqslant S(x) \leqslant x$;
(x) $1 - \dfrac{x^2}{2} \leqslant C(x) \leqslant 1 - \dfrac{x^2}{2} + \dfrac{x^4}{24}$.

Proof. It follows from Corollary 7.4.3 that $-1 \leqslant C(t) \leqslant 1$ for $t \in \boldsymbol{R}$ so that if $x \geqslant 0$, then

$$-x \leqslant \int_0^x C(t)\,dt \leqslant x,$$

whence we have (vii). If we integrate this inequality, we obtain

$$-\frac{x^2}{2} \leqslant \int_0^x S(t)\,dt \leqslant \frac{x^2}{2},$$

whence we have

$$-\frac{x^2}{2} \leqslant -C(x) + 1 \leqslant \frac{x^2}{2}.$$

Thus we have $1 - (x^2/2) \leqslant C(x)$, from which (viii) follows. Inequality (ix) follows by integrating (viii), and (x) follows by integrating (ix). Q.E.D.

The number π is obtained via the following lemma.

7.4.9 Lemma. *There exists a root γ of the cosine function in the interval $(\sqrt{2},\sqrt{3})$. Moreover $C(x) > 0$ for $x \in [0,\gamma)$. The number 2γ is the smallest strictly positive root of S.*

Proof. It follows from inequality (x) of Theorem 7.4.8 that C has a root between the positive root $\sqrt{2}$ of $x^2 - 2 = 0$ and the smallest positive root of $x^4 - 12x^2 + 24 = 0$, which is $\sqrt{6 - 2\sqrt{3}} < \sqrt{3}$. We let γ be the smallest such root of C.

It follows from the second formula in (vi) with $x = y$ that $S(2x) = 2S(x)C(x)$. This relation implies that $S(2\gamma) = 0$, so that 2γ is a strictly positive root of S. The same relation implies that if $2\delta > 0$ is the smallest strictly positive root of S, then $C(\delta) = 0$. Since γ is the smallest positive root of C, we have $\delta = \gamma$.

Q.E.D.

7.4.10 Definition. We let $\pi := 2\gamma$ denote the smallest strictly positive root of S.

Note. It follows from the inequality $\sqrt{2} < \gamma < \sqrt{6 - 2\sqrt{3}}$ that $2.829 < \pi = 2\gamma < 3.185$.

7.4.11 Theorem. *The functions C and S have period 2π in the sense that*

(xi) $C(x + 2\pi) = C(x)$ *and* $S(x + 2\pi) = S(x)$ *for* $x \in \boldsymbol{R}$.

Moreover we have

(xii) $S(x) = C(\frac{1}{2}\pi - x) = -C(x + \frac{1}{2}\pi)$, $C(x) = S(\frac{1}{2}\pi - x) = S(x + \frac{1}{2}\pi)$, *for all* $x \in \boldsymbol{R}$.

Proof. (xi) Since $S(2x) = 2S(x)C(x)$ and $S(\pi) = 0$, it follows that $S(2\pi) = 0$. Further, if $x = y$ in (vi), we obtain $C(2x) = (C(x))^2 - (S(x))^2 + 1$. Therefore $C(2\pi) = 1$. Hence it follows from (vi) with $y = 2\pi$ that

$$C(x + 2\pi) = C(x)C(2\pi) - S(x)S(2\pi) = C(x),$$

and

$$S(x + 2\pi) = S(x)C(2\pi) + C(x)S(2\pi) = S(x).$$

(xii) We note that $C(\frac{1}{2}\pi) = 0$, and it is an exercise to show that $S(\frac{1}{2}\pi) = 1$. If we employ these together with formulas (vi), the desired relations are obtained.

Q.E.D.

Exercises for Section 7.4

1. Calculate cos (.2), sin (.2) and cos 1, sin 1 correct to four decimal places.
2. Show that $|\sin x| \leq 1$ and $|\cos x| \leq 1$ for all $x \in \boldsymbol{R}$.
3. Show that property (viii) of Theorem 7.4.8 does not hold if $x < 0$, but that we have $|\sin x| \leq |x|$ for all $x \in \boldsymbol{R}$. Also show that $|\sin x - x| \leq |x|^3/6$ for all $x \in \boldsymbol{R}$.

4. Show that if $x > 0$ then

$$1 - \frac{x^2}{2} + \frac{x^4}{24} - \frac{x^6}{720} \leqslant \cos x \leqslant 1 - \frac{x^2}{2} + \frac{x^4}{24}.$$

Use this inequality to establish a lower bound for π.

5. Calculate π by approximating the smallest strictly positive zero of sin. (Either bisect intervals or use Newton's Method of Section 5.4.)

6. Define the sequences (c_n) and (s_n) inductively by $c_1(x) := 1$, $s_1(x) := x$, and

$$s_n(x) = \int_0^x c_n(t)\,dt, \qquad c_{n+1}(x) = 1 + \int_0^x s_n(t)\,dt$$

for all $n \in \boldsymbol{N}$, $x \in \boldsymbol{R}$. Reason as in the proof of Theorem 7.4.1 to conclude that there exist functions $c : \boldsymbol{R} \to \boldsymbol{R}$ and $s : \boldsymbol{R} \to \boldsymbol{R}$ such that (j) $c''(x) = c(x)$ and $s''(x) = s(x)$ for all $x \in \boldsymbol{R}$, and (jj) $c(0) = 1$, $c'(0) = 0$, and $s(0) = 0$, $s'(0) = 1$. Moreover, $c'(x) = s(x)$ and $s'(x) = c(x)$ for all $x \in \boldsymbol{R}$.

7. Show that the functions c, s in the preceding exercise have derivatives of all orders, that they satisfy the identity $(c(x))^2 - (s(x))^2 = 1$ for all $x \in \boldsymbol{R}$. Moreover, they are the unique functions satisfying (j) and (jj). (The functions c, s are called the **hyperbolic cosine** and **hyperbolic sine functions**, respectively.)

8. If $f : \boldsymbol{R} \to \boldsymbol{R}$ is such that $f''(x) = f(x)$ for all $x \in \boldsymbol{R}$, show that there exist real numbers α, β such that $f(x) = \alpha c(x) + \beta s(x)$ for all $x \in \boldsymbol{R}$. Apply this result to the functions $f_1(x) := e^x$ and $f_2(x) := e^{-x}$ for $x \in \boldsymbol{R}$. Show that $c(x) = \frac{1}{2}(e^x + e^{-x})$ and $s(x) = \frac{1}{2}(e^x - e^{-x})$ for $x \in \boldsymbol{R}$.

9. Show that the functions c, s in the preceding exercises are even and odd, respectively, and that

$$c(x + y) = c(x)c(y) + s(x)s(y), \qquad s(x + y) = s(x)c(y) + c(x)s(y),$$

for all $x, y \in \boldsymbol{R}$.

10. Show that $c(x) \geqslant 1$ for all $x \in \boldsymbol{R}$, that both c and s are strictly increasing on $(0,\infty)$, and that $\lim_{x\to\infty} c(x) = \lim_{x\to\infty} s(x) = \infty$.

CHAPTER EIGHT

INFINITE SERIES

In this chapter we are concerned with establishing the most important theorems in the theory of infinite series. Although a few peripheral results are included here, our attention is directed to the basic propositions. The reader is referred to more extensive treatises for advanced results and applications.

In the first section we shall present the main theorems concerning the convergence of infinite series in $\boldsymbol{R}$. We shall obtain some results of a general nature that serve to establish the convergence of series and justify certain manipulations with series.

In Section 8.2 we shall give some familiar "tests" for absolute convergence of series. In addition to guaranteeing the convergence of the series to which the tests are applicable, each of these tests yields a quantitative estimate concerning the *rapidity* of the convergence.

In Section 8.3 we introduce the study of series of *functions* and establish the basic properties of power series.

SECTION 8.1 Convergence of Infinite Series

In elementary texts, an infinite series is sometimes "defined" to be "an expression of the form

(1) $$x_1 + x_2 + \cdots + x_n + \cdots".$$

This "definition" lacks clarity, however, since there is no particular value that we can attach *a priori* to this array of symbols, which calls for an infinite number of additions to be performed. Although there are other definitions that are suitable, we shall take an infinite series to be the same as the sequence of partial sums.

8.1.1 Definition. If $X := (x_n)$ is a sequence in $\boldsymbol{R}$, then the **infinite series** (or simply the series) generated by X is the sequence $S := (s_k)$ defined by

$$
\begin{aligned}
s_1 &:= x_1, \\
s_2 &:= s_1 + x_2 \qquad (= x_1 + x_2), \\
&\cdots\cdots\cdots\cdots \\
s_k &:= s_{k-1} + x_k \qquad (= x_1 + x_2 + \cdots + x_k), \\
&\cdots\cdots\cdots\cdots\cdots
\end{aligned}
$$

If S converges, we refer to lim S as the **sum** of the infinite series. The elements x_n are called the **terms** and the elements s_k are called the **partial sums** of this infinite series.

It is conventional to use the expression (1) or one of the symbols

$$\sum (x_n), \qquad \sum_{n=1}^{\infty} (x_n), \qquad \sum_{n=1}^{\infty} x_n$$

both to denote the infinite series generated by the sequence $X = (x_n)$ and also to denote lim S in the case that this infinite series is convergent. In actual practice, the double use of these notations does not lead to confusion, provided it is understood that the convergence of the series must be established.

The reader should guard against confusing the words "sequence" and "series". In non-mathematical language, these words are interchangeable; in mathematics, however, they are not synonyms. According to our definition, an infinite series is a sequence S obtained from a given sequence X according to a special procedure that was stated above.

A final word on notational matters. Although we generally index the elements of the series by natural numbers, it is sometimes more convenient to start with $n = 0$, with $n = 5$, or with $n = k$. When such is the case, we shall denote the resulting series or their sums by notations such as

$$\sum_{n=0}^{\infty} x_n, \qquad \sum_{n=5}^{\infty} x_n, \qquad \sum_{n=k}^{\infty} x_n.$$

In Definition 3.1.3, we defined the sum and difference of two sequences X, Y in $\mathbf{R}$. Similarly, if c is a real number, we defined the sequence $cX = (cx_n)$. We now examine the series generated by these sequences.

8.1.2 Theorem. (a) *If the series $\Sigma(x_n)$ and $\Sigma(y_n)$ converge, then the series $\Sigma(x_n + y_n)$ converges and the sums are related by the formula*

$$\sum (x_n + y_n) = \sum (x_n) + \sum (y_n).$$

A similar result holds for the series generated by $X - Y$.

(b) *If the series $\Sigma(x_n)$ is convergent and c is a real number, then the series $\Sigma(cx_n)$ converges and*

$$\sum (cx_n) = c \sum (x_n).$$

Proof. This result follows directly from Theorem 3.2.3 and Definition 8.1.1. Q.E.D.

It might be expected that if the sequences $X = (x_n)$ and $Y = (y_n)$ generate convergent series, then the sequence $XY = (x_n y_n)$ also generates a convergent series. That this is not always true may be seen by taking the sequences $X = Y := ((-1)^n/\sqrt{n})$.

We now present a very simple necessary condition for convergence of a series. It is far from sufficient, however.

8.1.3 Lemma. *If* $\Sigma(x_n)$ *converges in* $\boldsymbol{R}$, *then* $\lim(x_n) = 0$.

Proof. By definition, the convergence of $\Sigma(x_n)$ means that $\lim(s_k)$ exists. But, since $x_k = s_k - s_{k-1}$, then $\lim(x_k) = \lim(s_k) - \lim(s_{k-1}) = 0$. Q.E.D.

The next result, although limited in scope, is of great importance.

8.1.4 Theorem. *Let* (x_n) *be a sequence of positive real numbers. Then* $\Sigma(x_n)$ *converges if and only if the sequence* $S = (s_k)$ *of partial sums is bounded. In this case,*

$$\sum x_n = \lim(s_k) = \sup\{s_k\}.$$

Proof. Since $x_n \geqslant 0$, the sequence of partial sums is monotone increasing:

$$s_1 \leqslant s_2 \leqslant \cdots \leqslant s_k \leqslant \cdots.$$

According to the Monotone Convergence Theorem 3.3.2, the sequence S converges if and only if it is bounded. Q.E.D.

Since the following Cauchy Criterion is precisely a reformulation of Theorem 3.5.4, we shall omit its proof.

8.1.5 Cauchy Criterion for Series. *The series* $\Sigma(x_n)$ *in* $\boldsymbol{R}$ *converges if and only if for each number* $\varepsilon > 0$ *there is a natural number* $M(\varepsilon)$ *such that if* $m \geqslant n \geqslant M(\varepsilon)$, *then*

$$|s_m - s_n| = |x_{n+1} + x_{n+2} + \cdots + x_m| < \varepsilon.$$

The notion of absolute convergence is often of great importance in treating series, as we shall show later.

8.1.6 Definition. Let $X := (x_n)$ be a sequence in $\boldsymbol{R}$. We say that the series $\Sigma(x_n)$ is **absolutely convergent** if the series $\Sigma(|x_n|)$ is convergent in $\boldsymbol{R}$. A series is said to be **conditionally convergent** if it is convergent but not absolutely convergent.

It is stressed that for series whose elements are *positive* real numbers, there is no distinction between ordinary convergence and absolute convergence. However, for other series there may be a difference.

8.1.7 Theorem. *If a series in* $\mathbf{R}$ *is absolutely convergent, then it is convergent.*

Proof. By hypothesis, the series $\Sigma(|x_n|)$ converges. Therefore, it follows from the necessity of the Cauchy Criterion 8.1.5 that given $\varepsilon > 0$ there is a natural number $M(\varepsilon)$ such that if $m \geqslant n \geqslant M(\varepsilon)$, then

$$|x_{n+1}| + |x_{n+2}| + \cdots + |x_m| < \varepsilon .$$

According to the Triangle Inequality, the left-hand side of this relation dominates

$$|x_{n+1} + x_{n+2} + \cdots + x_m| .$$

We apply the sufficiency of the Cauchy Criterion to conclude that the series $\Sigma(x_n)$ must converge. Q.E.D.

8.1.8 Examples. (a) We consider the real sequence $X := (a^n)$, which generates the **geometric series**:

$$a + a^2 + \cdots + a^n + \cdots . \tag{2}$$

A necessary condition for convergence is that $\lim (a^n) = 0$, which requires that $|a| < 1$. If $m \geqslant n$, then

$$a^{n+1} + a^{n+2} + \cdots + a^m = \frac{a^{n+1} - a^{m+1}}{1 - a}, \tag{3}$$

as can be verified by multiplying both sides by $1 - a$ and noticing the telescoping on the left side. Hence the partial sums satisfy

$$|s_m - s_n| = |a^{n+1} + \cdots + a^m| \leqslant \frac{|a^{n+1}| + |a^{m+1}|}{|1 - a|}, \qquad m \geqslant n .$$

If $|a| < 1$, then $|a^{n+1}| \to 0$ so the Cauchy Criterion implies that the geometric series (2) converges if and only if $|a| < 1$. Letting $n = 0$ in (3) and passing to the limit with respect to m we find that (2) converges to the limit $a/(1 - a)$ when $|a| < 1$.

(b) Consider the **harmonic series** $\Sigma(1/n)$, which is well-known to diverge. Since $\lim (1/n) = 0$, we cannot use Lemma 8.1.3 to establish this divergence, but must carry out a more delicate argument, which we shall base on Theorem 8.1.4. We shall show that a subsequence of the partial sums is not bounded. In fact, if $k_1 = 2$, then

$$s_{k_1} = \frac{1}{1} + \frac{1}{2},$$

and if $k_2 = 2^2$, then

$$s_{k_2} = \frac{1}{1} + \frac{1}{2} + \frac{1}{3} + \frac{1}{4} = s_{k_1} + \frac{1}{3} + \frac{1}{4} > s_{k_1} + 2\left(\frac{1}{4}\right) = 1 + \frac{2}{2}.$$

By mathematical induction, we establish that if $k_r = 2^r$, then

$$s_{k_r} > s_{k_{r-1}} + 2^{r-1}\left(\frac{1}{2^r}\right) = s_{k_{r-1}} + \frac{1}{2} \geqslant 1 + \frac{r}{2}.$$

Therefore, the subsequence (s_{k_n}) is not bounded and the harmonic series does not converge.

(c) We now treat the p-**series** $\Sigma(1/n^p)$ where $0 < p \leqslant 1$ and use the elementary inequality $n^p \leqslant n$, for $n \in \mathbf{N}$. From this it follows that, when $0 < p \leqslant 1$, then

$$\frac{1}{n} \leqslant \frac{1}{n^p}, \qquad \text{for} \quad n \in \mathbf{N}.$$

Since the partial sums of the harmonic series are not bounded, this inequality shows that the partial sums of $\Sigma(1/n^p)$ are not bounded for $0 < p \leqslant 1$. Hence the series diverges for these values of p.

(d) Consider the p-series for $p > 1$. Since the partial sums are monotone, it is sufficient to show that some subsequence remains bounded in order to establish the convergence of the series. If $k_1 = 2^1 - 1 = 1$, then $s_{k_1} = 1$. If $k_2 = 2^2 - 1 = 3$, we have

$$s_{k_2} = \frac{1}{1} + \left(\frac{1}{2^p} + \frac{1}{3^p}\right) < 1 + \frac{2}{2^p} = 1 + \frac{1}{2^{p-1}},$$

and if $k_3 = 2^3 - 1$, we have

$$s_{k_3} = s_{k_2} + \left(\frac{1}{4^p} + \frac{1}{5^p} + \frac{1}{6^p} + \frac{1}{7^p}\right) < s_{k_2} + \frac{4}{4^p} < 1 + \frac{1}{2^{p-1}} + \frac{1}{4^{p-1}}.$$

Let $a = 1/2^{p-1}$; since $p > 1$, it is seen that $0 < a < 1$. By mathematical induction, we find that if $k_r = 2^r - 1$, then

$$0 < s_{k_r} < 1 + a + a^2 + \cdots + a^{r-1}.$$

Hence the number $1/(1 - a)$ is an upper bound for the partial sums of the p-series when $1 < p$. From Theorem 8.1.4 it follows that for such values of p, the p-series converges.

(e) Consider the series $\Sigma\left(1/(n^2 + n)\right)$. By using partial fractions, we can write

$$\frac{1}{k^2 + k} = \frac{1}{k(k+1)} = \frac{1}{k} - \frac{1}{k+1}.$$

This expression shows that the partial sums are telescoping and hence

$$s_n = \frac{1}{1\cdot 2} + \frac{1}{2\cdot 3} + \cdots + \frac{1}{n(n+1)} = \frac{1}{1} - \frac{1}{n+1}.$$

It follows that the sequence (s_n) is convergent to 1.

Rearrangements of Series

Loosely speaking, a rearrangement of a series is another series which is obtained from the given one by using all of the terms exactly once, but scrambling the order in which the terms are taken. For example, the harmonic series

$$\frac{1}{1} + \frac{1}{2} + \frac{1}{3} + \cdots + \frac{1}{n} + \cdots$$

has rearrangements

$$\frac{1}{2} + \frac{1}{1} + \frac{1}{4} + \frac{1}{3} + \cdots + \frac{1}{2n} + \frac{1}{2n-1} + \cdots,$$

$$\frac{1}{1} + \frac{1}{2} + \frac{1}{4} + \frac{1}{3} + \frac{1}{5} + \frac{1}{7} + \cdots$$

The first rearrangement is obtained by interchanging the first and second terms, the third and fourth terms, and so forth. The second rearrangement is obtained from the harmonic series by taking one "odd term", two "even terms", three "odd terms", and so on. It is evident that there are infinitely many other possible rearrangements of the harmonic series.

8.1.9 Definition. A series $\Sigma(y_m)$ in $\boldsymbol{R}$ is a **rearrangement** of a series $\Sigma(x_n)$ if there exists a bijection f of $\boldsymbol{N}$ onto $\boldsymbol{N}$ such that $y_m = x_{f(m)}$ for all $m \in \boldsymbol{N}$.

There is a remarkable observation due to Riemann, that if $\Sigma(x_n)$ is a series in $\boldsymbol{R}$ that is conditionally convergent (that is, it is convergent but not absolutely convergent) and if c is an arbitrary real number, then there exists a rearrangement of $\Sigma(x_n)$ which converges to c. The idea of the proof of this assertion is very elementary: we take positive terms until we obtain a partial sum exceeding c, then we take negative terms from the given series until we obtain a partial sum of terms less than c, etc. Since $\lim (x_n) = 0$, it is not difficult to see that a rearrangement which converges to c can be constructed.

In our manipulations with series, we generally find it convenient to be sure that rearrangements will not affect the convergence or the value of the limit.

8.1.10 Rearrangement Theorem. *Let $\Sigma(x_n)$ be an absolutely convergent series in $\boldsymbol{R}$. Then any rearrangement of $\Sigma(x_n)$ converges absolutely to the same value.*

Proof. Let $x = \Sigma(x_n)$, let $\Sigma(y_m)$ be a rearrangement of $\Sigma(x_n)$, and let K be an upper bound for the partial sums of $\Sigma(|x_n|)$. Clearly, if $t_r = y_1 + \cdots + y_r$ is a partial sum of $\Sigma(y_m)$, then

$$|y_1| + \cdots + |y_r| \leqslant K,$$

whence it follows that $\Sigma(y_m)$ is absolutely convergent to some element y of $\boldsymbol{R}$. We wish to show that $x = y$. If $\varepsilon > 0$, let $N(\varepsilon)$ be such that if $m > n \geqslant N(\varepsilon)$, and $s_n = x_1 + \cdots + x_n$, then $|x - s_n| < \varepsilon$ and

$$\sum_{k=n+1}^{m} |x_k| < \varepsilon .$$

Choose a partial sum t_r of $\Sigma(y_m)$ such that $|y - t_r| < \varepsilon$ and such that each $x_1, x_2, \ldots, x_n$ occurs in t_r. Having done this, choose $m > n$ so large that every y_k appearing in t_r also appears in s_m. Therefore

$$|x - y| \leqslant |x - s_m| + |s_m - t_r| + |t_r - y| < \varepsilon + \sum_{n+1}^{m} |x_k| + \varepsilon < 3\varepsilon .$$

Since $\varepsilon > 0$ is arbitrary, we infer that $x = y$. Q.E.D.

Exercises for Section 8.1

1. Let $\Sigma(a_n)$ be a given series and let $\Sigma(b_n)$ be one in which the terms are the same as those in $\Sigma(a_n)$, except those for which $a_n = 0$ have been omitted. Show that $\Sigma(a_n)$ converges to a number A if and only if $\Sigma(b_n)$ converges to A.
2. Show that the convergence of a series is not affected by changing a finite number of its terms. (Of course, the sum may well be changed.)
3. Show that grouping the terms of a convergent series by introducing parentheses containing a finite number of terms does not destroy the convergence or the value of the limit. However, grouping terms in a divergent series can produce convergence.
4. Show that if a convergent series contains only a finite number of negative terms, then it is absolutely convergent.

5. Show that if a series is conditionally convergent, then the series of positive terms is divergent and the series of negative terms is divergent.
6. By using partial fractions, show that

$$\text{(a)} \sum_{n=0}^{\infty} \frac{1}{(\alpha + n)(\alpha + n + 1)} = \frac{1}{\alpha} \quad \text{if } \alpha > 0,$$

$$\text{(b)} \sum_{n=1}^{\infty} \frac{1}{n(n+1)(n+2)} = \frac{1}{4}.$$

7. If $\Sigma(a_n)$ is a convergent series, then is $\Sigma(a_n^2)$ always convergent? If $a_n \geqslant 0$, then is it true that $\Sigma(\sqrt{a_n})$ is always convergent?
8. If $\Sigma(a_n)$ is convergent and $a_n \geqslant 0$, then is $\Sigma(\sqrt{a_n a_{n+1}})$ convergent?
9. Let $\Sigma(a_n)$ be a series of strictly positive numbers and let b_n, $n \in \mathbf{N}$, be defined to be $b_n := (a_1 + a_2 + \cdots + a_n)/n$. Show that $\Sigma(b_n)$ always diverges.
10. Let $\Sigma(a_n)$ be convergent and let c_n, $n \in \mathbf{N}$, be defined to be the weighted means

$$c_n := \frac{a_1 + 2a_2 + \cdots + na_n}{n(n+1)}.$$

Then $\Sigma(c_n)$ converges and equals $\Sigma(a_n)$.
11. Let $\Sigma(a_n)$ be a series of monotone decreasing positive numbers. Prove that $\Sigma_{n=1}^{\infty}(a_n)$ converges if and only if the series

$$\sum_{n=1}^{\infty} 2^n a_{2^n}$$

converges. This result is often called the **Cauchy Condensation Test**. [*Hint*: group the terms into blocks as in Examples 8.1.8(b, d).]
12. Use the Cauchy Condensation Test to discuss the convergence of the p-series $\Sigma(1/n^p)$.
13. Use the Cauchy Condensation Test to establish the divergence of the series:

$$\sum \frac{1}{n \log n}, \qquad \sum \frac{1}{n(\log n)(\log \log n)},$$

$$\sum \frac{1}{n(\log n)(\log \log n)(\log \log \log n)}.$$

14. Show that if $c > 1$, the following series are convergent:

$$\sum \frac{1}{n(\log n)^c}, \qquad \sum \frac{1}{n(\log n)(\log \log n)^c}.$$

SECTION 8.2 Tests for Convergence

In the preceding section we obtained some results concerning the manipulation of infinite series, especially in the important case where the series are absolutely convergent. However, except for the Cauchy Criterion and the fact that the terms of a convergent series converge to zero, we did not establish any necessary or sufficient conditions for convergence of infinite series. We shall now give some results which can be used to establish the convergence or divergence of infinite series. In view of its importance, we shall pay special attention to absolute convergence.

Our first test shows that if the terms of a positive real series are dominated by the corresponding terms of a convergent series, then the first series is convergent. It yields a test for absolute convergence that the reader should formulate.

8.2.1 Comparison Test. *Let $X := (x_n)$ and $Y := (y_n)$ be positive real sequences and suppose that for some natural number K,*

$$(1) \qquad x_n \leqslant y_n \quad \text{for} \quad n \geqslant K,$$

Then the convergence of $\Sigma(y_n)$ implies the convergence of $\Sigma(x_n)$.

Proof. If $m \geqslant n \geqslant \sup\{K, M(\varepsilon)\}$, then

$$x_{n+1} + \cdots + x_m \leqslant y_{n+1} + \cdots + y_m < \varepsilon,$$

from which the assertion is evident. Q.E.D.

8.2.2 Limit Comparison Test. *Suppose that $X := (x_n)$ and $Y := (y_n)$ are positive real sequences.*

(a) *If the relation*

$$(2) \qquad \lim (x_n/y_n) \neq 0$$

holds, then $\Sigma(x_n)$ is convergent if and only if $\Sigma(y_n)$ is convergent.

(b) *If the limit in (2) is zero and $\Sigma(y_n)$ is convergent, then $\Sigma(x_n)$ is convergent.*

Proof. It follows from (2) that for some real number $c > 1$ and some natural number K, then

$$(1/c)y_n \leqslant x_n \leqslant cy_n \quad \text{for} \quad n \geqslant K.$$

If we apply the Comparison Test 8.2.1 twice, we obtain the assertion in part (a). The proof of (b) is similar and will be omitted. Q.E.D.

The Root and Ratio Tests

We now give an important test due to Cauchy.

8.2.3 Root Test. (a) *If $X := (x_n)$ is a sequence in* **R** *and there exists a positive number $r < 1$ and a natural number K such that*

$$(3) \qquad |x_n|^{1/n} \leqslant r \quad \textit{for} \quad n \geqslant K,$$

then the series $\Sigma(x_n)$ is absolutely convergent.

(b) *If there exists a number $r > 1$ and a natural number K such that*

$$(4) \qquad |x_n|^{1/n} \geqslant r \quad \textit{for} \quad n \geqslant K,$$

then the series $\Sigma(x_n)$ is divergent.

Proof. (a) If (3) holds, then we have $|x_n| \leqslant r^n$. Now for $0 \leqslant r \leqslant 1$, the series $\Sigma(r^n)$ is convergent, as was seen in Example 8.1.8(a). Hence it follows from the Comparison Test that $\Sigma(x_n)$ is absolutely convergent.

(b) If (4) holds, then $|x_n| \geqslant r^n$. However, since $r \geqslant 1$, it is false that $\lim(|x_n|) = 0$. Q.E.D.

In addition to establishing the convergence of $\Sigma(x_n)$, the root test can be used to obtain an estimate of the rapidity of convergence. This estimate is useful in numerical computations and in some theoretical estimates as well.

8.2.4 Corollary. *If r satisfies $0 < r < 1$ and if the sequence $X := (x_n)$ satisfies (3), then the partial sums s_n, $n \geqslant K$, approximate the sum $s = \Sigma(x_n)$ according to the estimate*

$$(5) \qquad |s - s_n| \leqslant \frac{r^{n+1}}{1 - r} \quad \textit{for} \quad n \geqslant K.$$

Proof. If $m \geqslant n \geqslant K$, we have

$$|s_m - s_n| = |x_{n+1} + \cdots + x_m| \leqslant |x_{n+1}| + \cdots + |x_m|$$
$$\leqslant r^{n+1} + \cdots + r^m < \frac{r^{n+1}}{1 - r}.$$

Now take the limit with respect to m to obtain (5). Q.E.D.

It is often convenient to make use of the following varriant of the Root Test.

8.2.5 Corollary. *Let $X := (x_n)$ be a sequence in* **R** *and set*

$$(6) \qquad r := \lim(|x_n|^{1/n}),$$

whenever this limit exists. Then $\Sigma(x_n)$ *is absolutely convergent when* $r < 1$ *and is divergent when* $r > 1$.

Proof. It follows that if the limit in (6) exists and is less than 1, then there is a real number r_1 with $r < r_1 < 1$ and a natural number K such that $|x_n|^{1/n} \leqslant r_1$ for $n \geqslant K$. In this case the series is absolutely convergent.

If this limit exceeds 1, then there is a real number $r_2 > 1$ and a natural number K such that $|x_n|^{1/n} \geqslant r_2$ for $n \geqslant K$, in which case the series is divergent. Q.E.D.

The next test is due to D'Alembert.

8.2.6 Ratio Test. (a) *If* $X := (x_n)$ *is a sequence of non-zero elements of* $\mathbf{R}$ *and there is a positive number* $r < 1$ *and a natural number* K *such that*

$$\frac{|x_{n+1}|}{|x_n|} \leqslant r \qquad \text{for} \quad n \geqslant K, \tag{7}$$

then the series $\Sigma(x_n)$ *is absolutely convergent.*

(b) *If there exists a number* $r \geqslant 1$ *and a natural number* K *such that*

$$\frac{|x_{n+1}|}{|x_n|} \geqslant r \qquad \text{for} \quad n \geqslant K, \tag{8}$$

then the series $\Sigma(x_n)$ *is divergent.*

Proof. (a) If (7) holds, then an elementary induction argument shows that $|x_{K+m}| \leqslant r^m |x_K|$ for $m \geqslant 1$. It follows that for $n \geqslant K$ the terms of $\Sigma(x_n)$ are dominated by a fixed multiple of the terms of the geometric series $\Sigma(r^n)$ with $0 \leqslant r < 1$. From the Comparison Test 8.2.1, we infer that $\Sigma(x_n)$ is absolutely convergent.

(b) If (8) holds, then an elementary induction argument shows that $|x_{K+m}| \geqslant r^m|x_K|$ *for* $m \geqslant 1$. Since $r \geqslant 1$, it is impossible to have $\lim(|x_n|) = 0$, so the series cannot converge. Q.E.D.

8.2.7 Corollary. *If* r *satisfies* $0 \leqslant r < 1$ *and if the sequence* $X := (x_n)$ *satisfies* (7) *for* $n \geqslant K$, *then the partial sums approximate the sum* $s = \Sigma(x_n)$ *according to the estimate*

$$|s - s_n| \leqslant \frac{r}{1 - r}|x_n| \qquad \text{for} \quad n \geqslant K. \tag{9}$$

Proof. The relation (7) implies that $|x_{n+k}| \leqslant r^k |x_n|$ when $n \geqslant K$. Therefore, if $m \geqslant n \geqslant K$, we have

$$|s_m - s_n| = |x_{n+1} + \cdots + x_m| \leqslant |x_{n+1}| + \cdots + |x_m|$$
$$\leqslant (r + r^2 + \cdots + r^{m-n})\,|x_n| < \frac{r}{1-r}|x_n|.$$

Again we take the limit with respect to m to obtain (9). Q.E.D.

8.2.8 Corollary. *Let $X := (x_n)$ be a sequence in $\mathbf{R}$ and set $r := \lim(|x_{n+1}|/|x_n|)$, whenever the limit exists. Then the series $\Sigma(x_n)$ is absolutely convergent when $r < 1$ and divergent when $r > 1$.*

Proof. Suppose that the limit exists and $r < 1$. If r_1 satisfies $r < r_1 < 1$, then there is a natural number K such that $|x_{n+1}|/|x_n| < r$ for $n \geqslant K$. In this case Theorem 8.2.6 establishes the absolute convergence of the series. If $r > 1$, and if r_2 satisfies $1 < r_2 < r$, then there is a natural number K such that $|x_{n+1}|/|x_n| > r_2$ for $n \geqslant K$, and in this case there is divergence. Q.E.D.

The Integral Test

The next test for convergence—a very powerful one—uses the notion of the improper integral that was introduced in Section 6.4.

8.2.9 Integral Test. *Let f be a positive, decreasing, continuous function on $\{t : t \geqslant 1\}$. Then the series $\Sigma(f(n))$ converges if and only if the improper integral*

$$\int_1^{+\infty} f(t)\,dt = \lim_n \left(\int_1^n f(t)\,dt\right)$$

exists. In the case of convergence, the partial sum $s_n = \Sigma_{k=1}^n (f(k))$ and the sum $s = \Sigma_{k=1}^\infty (f(k))$ satisfy the estimate

$$\int_{n+1}^{+\infty} f(t)\,dt \leqslant s - s_n \leqslant \int_n^{+\infty} f(t)\,dt. \tag{10}$$

Proof. Since f is positive, continuous, and decreasing on the interval $[k-1,k]$, it follows that

$$f(k) \leqslant \int_{k-1}^k f(t)\,dt \leqslant f(k-1). \tag{11}$$

By summing this inequality for $k = 2,3,\ldots,n$, we obtain the relation

$$s_n - f(1) \leqslant \int_1^n f(t)\,dt \leqslant s_{n-1},$$

which shows that both or neither of the limits

$$\lim (s_n), \qquad \lim \left(\int_1^n f(t)\, dt \right)$$

exist. If they exist, we obtain on summing relation (11) for $k = n + 1, \ldots, m$, that

$$s_m - s_n \leqslant \int_n^m f(t)\, dt \leqslant s_{m-1} - s_{n-1},$$

whence it follows that

$$\int_{n+1}^{m+1} f(t)\, dt \leqslant s_m - s_n \leqslant \int_n^m f(t)\, dt.$$

If we take the limit with respect to m in this last inequality, we obtain the inequality (10). Q.E.D.

We shall show how the results in Theorems 8.2.1 through 8.2.9 can be applied to the p-series, which were introduced in Example 8.1.8(c).

8.2.10 Examples. (a) First we shall apply the Comparison Test. Knowing that the harmonic series $\Sigma(1/n)$ diverges, it is seen that if $p \leqslant 1$, then $n^p \leqslant n$ and hence $1/n \leqslant 1/n^p$. After using the Comparison Test 8.2.1, we conclude that the p-series $\Sigma(1/n^p)$ diverges for $p \leqslant 1$.

(b) Now consider the case $p = 2$; that is, the series $\Sigma(1/n^2)$. We compare the series with the convergent series $\Sigma(1/n(n+1))$ of Example 8.1.8(e). Since the relation

$$\frac{1}{n(n+1)} < \frac{1}{n^2}$$

holds and the terms on the left form a convergent series, we cannot apply the Comparison Theorem directly. However, we could apply this theorem if we compared the nth term of $\Sigma(1/n(n+1))$ with the $(n+1)$st term of $\Sigma(1/n^2)$. Instead, we choose to apply the Limit Comparison Test 8.2.2 and note that

$$\frac{1}{n(n+1)} \div \frac{1}{n^2} = \frac{n^2}{n(n+1)} = \frac{n}{n+1}.$$

Since the limit of this quotient is 1 and $\Sigma(1/n(n+1))$ converges, then so does the series $\Sigma(1/n^2)$.

(c) Now consider the case $p \geqslant 2$. If we note that $n^p \geqslant n^2$ for $p \geqslant 2$, then $1/n^p \leqslant 1/n^2$. A direct application of the Comparison Test assures that $\Sigma(1/n^p)$ converges for $p \geqslant 2$. Alternatively, we could apply the Limit Comparison Test and note that

$$\frac{1}{n^p} \div \frac{1}{n^2} = \frac{n^2}{n^p} = \frac{1}{n^{p-2}}.$$

If $p > 2$, this expression converges to 0, whence it follows from Corollary 8.2.2(b) that the series $\Sigma(1/n^p)$ converges for $p \geqslant 2$.

By using the Comparison Test, we cannot gain any information concerning the p-series for $1 < p < 2$ unless we can find a series whose convergence character is known and which can be compared to the series in this range.

(d) We demonstrate the Root and the Ratio Tests as applied to the p-series. Note that

$$\left(\frac{1}{n^p}\right)^{1/n} = (n^{-p})^{1/n} = (n^{1/n})^{-p}.$$

Now it is known (see Example 3.1.11(e)) that the sequence $(n^{1/n})$ converges to 1. Hence we have

$$\lim \left(\left(\frac{1}{n^p}\right)^{1/n}\right) = 1,$$

so that the Root Test (in the form of Corollary 8.2.5) does not apply.

In the same way, since

$$\frac{1}{(n+1)^p} \div \frac{1}{n^p} = \frac{1}{(1+1/n)^p},$$

and since the sequence $((1 + 1/n)^p)$ converges to 1, the Ratio Test (in the form of Corollary 8.2.8) does not apply.

(e) Finally, we apply the Integral Test to the p-series. Let $f(t) := t^{-p}$ and recall that

$$\int_1^n \frac{1}{t}\,dt = \log(n) - \log(1),$$

$$\int_1^n \frac{1}{t^p}\,dt = \frac{1}{1-p}(n^{1-p} - 1) \qquad \text{for} \quad p \neq 1.$$

From these relations we see that the p-series converges if $p > 1$ and diverges if $p \leqslant 1$.

Alternating Series

The tests considered thus far all have the character that, if certain hypotheses are satisfied, then the series $\Sigma(x_n)$ is absolutely convergent. However, there are many series, such as $\Sigma(-1)^n/n$, that are convergent, but not absolutely convergent. The ones that occur most frequently are series that are "alternating" in the sense of the next definition.

8.2.11 Definition. A sequence $X := (x_n)$ of non-zero real numbers is **alternating** if the terms $(-1)^{n+1}x_n$, $n = 1, 2, \ldots$, are all positive (or all negative) real numbers. If a sequence $X = (x_n)$ is alternating, we say that the series $\Sigma(x_n)$ it generates is an **alternating series**.

It is useful to set $x_n := (-1)^{n+1}z_n$ and require that $z_n > 0$ (or $z_n < 0$) for all $n = 1, 2, \ldots$. The convergence of alternating series is easily treated when the next result, proved by Leibniz, can be applied.

8.2.12 Alternating Series Test. *Let $Z := (z_n)$ be a decreasing sequence of strictly positive numbers with $\lim (z_n) = 0$. Then the alternating series $\Sigma((-1)^{n+1}z_n)$ is convergent. Moreover, if s is the sum of this series and s_n is the nth partial sum, then we have the estimate*

$$|s - s_n| \leqslant z_{n+1} \tag{12}$$

for the rapidity of convergence.

Proof. Since we have

$$s_{2n} = (z_1 - z_2) + (z_3 - z_4) + \cdots + (z_{2n-1} - z_{2n}),$$

and since $z_k - z_{k+1} \geqslant 0$, it follows that the subsequence (s_{2n}) of partial sums is increasing. Since

$$s_{2n} = z_1 - (z_2 - z_3) - \cdots - (z_{2n-2} - z_{2n-1}) - z_{2n},$$

it also follows that $s_{2n} \leqslant z_1$ for all $n \in \boldsymbol{N}$. It follows from the Monotone Convergence Theorem 3.3.2 that the subsequence (s_{2n}) converges to some number $s \in \boldsymbol{R}$.

We now show that the entire sequence (s_n) converges to s. Indeed, if $\varepsilon > 0$, let n_ε be such that if $n \geqslant n_\varepsilon$ then

$$|s_{2n} - s| \leqslant \tfrac{1}{2}\varepsilon \quad \text{and} \quad |z_{2n+1}| \leqslant \tfrac{1}{2}\varepsilon.$$

It follows that if $n \geqslant n_\varepsilon$ then

$$\begin{aligned} |s_{2n+1} - s| &= |s_{2n} + z_{2n+1} - s| \\ &\leqslant |s_{2n} - s| + |z_{2n+1}| \\ &\leqslant \tfrac{1}{2}\varepsilon + \tfrac{1}{2}\varepsilon = \varepsilon. \end{aligned}$$

Therefore every partial sum of an odd number of terms is also within ε of s if n is large enough. Since $\varepsilon > 0$ is arbitrary, the convergence of (s_n) and hence of $\Sigma(-1)^{n+1}z_n$ is established.

To obtain (12), suppose that n is even. If m is even and $m \geqslant n$, then

$$0 \leqslant s_m - s_n = z_{n+1} - (z_{n+2} - z_{n+3}) - \cdots - z_m \leqslant z_{n+1}.$$

Since $s = \lim (s_m)$ it follows that $0 \leqslant s - s_n \leqslant z_{n+1}$, when n is even. The case where n is odd is handled similarly and is left as an exercise. Q.E.D.

Exercises for Section 8.2

1. Establish the convergence or the divergence of the series whose nth term is given by

   (a) $\dfrac{1}{(n+1)(n+2)}$, (b) $\dfrac{n}{(n+1)(n+2)}$,

   (c) $2^{-1/n}$, (d) $n/2^n$,

   (e) $(n(n+1))^{-1/2}$, (f) $(n^2(n+1))^{-1/2}$,

   (g) $n!/n^n$, (h) $(-1)^n n/(n+1)$.

2. For each of the series in Exercise 1 which converge, estimate the remainder if only four terms are taken. If we wish to determine the sum within 1/1000, how many terms should we take?
3. Discuss the convergence or the divergence of the series with nth term (for sufficiently large n) given by

   (a) $(\log n)^{-p}$, (b) $(\log n)^{-n}$,
   (c) $(\log n)^{-\log n}$, (d) $(\log n)^{-\log \log n}$,
   (e) $(n \log n)^{-1}$, (f) $(n(\log n)(\log \log n)^2)^{-1}$.
4. Discuss the convergence or the divergence of the series with nth term

   (a) $2^n e^{-n}$, (b) $n^n e^{-n}$,
   (c) $e^{-\log n}$, (d) $(\log n)\, e^{-\sqrt{n}}$,
   (e) $n!e^{-n}$, (f) $n!e^{-n^2}$.
5. Show that the series

$$\frac{1}{1^2} + \frac{1}{2^3} + \frac{1}{3^2} + \frac{1}{4^3} + \cdots$$

   is convergent, but that both the Ratio and the Root Tests fail to apply.

6. If a and b are positive numbers, then $\Sigma\,(an + b)^{-p}$ converges if $p > 1$ and diverges if $p \leq 1$.
7. Discuss the series whose nth term is

(a) $\dfrac{n!}{3\cdot 5\cdot 7\cdots(2n+1)}$, (b) $\dfrac{(n!)^2}{(2n)!}$,

(c) $\dfrac{2\cdot 4\cdots(2n)}{3\cdot 5\cdots(2n+1)}$, (d) $\dfrac{2\cdot 4\cdots(2n)}{5\cdot 7\cdots(2n+3)}$.

8. The series given by

$$\left(\frac{1}{2}\right)^p + \left(\frac{1\cdot 3}{2\cdot 4}\right)^p + \left(\frac{1\cdot 3\cdot 5}{2\cdot 4\cdot 6}\right)^p + \cdots$$

converges for $p > 2$ and diverges for $p \leq 2$.
9. Consider the series

$$1 - \frac{1}{2} - \frac{1}{3} + \frac{1}{4} + \frac{1}{5} - \frac{1}{6} - \frac{1}{7} + + - - \cdots,$$

where the signs come in pairs. Does it converge?
10. Let $a_n \in \boldsymbol{R}$ for $n \in \mathbf{N}$ and let $p < q$. If the series $\Sigma(a_n/n^p)$ is convergent, then the series $\Sigma(a_n/n^q)$ is also convergent.
11. If p and q are strictly positive numbers, then $\Sigma(-1)^n\,(\log n)^p/n^q$ is a convergent series.
12. Discuss the series whose nth term is

(a) $(-1)^n \dfrac{n^n}{(n+1)^{n+1}}$, (b) $\dfrac{n^n}{(n+1)^{n+1}}$,

(c) $(-1)^n \dfrac{(n+1)^n}{n^n}$, (d) $\dfrac{(n+1)^n}{n^{n+1}}$.

13. Suppose that $\Sigma(a_n)$ is a convergent series of real numbers. Either prove that $\Sigma(b_n)$ converges or give a counter-example, when we define b_n by
(a) a_n/n, (b) $\sqrt{a_n}/n$ $(a_n \geq 0)$,
(c) $a_n \sin n$, (d) $\sqrt{a_n/n}$ $(a_n \geq 0)$,
(e) $n^{1/n}a_n$, (f) $a_n/(1 + |a_n|)$.
14. Show that the series

$$1 + \frac{1}{2} - \frac{1}{3} + \frac{1}{4} + \frac{1}{5} - \frac{1}{6} + + - \cdots$$

is divergent.
15. If the hypothesis that (z_n) is decreasing is dropped, show that the Alternating Series Test 8.2.11 may fail.
16. For $n \in \mathbf{N}$, let c_n be defined by

$$c_n := \frac{1}{1} + \frac{1}{2} + \cdots + \frac{1}{n} - \log n.$$

Show that (c_n) is a decreasing sequence of positive numbers. The limit C of this sequence is called **Euler's Constant** and is approximately equal to 0.577. Show that if we put

$$b_n := \frac{1}{1} - \frac{1}{2} + \frac{1}{3} - \cdots - \frac{1}{2n},$$

then the sequence (b_n) converges to $\log 2$. [*Hint*: $b_n = c_{2n} - c_n + \log 2$.]

17. Let $0 < a < 1$ and consider the series

$$a^2 + a + a^4 + a^3 + \cdots + a^{2n} + a^{2n-1} + \cdots.$$

Show that the Root Test applies, but that the Ratio Test does not apply.

18. (Raabe's Test) Let $c_n > 0$ for $n \in \boldsymbol{N}$.
    (a) Suppose there exists $a > 1$ and $K > 0$ such that $c_{n+1}/c_n \leqslant 1 - a_n/n$ for all $n \in \boldsymbol{N}$, $n \geqslant K$. Show that $\Sigma(c_n)$ converges to some $c \in \boldsymbol{R}$. [*Hint*: Show that $(k-1)c_k - kc_{k+1} \geqslant (a-1)c_k$ for all $k \in \boldsymbol{N}$, $k \geqslant K$.]
    (b) Suppose there exists $a \leqslant 1$ and $K > 0$ such that $c_{n+1}/c_n \geqslant 1 - a/n$ for all $n \in \boldsymbol{N}$, $n \geqslant K$. Show that $\Sigma(c_n)$ diverges. [*Hint*: Show that (nc_{n+1}) is increasing for $n \geqslant K$.]
19. If the condition in part (a) of the preceding exercise is satisfied, show that the partial sums (s_n) of $\Sigma(c_n)$ satisfy the estimate $|c - s_n| \leqslant nc_n/(a-1)$ for $n \geqslant K$.
20. Use the result in Exercise 18 to note that if (x_n) is a sequence of non-zero real numbers and if the limit

$$a := \lim\left(n(1 - |x_{n+1}/x_n|)\right)$$

exists, then $\Sigma(x_n)$ is absolutely convergent when $a > 1$ and is not absolutely convergent when $a < 1$.

21. If $p > 0$, $q > 0$, show that the series

$$\sum \frac{(p+1)(p+2)\cdots(p+n)}{(q+1)(q+2)\cdots(q+n)}$$

converges for $q > p + 1$ and diverges for $q \leqslant p + 1$.

22. Suppose that none of the numbers a, b, c is a negative integer or zero. Prove that the **hypergeometric series**

$$\frac{ab}{1!c} + \frac{a(a+1)b(b+1)}{2!c(c+1)} + \frac{a(a+1)(a+2)b(b+1)(b+2)}{3!c(c+1)(c+2)} + \cdots$$

is absolutely convergent for $c > a + b$ and divergent for $c \leqslant a + b$.

23. Let $a_n > 0$ and suppose that $\Sigma(a_n)$ converges. Construct a convergent series $\Sigma(b_n)$ with $b_n > 0$ such that $\lim(a_n/b_n) = 0$; hence $\Sigma(b_n)$ converges less rapidly than $\Sigma(a_n)$. [*Hint*: Let (A_n) be the partial sums of $\Sigma(a_n)$ and A its limit. Define $r_0 := A$, $r_n := A - A_n$ and $b_n := \sqrt{r_{n-1}} - \sqrt{r_n}$.]
24. Let $a_n > 0$ and suppose that $\Sigma(a_n)$ diverges. Construct a divergent series $\Sigma(b_n)$ with $b_n > 0$ such that $\lim(b_n/a_n) = 0$; hence $\Sigma(b_n)$ diverges less rapidly than $\Sigma(a_n)$. [*Hint*: Let $b_1 := \sqrt{a_1}$ and $b_n := \sqrt{a_{n-1}} - \sqrt{a_n}$, $n > 1$.]

25. Let $\{n_1, n_2, \ldots\}$ denote the collection of natural numbers that do not use the digit 6 in their decimal expansion. Show that the series $\Sigma(1/n_k)$ converges to a number less than 90. If $\{m_1, m_2, \ldots\}$ is the collection that ends in 6, then $\Sigma(1/m_k)$ diverges.

SECTION 8.3 Series of Functions

Because of their frequent appearance and importance, we now present a discussion of infinite series of functions. Since the convergence of an infinite series is handled by examining the sequence of partial sums, questions concerning series of functions are answered by examining corresponding questions for sequences of functions. For this reason, a portion of the present section is merely a translation of facts already established for sequences of functions into series terminology. This is the case, for example, for the portion of the section dealing with series of general functions. However, in the second part of the section, where we discuss power series, some new features arise merely because of the special character of the functions involved.

8.3.1 Definition. If (f_n) is a sequence of functions defined on a subset D of $\boldsymbol{R}$ with values in $\boldsymbol{R}$, the sequence of **partial sums** (s_n) of the infinite series $\Sigma(f_n)$ is defined for x in D by

$$s_1(x) := f_1(x),$$

$$s_2(x) := s_1(x) + f_2(x) \quad (= f_1(x) + f_2(x)),$$

$$\cdots\cdots\cdots\cdots\cdots\cdots\cdots\cdots\cdots\cdots$$

$$s_{n+1}(x) := s_n(x) + f_{n+1}(x) \quad (= f_1(x) + \cdots + f_n(x) + f_{n+1}(x)),$$

$$\cdots\cdots\cdots\cdots\cdots\cdots\cdots\cdots\cdots\cdots$$

In case the sequence (s_n) converges on D to a function f, we say that the infinite series of functions $\Sigma(f_n)$ **converges** to f on D. We shall often write

$$\Sigma(f_n), \quad \sum_{n=1}^{\infty}(f_n), \quad \text{or} \quad \sum_{n=1}^{\infty} f_n$$

to denote either the series or the limit function, when it exists.

If the series $\Sigma(|f_n(x)|)$ converges for each x in D, then we say that $\Sigma(f_n)$ is **absolutely convergent** on D. If the sequence (s_n) is uniformly convergent on D to f, then we say that $\Sigma(f_n)$ is **uniformly convergent** on D, or that it **converges to f uniformly on D**.

One of the main reasons for the interest in uniformly convergent series of functions is the validity of the following results which give conditions justifying the change of order of the summation and other limiting operations.

8.3.2 Theorem. *If f_n is continuous on $D \subseteq \mathbf{R}$ to $\mathbf{R}$ for each $n \in \mathbf{N}$ and if $\Sigma(f_n)$ converges to f uniformly on D, then f is continuous on D.*

This is a direct translation of Theorem 7.2.2 for series. The next result is a translation of Theorem 7.2.4.

8.3.3 Theorem. *Suppose that the real-valued functions f_n, $n \in \mathbf{N}$, are integrable on the interval $J := [a,b]$. If the series $\Sigma(f_n)$ converges to f uniformly on J, then f is integrable and*

$$(1) \qquad \int_a^b f = \sum_{n=1}^{\infty} \int_a^b f_n .$$

Next we turn to the corresponding theorem pertaining to differentiation. Here we assume the uniform convergence of the series obtained after term-by-term differentiation of the given series. This result is an immediate consequence of Theorem 7.2.3.

8.3.4 Theorem. *For each $n \in \mathbf{N}$, let f_n be a real-valued function on $J := [a,b]$ which has a derivative f_n' on J. Suppose that the series $\Sigma(f_n)$ converges for at least one point of J and that the series of derivatives $\Sigma(f_n')$ converges uniformly on J.*

Then there exists a real-valued function f on J such that $\Sigma(f_n)$ converges uniformly on J to f. In addition, f has a derivative on J and $f' = \Sigma f_n'$.

Tests for Uniform Convergence

Since we have stated some consequences of uniform convergence of series, we shall now present a few tests that can be used to establish uniform convergence.

8.3.5 Cauchy Criterion. *Let (f_n) be a sequence of functions on $D \subseteq \mathbf{R}$ to $\mathbf{R}$. The infinite series $\Sigma(f_n)$ is uniformly convergent on D if and only if for every $\varepsilon > 0$ there exists an $M(\varepsilon)$ such that if $m \geq n \geq M(\varepsilon)$, then*

$$|f_n(x) + \cdots + f_m(x)| < \varepsilon,$$

for all $x \in D$.

8.3.6 Weierstrass *M*-Test. *Let (M_n) be a sequence of positive real numbers such that $|f_n(x)| \leq M_n$ for $x \in D$, $n \in \mathbf{N}$. If the series $\Sigma(M_n)$ is convergent, then $\Sigma(f_n)$ is uniformly convergent on D.*

Proof. If $m > n$, we have the relation

$$|f_n(x) + \cdots + f_n(x)| \leq M_n + \cdots + M_m \qquad \text{for} \quad x \in D.$$

Now apply 8.1.5, 8.3.5, and the convergence of $\Sigma(M_n)$. Q.E.D.

Power Series

We shall now turn to a discussion of power series. This is an important class of series of functions and enjoys properties that are not valid for general series of functions.

8.3.7 Definition. A series of real functions $\Sigma(f_n)$ is said to be a **power series around** $x = c$ if the function f_n has the form

$$f_n(x) = a_n(x - c)^n,$$

where a_n and c belong to $\boldsymbol{R}$ and where $n = 0, 1, 2, \ldots$.

For the sake of simplicity of our notation, we shall treat only the case where $c = 0$. This is no loss of generality, however, since the translation $x' = x - c$ reduces a power series around c to a power series around 0. Thus, whenever we refer to a power series, we shall mean a series of the form

$$\text{(2)} \qquad \sum_{n=0}^{\infty} a_n x^n = a_0 + a_1 x + \cdots + a_n x^n + \cdots.$$

Even though the functions appearing in (2) are defined over all of $\boldsymbol{R}$, it is not to be expected that the series (2) will converge for all x in $\boldsymbol{R}$. For example, by using the Ratio Test 8.2.6, we can show that the series

$$\sum_{n=0}^{\infty} n!x^n, \qquad \sum_{n=0}^{\infty} x^n, \qquad \sum_{n=0}^{\infty} x^n/n!,$$

converge for x in the sets

$$\{0\}, \qquad \{x \in \boldsymbol{R} : |x| < 1\}, \qquad \boldsymbol{R},$$

respectively. Thus, the set on which a power series converges may be small, medium, or large. However, an arbitrary subset of $\boldsymbol{R}$ cannot be the precise set on which a power series converges, as we shall show.

If (b_n) is a bounded sequence of non-negative real numbers, then we define the **limit superior** of (b_n) to be the infimum of those numbers v such that $b_n \leqslant v$ for all sufficiently large $n \in \boldsymbol{N}$. This infimum is uniquely determined and is denoted by $\limsup (b_n)$. The only thing we need to know is (i) that if $v > \limsup (b_n)$, then $b_n \leqslant v$ for all sufficiently large $n \in \boldsymbol{N}$, and (ii) that if $w < \limsup (b_n)$, then $w \leqslant b_n$ for infinitely many $n \in \boldsymbol{N}$.

8.3.8 Definition. Let $\Sigma(a_n x^n)$ be a power series. If the sequence $(|a_n|^{1/n})$ is bounded, we set $\rho := \limsup (|a_n|^{1/n})$; if this sequence is not bounded we set

$\rho = +\infty$. We define the **radius of convergence** of $\Sigma(a_n x^n)$ to be given by

$$\begin{aligned} R &:= 0, && \text{if} \quad \rho = +\infty, \\ &:= 1/\rho, && \text{if} \quad 0 < \rho < +\infty, \\ &:= +\infty, && \text{if} \quad \rho = 0. \end{aligned}$$

The **interval of convergence** is the open interval $(-R, R)$.

We shall now justify the term "radius of convergence".

8.3.9 Cauchy-Hadamard Theorem. *If R is the radius of convergence of the power series $\Sigma(a_n x^n)$, then the series is absolutely convergent if $|x| < R$ and divergent if $|x| > R$.*

Proof. We shall treat only the case where $0 < R < +\infty$, leaving the cases $R = 0$, $R = +\infty$, as exercises. If $0 < |x| < R$, then there exists a positive number $c < 1$ such that $|x| < cR$. Therefore $\rho < c/|x|$ and so it follows that if n is sufficiently large, then $|a_n|^{1/n} \leq c/|x|$. This is equivalent to the statement that

$$(3) \qquad |a_n x^n| \leq c^n$$

for all sufficiently large n. Since $c < 1$, the absolute convergence of $\Sigma(a_n x^n)$ follows from the Comparison Test 8.2.1.

If $|x| > R = 1/\rho$, then there are infinitely many $n \in \mathbf{N}$ for which we have $|a_n|^{1/n} > 1/|x|$. Therefore, $|a_n x^n| > 1$ for infinitely many n, so that the sequence $(a_n x^n)$ does not converge to zero. Q.E.D.

It will be noted that the Cauchy-Hadamard Theorem makes no statement as to whether the power series converges when $|x| = R$. Indeed, anything can happen, as the examples

$$\Sigma\, x^n, \qquad \Sigma\, \frac{1}{n} x^n, \qquad \Sigma\, \frac{1}{n^2} x^n,$$

show. Since $\lim (n^{1/n}) = 1$, each of these power series has radius of convergence equal to 1. The first power series converges at neither of the points $x = -1$ and $x = +1$; the second series converges at $x = -1$ but diverges at $x = +1$; and the third power series converges at both $x = -1$ and $x = +1$. (Find a power series with $R = 1$ which converges at $x = +1$ but diverges at $x = -1$.)

It is an exercise to show that the radius of convergence of $\Sigma(a_n x^n)$ is also given by

$$(3) \qquad \lim \left(\frac{|a_n|}{|a_{n+1}|} \right),$$

provided this limit exists. Frequently, it is more convenient to use (3) than Definition 8.3.8.

The argument used in the proof of the Cauchy-Hadamard Theorem yields the uniform convergence of the power series on any fixed compact subset in the interval of convergence $(-R, R)$.

8.3.10 Theorem. *Let R be the radius of convergence of $\Sigma(a_n x^n)$ and let K be a compact subset of the interval of convergence $(-R, R)$. Then the power series converges uniformly on K.*

Proof. The compactness of $K \subseteq (-R, R)$ implies that there exists a positive constant $c < 1$ such that $|x| < cR$ for all $x \in K$. (Why?) By the argument in 8.3.9, we infer that for sufficiently large n, the estimate (3) holds for all $x \in K$. Since $c < 1$, the uniform convergence of $\Sigma(a_n x^n)$ on K is a direct consequence of the Weierstrass M-test with $M_n := c^n$. Q.E.D.

8.3.11 Theorem. *The limit of a power series is continuous on the interval of convergence. A power series can be integrated term-by-term over any compact interval contained in the interval of convergence.*

Proof. If $|x_0| < R$, then the preceding result asserts that $\Sigma(a_n x^n)$ converges uniformly on any compact neighborhood of x_0 contained in $(-R, R)$. The continuity at x_0 then follows from Theorem 8.3.2, and the term-by-term integration is justified by Theorem 8.3.3. Q.E.D.

We now show that a power series can be differentiated term-by-term. Unlike the situation for general series, we do not need to assume that the differentiated series is uniformly convergent. Hence this result is stronger than the corresponding result for the differentiation of infinite series.

8.3.12 Differentiation Theorem. *A power series can be differentiated term-by-term within the interval of convergence. In fact, if*

$$f(x) = \sum_{n=0}^{\infty}(a_n x^n), \qquad \textit{then} \qquad f'(x) = \sum_{n=1}^{\infty}(n a_n x^{n-1}).$$

Both series have the same radius of convergence.

Proof. Since $\lim(n^{1/n}) = 1$, the sequence $(|na_n|^{1/n})$ is bounded if and only if the sequence $(|a_n|^{1/n})$ is bounded. Moreover, it is easily seen that

$$\limsup(|na_n|^{1/n}) = \limsup(|a_n|^{1/n}).$$

Therefore, the radius of convergence of the two series is the same, so the formally differentiated series is uniformly convergent on each compact subset of the interval of convergence. We can then apply Theorem 8.3.4 to conclude that the formally differentiated series converges to the derivative of the given series. Q.E.D.

It is to be observed that the theorem makes no assertion about the end points of the interval of convergence. If a series is convergent at an end point, then the differentiated series may or may not be convergent at this point. For example, the series $\Sigma_{n=0}^{\infty}(x^n/n^2)$ converges at both end points $x = -1$ and $x = +1$. However, the differentiated series

$$\sum_{n=1}^{\infty} \frac{x^{n-1}}{n} = \sum_{m=0}^{\infty} \frac{x^m}{m+1}$$

converges at $x = -1$ but diverges at $x = +1$.

By repeated application of the preceding result, we conclude that if k is any natural number, then the power series $\Sigma_{n=0}^{\infty}(a_n x^n)$ can be differentiated term-by-term k times to obtain

$$(4) \qquad \sum_{n=k}^{\infty} \frac{n!}{(n-k)!} a_n x^{n-k}.$$

Moreover, this series converges absolutely to $f^{(k)}(x)$ for $|x| < R$ and uniformly over any compact subset of the interval of convergence.

If we substitute $x = 0$ in (4), we obtain the important formula

$$(5) \qquad f^{(k)}(0) = k! a_k.$$

8.3.13 Uniqueness Theorem. *If $\Sigma(a_n x^n)$ and $\Sigma(b_n x^n)$ converge on some interval $(-r, r)$, $r > 0$, to the same function f, then*

$$a_n = b_n \qquad \textit{for all} \quad n \in \mathbf{N}.$$

Proof. Our preceding remarks show that $n! a_n = f^{(n)}(0) = n! b_n$ for all $n \in \mathbf{N}$. Q.E.D.

Taylor Series

If a function f has derivatives of all order at a point c in $\mathbf{R}$, then we can calculate the Taylor coefficients $a_n = f^{(n)}(c)/n!$ for $n \in \mathbf{N}$ and in this way obtain a power series with these coefficients. However, it is not necessarily true that the resulting power series converges to the function f in an interval about c. (See Exercise 8.3.12 for an example.) The issue of convergence is resolved by the remainder term R_n in Taylor's Theorem 5.4.1. We shall write

$$(6) \qquad f(x) = \sum_{n=0}^{\infty} \frac{f^{(n)}(c)}{n!}(x-c)^n$$

for $|x - c| < R$ if and only if the sequence $(R_n(x))$ of remainders converges to 0 for each x in some interval $\{x : |x - c| < R\}$. In this case we say that the power series (6) is the **Taylor expansion** of f at c. We observe that the Taylor polynomials for f discussed in Section 5.4 are just the partial sums of the Taylor expansion (6) of f.

8.3.14 Examples. (a) If $f(x) := \sin x$, $x \in \mathbf{R}$, we have $f^{(2n)}(x) = (-1)^n \cos x$ and $f^{(2n+1)}(x) = (-1)^n \sin x$ for $n \in \mathbf{N}$, $x \in \mathbf{R}$. Evaluating at $c = 0$, we get the Taylor coefficients $a_{2n} = 0$ and $a_{2n+1} = (-1)^n/(2n+1)!$ for $n \in \mathbf{N}$. Since $|\sin x| \leqslant 1$ and $|\cos x| \leqslant 1$ for all x, it follows that $|R_n(x)| \leqslant |x|^n/n!$ for $n \in \mathbf{N}$ and $x \in \mathbf{R}$. Since $\lim (R_n(x)) = 0$ for each $x \in \mathbf{R}$, we obtain the Taylor expansion

$$(7) \qquad \sin x = \sum_{n=0}^{\infty} \frac{(-1)^n}{(2n+1)!} x^{2n+1}, \qquad x \in \mathbf{R}.$$

An application of Theorem 8.3.12 gives us the Taylor expansion

$$(8) \qquad \cos x = \sum_{n=0}^{\infty} \frac{(-1)^n}{(2n)!} x^{2n}, \qquad x \in \mathbf{R}.$$

(b) If $g(x) := e^x$, $x \in \mathbf{R}$, then $g^{(n)}(x) = e^x$ for all $n \in \mathbf{N}$, and hence the Taylor coefficients are given by $a_n = 1/n!$ for $n \in \mathbf{N}$. For a given $x \in \mathbf{R}$, we have $|R_n(x)| \leqslant e^{|x|}|x|^n/n!$ and therefore $(R_n(x))$ tends to 0 as $n \to \infty$. Therefore, we obtain the Taylor expansion

$$(9) \qquad e^x = \sum_{n=0}^{\infty} \frac{1}{n!} x^n, \qquad x \in \mathbf{R}.$$

We can obtain the Taylor expansion at an arbitrary $c \in \mathbf{R}$ by the device of replacing x by $x - c$ in (9) and noting that

$$(10) \qquad e^x = e^c \cdot e^{x-c} = e^c \sum_{n=0}^{\infty} \frac{1}{n!}(x-c)^n = \sum_{n=0}^{\infty} \frac{e^c}{n!}(x-c)^n$$

for all $x \in \mathbf{R}$.

Exercises for Section 8.3

1. Discuss the convergence and the uniform convergence of the series $\Sigma(f_n)$, where $f_n(x)$ is given by:
   (a) $(x^2 + n^2)^{-1}$,
   (b) $(nx)^{-2}$, $x \neq 0$,
   (c) $\sin (x/n^2)$,
   (d) $(x^n + 1)^{-1}$, $x \geqslant 0$,
   (e) $x^n(x^n + 1)^{-1}$, $x \geqslant 0$,
   (f) $(-1)^n(n + x)^{-1}$, $x \geqslant 0$.
2. If $\Sigma(a_n)$ is an absolutely convergent series, then the series $\Sigma(a_n \sin nx)$ is absolutely and uniformly convergent.
3. Let (c_n) be a decreasing sequence of positive numbers. If $\Sigma(c_n \sin nx)$ is uniformly convergent, then $\lim (nc_n) = 0$.

4. Discuss the cases $R = 0$, $R = +\infty$ in the Cauchy-Hadamard Theorem 8.3.9.
5. Show that the radius of convergence R of the power series $\Sigma(a_n x^n)$ is given by $\lim (|a_n|/|a_{n+1}|)$ whenever this limit exists. Give an example of a power series where this limit does not exist.
6. Determine the radius of convergence of the series $\Sigma(a_n x^n)$, where a_n is given by:
   (a) $1/n^n$, (b) $n^\alpha/n!$,
   (c) $n^n/n!$, (d) $\log(n)^{-1}$, $n \geq 2$.
   (e) $(n!)^2/(2n)!$, (f) $n^{-\sqrt{n}}$.
7. If $a_n = 1$ when n is the square of a natural number and $a_n = 0$ otherwise, find the radius of convergence of $\Sigma(a_n x^n)$. If $b_n = 1$ when $n = m!$ for $m \in \boldsymbol{N}$ and $b_n = 0$ otherwise, find the radius of convergence of $\Sigma(b_n x^n)$.
8. Prove in detail that $\lim\sup(|na_n|^{1/n}) = \lim\sup(|a_n|^{1/n})$.
9. If $0 < p \leq |a_n| \leq q$ for all $n \in \boldsymbol{N}$, find the radius of convergence of $\Sigma(a_n x^n)$.
10. Let $f(x) = \Sigma(a_n x^n)$ for $|x| < R$. If $f(x) = f(-x)$ for all $|x| < R$, show that $a_n = 0$ for all odd n.
11. Prove that if f is defined for $|x| < r$ and if there exists a constant B such that $|f^{(n)}(x)| \leq B$ for all $|x| < r$ and $n \in \boldsymbol{N}$, then the Taylor series expansion

$$\sum_{n=0}^{\infty} \frac{f^{(n)}(0)}{n!} x^n$$

converges to $f(x)$ for $|x| < r$.
12. Prove by induction that the function given by $f(x) := e^{-x^2}$ for $x \neq 0$, $f(0) := 0$, has derivatives of all orders at every point and that all of these derivatives vanish at $x = 0$. Hence this function is not given by its Taylor expansion about $x = 0$.
13. Give an example of a function which is equal to its Taylor series expansion about $x = 0$ for $x \geq 0$, but which is not equal to this expansion for $x < 0$.
14. Use the Lagrange form of the remainder to justify the general Binomial Expansion

$$(1 + x)^m = \sum_{n=0}^{\infty} \binom{m}{n} x^n$$

when x is in the interval $0 \leq x < 1$.
15. (Geometric series) Show directly that if $|x| < 1$, then

$$\frac{1}{1 - x} = \sum_{n=0}^{\infty} x^n.$$

16. Show by integrating the series for $1/(1 + x)$ that if $|x| < 1$, then

$$\log(1 + x) = \sum_{n=0}^{\infty} \frac{(-1)^{n+1}}{n} x^n.$$

17. Show that if $|x| < 1$, then

$$\text{Arctan } x = \sum_{n=0}^{\infty} \frac{(-1)^n}{2n + 1} x^{2n+1}.$$

18. Show that if $|x| < 1$, then

$$\text{Arcsin } x = \sum_{n=0}^{\infty} \frac{1 \cdot 3 \cdots (2n-1)}{2 \cdot 4 \cdots 2n} \cdot \frac{x^{2n+1}}{2n+1}.$$

19. Find a series expansion for

$$\int_0^x e^{-t^2}\, dt \qquad \text{for} \quad x \in \mathbf{R}.$$

20. If $\alpha \in R$ and $|k| < 1$, the integral $F(\alpha,h) := \int_1^\alpha (1 - k^2 (\sin x)^2)^{-1/2}\, dx$ is called an **elliptic integral of the first kind**. Show that

$$F\left(\frac{\pi}{2}, k\right) = \frac{\pi}{2} \sum_{n=0}^{\infty} \left(\frac{1 \cdot 3 \cdots (2n-1)}{2 \cdot 4 \cdots 2n}\right)^2 k^{2n}$$

for $|k| < 1$.

References

Apostol, T.M., *Mathematical Analysis*, Second Edition, Addison-Wesley, Reading, Mass., 1974.

Bartle, R.G., *The Elements of Integration*, Wiley, New York, 1966.

Bartle, R.G., *The Elements of Real Analysis*, Second Edition, Wiley, 1976.

Boas, R.P., Jr., *A Primer of Real Functions*, Carus Monograph Number 13, Math. Assoc. Amer., 1960.

Gelbaum, B.R., and J.M.H. Olmsted, *Counterexamples in Analysis*, Holden-Day, San Francisco, 1964.

Halmos, P.R., *Naive Set Theory*, Springer-Verlag, New York, 1974.

Hamilton, N., and J. Landin, *Set Theory*, Allyn-Bacon, Boston, 1961.

Titchmarsh, E.C., *The Theory of Functions*, Second Edition, Oxford Univ. Press, London, 1939.

Wilder, R.L., *The Foundations of Mathematics*, Wiley, New York, 1952.

HINTS FOR SELECTED EXERCISES

Reader: Do not look at these hints unless you are stymied. However, sometimes just a little hint is all that one needs. Many of the exercises call for proofs, and there is no single way that is correct, so even if you have a totally different argument, yours may well be entirely correct. Few of the following arguments are complete, although more detail is presented for the early material.

SECTION 1.1

4. By definition $A \cap B \subseteq A$. If $A \subseteq B$, then $A \cap B \supseteq A$ so that $A \cap B = A$. Conversely, if $A \cap B = A$, then $A \cap B \supseteq A$ and it follows that $A \subseteq B$.

5,6. The set D is the union of $\{x : x \in A \text{ and } x \notin B\}$ and $\{x : x \notin A \text{ and } x \in B\}$.

8. $(A \cap B) \cap (A \setminus B) = A \cap (B \setminus B) = A \cap \emptyset = \emptyset$

10. If $x \in E \cap \bigcup A_j$, then $x \in E$ and $x \in \bigcup A_j$. Therefore, $x \in E$ and $x \in A_j$ for at least one j. This implies that $x \in E \cap A_j$ for at least one j, so that $E \cap \bigcup A_j \subseteq \bigcup (E \cap A_j)$. The opposite inclusion is similar.

13. If $x \notin \bigcap \{A_j : j \in J\}$, then there exists $k \in J$ such that $x \notin A_k$. (Why?) Therefore $x \in \mathscr{C}(A_k)$ and hence $x \in \bigcup \{\mathscr{C}(A_j) : j \in J\}$. Therefore $\mathscr{C}(\bigcap A_j) \subseteq \bigcup \mathscr{C}(A_j)$. The proof of the opposite inclusion is similar.

SECTION 1.2

1. No. Both $(0,1)$ and $(0,-1)$ belong to C.
2. Let $f(x) := 2x$ and $g(x) := 3x$.
3. If (b,a) and (b,a') belong to f^{-1}, then (a,b) and (a',b) belong to f. Since f is injective, then $a = a'$. Hence f^{-1} is a function.
5. If $f(x_1) = f(x_2)$, then $x_1 = g \circ f(x_1) = g \circ f(x_2) = x_2$, whence f is injective.
6. Apply Exercise 5 twice.
11. If $y \in f(E \cap F)$, then there exists $x \in E \cap F$ such that $y = f(x)$. Since $x \in E$, then $y \in f(E)$; since $x \in F$, then $y \in f(F)$. Therefore $y \in f(E) \cap f(F)$. This proves that $f(E \cap F) \subseteq f(E) \cap f(F)$.

SECTION 1.3

1. The assertion is true for $n = 1$ since $(\frac{1}{2} \cdot 1 \cdot 2)^2 = 1 = 1^3$. To complete the proof use the fact that $(\frac{1}{2} k(k+1))^2 + (k+1)^3 = (\frac{1}{2}(k+1)(k+2))^2$.

2. If $k < 2^k$, then $k + 1 < 2^k + 1 < 2^k + 2^k = 2^{k+1}$.
3. If $k^3 + (k + 1)^3 + (k + 2)^3$ is divisible by 9, then $(k + 1)^3 + (k + 2)^3 + (k + 3)^3 = k^3 + (k + 1)^3 + (k + 2)^3 + 9(k^2 + k + 3)$ is also divisible by 9.
4. Note that $1 + 3 + 5 + \cdots + (2n - 1) = n^2$.
6. If $n = 4$, then $2^4 = 16 < 24 = 4!$. If $2^k < k!$, then $2^{k+1} = 2 \cdot 2^k < 2 \cdot k! < (k + 1)k! = (k + 1)!$
7. It is true for $n = 1$ and $n \geq 5$, but not for $n = 2,3,4$.

SECTION 1.4

1. Let $f(n) := n/2$ for $n \in E$.
2. Let $f(n) := (n + 1)/2$ for $n \in O$.
3. Let $f(n) := n + 1$ for $n \in \mathbf{N}$.
5. Let $A_n := \{n\}$ for $n \in \mathbf{N}$; then each set A_n has a single point but $\mathbf{N} = \bigcup A_n$ is infinite.
6. If A is infinite and $B = \{b_n : n \in \mathbf{N}\}$ is a subset of A, then the function defined on A by $f(x) := b_{n+1}$ for $x = b_n \in B$ and $f(x) := x$ for $x \in A \setminus B$, is injective and maps A onto $A \setminus \{b_1\}$.

SECTION 2.1

4. (a) $-(a + b) = (-1)(a + b) = (-1)a + (-1)b = (-a) + (-b)$.
5. Evidently $a = 0$ satisfies $a \cdot a = a$. If $a \neq 0$ and $a \cdot a = a$, then $(a \cdot a)(1/a) = a(1/a)$, whence $a = a(a(1/a)) = a(1/a) = 1$.
7. Note that if $q \in \mathbf{Z}$ and if $3q^2$ is even, then q is even. (Why?) Hence if $(p/q)^2 = 6$, we must have p even, say $p = 2m$, whence $2m^2 = 3q^2$ so that q is also even.
8. If $p \in \mathbf{Z}$, then there are three possibilities: for some $m \in \mathbf{Z}$, (i) $p = 3m$, (ii) $p = 3m + 1$, or (iii) $p = 3m + 2$. In either case (ii) or (iii), we have $p^2 = 3h^2 + 1$ for some $h \in \mathbf{N}$.
9. If $s := r + \xi \in \mathbf{Q}$, then $\xi = s - r \in \mathbf{Q}$, which is a contradiction. If $t := r\xi \in \mathbf{Q}$ and $r \neq 0$, then $\xi = t/r \in \mathbf{Q}$, which is a contradiction.
11. Take $x = -y = \sqrt{2}$.
12. (a) Commutative, not associative, no identity
    (b) Commutative, associative, $e = 2$
    (c) Not commutative, not associative, no identity
13. (a) Not distributive (b) Distributive

SECTION 2.2

INFINITE SERIES

1. (a) If $a < b$, then 2.2.6(a) implies that $a + c < b + c$. If $a = b$, then $a + c = b + c$. Hence, if $a \leq b$, then $a + c \leq b + c$. If $c < d$, then 2.2.6(a) implies that $b + c < b + d$, whence $a + c < b + d$.
3. Since $b - a$ and $d - c$ are in P, it follows from 2.2.1(ii) that $(bd + ac) - (ad + bc) = (b - a)(d - c) \in P$.
5. If $a = b = 0$, then $a^2 + b^2 = 0 + 0 = 0$. Suppose that $a \neq 0$; then 2.2.5(a) implies that $a^2 > 0$. Since $b^2 \geq 0$, it follows that $a^2 + b^2 > 0$.

7. If $0 < a < b$, it follows from the preceding exercise that $0 < a^2 < ab < b^2$. Hence, by the second part of Example 2.2.13, we infer that $a = \sqrt{a^2} < \sqrt{ab} < \sqrt{b^2} = b$.
9. (a) $\{x \in \mathbf{R} : x < -1 \text{ or } x > 4\}$
   (b) $\{x \in \mathbf{R} : 1 < x < 2 \text{ or } -2 < x < -1\}$
   (c) $\{x \in \mathbf{R} : -1 < x < 0 \text{ or } 1 < x\}$
10. (a) If, on the contrary, $a > b$, then $\varepsilon_0 := a - b > 0$ and $a - \varepsilon_0 = b$, which provides a contradiction.
12. (a) If $0 < c < 1$, then 2.2.6(c) implies that $0 < c^2 < c$, whence $0 < c^2 < c < 1$.
    (b) If $1 < c$, then 2.2.6(c) implies that $c < c^2$, whence $1 < c < c^2$.
13. Let $c := 1 + x$ with $x > 0$. By Bernoulli's Inequality 2.2.13(c), we infer that $c^n = (1 + x)^n \geqslant 1 + nx \geqslant 1 + x = c$ for all $n \in \mathbf{N}$.
14. Use Exercise 13 and Exercise 2.1.14.
16. Let $b := 1/c$ and use Exercise 14.
17. Let $c := b/a$ and use Exercise 13.
18. Take $a_j := 1/\sqrt{c_j}$ and $b_j := \sqrt{c_j}$ in Cauchy's Inequality 2.2.13(d).
19. Take $a_j := c_j$ and $b_j := 1$ in Cauchy's Inequality. Also note that $(c_1 + \ldots + c_n)^2 > c_1^2 + \ldots + c_n^2$.

SECTION 2.3

1. (a) If $a \geqslant 0$, then $|a| = a = \sqrt{a^2}$; if $a < 0$, then $|a| = -a = \sqrt{a^2}$.
   (b) $a^2 \geqslant 0$ for all $a \in \mathbf{R}$
3. Show that $ab \geqslant 0 \Leftrightarrow |ab| = ab$. Then show that $(|a| + |b|)^2 = (a + b)^2 \Leftrightarrow |ab| = ab$.
5. (a) $\{x \in \mathbf{R} : -2 \leqslant x \leqslant 9/2\}$
   (b) $\{x \in \mathbf{R} : -2 \leqslant x \leqslant 2\}$
   (c) $\{x \in \mathbf{R} : x < 0\}$
8. (a) $\{(x,y) : y = \pm x\}$
   (b) The point (x,y) lies on the square with vertices $(\pm 1,0)$ and $(0, \pm 1)$.
9. (b) The point (x,y) lies inside or on the square with vertices $(\pm 1,0)$ and $(0, \pm 1)$.
   (c) The point (x,y) lies on or between the hyperbolas $y = \pm 2/x$ and the coordinate axes.
11. Let $\varepsilon := |a - b|/2$ and let $U := \{x \in \mathbf{R} : |x - a| < \varepsilon\}$ and $V := \{x \in \mathbf{R} : |x - b| < \varepsilon\}$.

SECTION 2.4

2. Zero is a lower bound for S_1. If $y \in \mathbf{R}$ is an upper bound for S_1, then the Trichotomy Property implies that either (i) $y < 0$, or (ii) $y = 0$, or (iii) $y > 0$. Cases (i) and (ii) are easily excluded. If $y > 0$, then since $y + 1 \in S_1$, we have a contradiction.
3. If $u \geqslant s^*$, then $u = \sup (S \cup \{u\})$. If $u < s^*$, then there exists $s \in S$ with $u < s < s^*$ so that $s^* = \sup (S \cup \{u\})$.
5. Let $u \in S$ be an upper bound of S. If v is another upper bound of S, then we must have $u \leqslant v$. Hence $u = \sup S$.
7. Let $u = \sup S$. If $n \in \mathbf{N}$, then since u is an upper bound of S, so is $u + 1/n$. Since u is the supremum of S and $u - 1/n < u$, then $u - 1/n$ is not an upper bound of S. Conversely, suppose $v \in \mathbf{R}$ is not the supremum of the set S. Then either (i) v is not

an upper bound of S (so that there exists $s_1 \in S$ with $v < s_1$, whence we take $n \in \mathbf{N}$ with $1/n < s_1 - v$ to show that $v + 1/n$ is not an upper bound of S), or (ii) there exists an upper bound v_1 of S with $v_1 < v$ (in which case we take $1/n < v - v_1$ to show that $v - 1/n$ is not an upper bound of S).

10. Since $\sup S$ is an upper bound for S, it is an upper bound for S_0 and hence $\sup S_0 \leqslant \sup S$.
12. Let $u := \sup f(X)$; then $f(x) \leqslant u$ for all $x \in X$, and so $a + f(x) \leqslant a + u$ for all $x \in X$, whence it follows that $\sup \{a + f(x) : x \in X\} \leqslant a + u$. If $w < a + u$, then $w - a < u$ so that there exists $x_w \in X$ with $w - a < f(x_w)$, whence $w < a + f(x_w)$. Therefore w is not an upper bound for $\{a + f(x) : x \in X\}$.
14. If $u := \sup f(X)$ and $v := \sup g(X)$, then $f(x) \leqslant u$ and $g(x) \leqslant v$ for all $x \in X$, whence $f(x) + g(x) \leqslant u + v$ for all $x \in X$. Thus $u + v$ is an upper bound for the set $\{f(x) + g(x) : x \in X\}$, whence it follows that $\sup \{f(x) + g(x) : x \in X\} \leqslant u + v$.
18. We have $h(x,y) \leqslant F(x)$ for all $x \in X$, $y \in Y$. Thus $\sup \{h(x,y) : x \in X, y \in Y\} \leqslant \sup \{F(x) : x \in X\}$. If $w < \sup \{F(x) : x \in X\}$, then there exists $x_0 \in X$ with $w < F(x_0) = \sup \{h(x_0,y) : y \in Y\}$, whence there exists $y_0 \in Y$ with $w < h(x_0,y_0)$. Thus w is not an upper bound for $\{h(x,y) : x \in X, y \in Y\}$ and so $w < \sup \{h(x,y) : x \in X, y \in Y\}$. Since this is true for any w with $w < \sup \{F(x) : x \in X\}$, we infer that $\sup \{F(x) : x \in X\} \leqslant \sup \{h(x,y) : x \in X, y \in Y\}$.
19. If $x \in \mathbf{Z}$, take $n := x + 1$. If $x \notin \mathbf{Z}$, we have two cases: (i) $x > 0$ (which is covered by Corollary 2.4.8), and (ii) $x < 0$. In case (ii), consider $z := -x$ and use 2.4.8.
23. Consider $T := \{t \in \mathbf{R} : 0 \leqslant t, t^3 < 2\}$. Note that if $t > 2$, then $t^3 > 2$ so that $t \notin T$. Hence $y := \sup T \geqslant 0$ exists. If $y^3 < 2$, choose $1/n < (2 - y^3)/(3y^2 + 3y + 1)$ and show that $(y + 1/n)^3 < 2$, a contradiction. Etc.

SECTION 2.5

1. Note that $[a',b'] \subseteq [a,b]$ if and only if $a \leqslant a' \leqslant b' \leqslant b$.
3. Since $\inf S$ is a lower bound for S and $\sup S$ is an upper bound for S, then $S \subseteq I_S$. Conversely, if $S \subseteq [a,b]$ then a is a lower bound for S and b is an upper bound for S, and so $[a,b] \subseteq I_S$.
5. If $x > 0$, then it follows from Corollary 2.4.8(b) that there exists $n \in \mathbf{N}$ with $1/n < x$, whence $x \notin [0,1/n]$.
7. If $x \leqslant 0$, then $x \notin K_n$ for any $n \in \mathbf{N}$. If $x > 0$, then it follows from the Archimedean Property 2.4.7 that there exists $n_x \in \mathbf{N}$ with $x \notin (n_x, \infty)$ whence $x \notin K_{n_x}$.
11. Note that $x - \varepsilon > 0$. Hence, by the Archimedean Property there exists $n_0 \in \mathbf{N}$ such that if $n \geqslant n_0$ then $1/n < x - \varepsilon$. Hence, if $n \geqslant n_0$, we have $1/n \notin (x - \varepsilon, x + \varepsilon)$.

SECTION 2.6

1. If $0 < x < 1$ and if $|u - x| < \inf \{x, 1 - x\}$, then (i) $x - u < x$ which implies that $0 < u$, and (ii) $u - x < 1 - x$ which implies that $u < 1$.
3. Clearly, if $x \in (0,1]$, then $x \in (0,1 + 1/n)$ for all $n \in \mathbf{N}$. Conversely, if $x > 1$, then $x - 1 > 0$ so that there exists $n_x \in \mathbf{N}$ such that $x - 1 > 1/n_x$, whence $x \notin (0,1 + 1/n_x)$.

5. If $a_x \in G$, then since G is open there exists $\varepsilon > 0$ with $(a_x - \varepsilon, a_x + \varepsilon) \subseteq G$. In this case $(a_x - \varepsilon, x] \subseteq G$, contradicting the definition of a_x.
7. If $z \in I_x \cap I_y$, then $a_x < z < b_y$ and $a_y < z < b_x$. If $b_x < b_y$, then $b_x \in (a_y, b_y) \subseteq G$ which contradicts the fact that $b_x \notin G$. Similarly if $b_y < b_x$. Hence $b_x = b_y$.
9. Note that : x is a boundary point of $A \Leftrightarrow$ every neighborhood V of x contains points in A and points in $\mathscr{C}(A) \Leftrightarrow x$ is a boundary point of $\mathscr{C}(A)$.
11. Let $F \in \mathbf{R}$ be closed and let x be a boundary point of F. If $x \notin F$, then $x \in \mathscr{C}(F)$; since $\mathscr{C}(F)$ is open, there exists a neighborhood V of x such that $V \subseteq \mathscr{C}(F)$. But this contradicts the hypothesis that x is a boundary point of F. Conversely, if $F \subseteq \mathbf{R}$ is a set that contains all of its boundary points, then if $y \notin F$, it follows that y is not a boundary point of F. Hence there exists a neighborhood V of y such that V does not contain any points of F. (Why?) Thus, $V \subseteq \mathscr{C}(F)$; since $y \in \mathscr{C}(F)$ is arbitrary this shows that $\mathscr{C}(F)$ is open. Hence F is closed.
13. Since A° is the union of sets contained in A, then $A^\circ \subseteq A$. Consequently (why?) we have $(A^\circ)^\circ \subseteq A^\circ$. Since A° is an open set (why?) contained in A° and $(A^\circ)^\circ$ is the union of all open sets contained in A°, then $A^\circ \subseteq (A^\circ)^\circ$. We conclude that $A^\circ = (A^\circ)^\circ$. Finally, note that if $A := [0,1]$ and $B := [1,2]$, then $A^\circ = (0,1)$ and $B^\circ = (1,2)$ while $(A \cup B)^\circ = (0,2)$.
15. Since A^- is the intersection of sets containing A, then $A \subseteq A^-$. Therefore (why?) $A^- \subseteq (A^-)^-$. Since A^- is a closed set (why?) containing A^- and $(A^-)^-$ is the intersection of all closed sets containing A^-, then $(A^-)^- \subseteq A^-$. We conclude that $(A^-)^- = A^-$. Finally, take $A := (0,1)$ and $B := (1,2)$ so that $(A \cap B)^- = \emptyset$, while $A^- \cap B^- = [0,1] \cap [1,2] = \{1\}$.

SECTION 3.1

1. (a) 0,2,0,2,0 (b) $-1, 1/2, -1/3, 1/4, -1/5$
3. (a) 1,4,13,40,121 (c) 1,2,3,5,4
6. (a) $0 < 1/\sqrt{n+7} < 1/\sqrt{n}$ (b) $|2n/(n+2) - 2| = 4/(n+2)$
   (c) $\sqrt{n}/(n+1) < 1/\sqrt{n}$. (d) $|(-1)^n n/(n^2+1)| \leqslant 1/n$
7. Consider $((-1)^n)$.
9. Let $V := (0,\infty)$ so that V is a neighborhood of x; hence there exists $K(V)$ such that if $n \geqslant K(V)$, then $x_n > 0$.
13. Use the argument in 3.1.11(e). If $(2n)^{1/n} = 1 + k_n$, show that $k_n^2 \leqslant 2(2n-1)/n(n-1) < 4/(n-1)$.

SECTION 3.2

1. (a) $\lim (x_n) = 1$ (b) Divergence
   (c) Divergence (d) $\lim (x_n) = 2$
3. $X = Y = ((-1)^n)$
5. If $z_n := x_n y_n$ and $\lim (x_n) = x \neq 0$, then ultimately $x_n \neq 0$ so that $y_n = z_n/x_n$.
7. It is not bounded.
9. In (3) the exponent $k \in \mathbf{N}$ is fixed, whereas in $(1 + 1/n)^n$ the exponent varies.
10. Note that $y_n = 1/(\sqrt{n+1} + \sqrt{n})$.

11. Note that $b \leqslant y_n \leqslant 2^{1/n}b$.
13. $(1/n)$, (n)
15. (a,c,d) Converges to 0.
(b) Diverges
17. $(1/n)$, (n).

SECTION 3.3

1. Show by induction that $0 \leqslant x_n \leqslant 2$ for $n \geqslant 2$, and that (x_n) is monotone. In fact, (x_n) is decreasing, for if $x_1 > x_2$ then we would have $(x_1 - 1)^2 < 0$.
3. Show that the sequence (z_n) is monotone. Let z^* be the positive root of the equation $z^2 - z - a = 0$ so that $z^* = (1 + \sqrt{1 + 4a})/2$. Show that if $0 < z_1 < z^*$, then $z_1^2 - z_1 - a < 0$ and the sequence increases to z^*. If $z^* < z_1$, then the sequence decreases to z^*.
5. Show that (s_n) is decreasing and bounded, and that (t_n) is increasing and bounded. Also $t_n \leqslant x_n \leqslant s_n$ for $n \in \mathbf{N}$.
8. Show that $y_{n+1} - y_n > 0$ and that $0 \leqslant y_n \leqslant n/2n = 1/2$.
10. (a) e (b) e^2
(c) e (d) Note that $1 - 1/n = (1 + 1/(n - 1))^{-1}$.
12. Note that $0 \leqslant s_n - \sqrt{5} \leqslant (s_n^2 - 5)/\sqrt{5} \leqslant (s_n^2 - 5)/2$.

SECTION 3.4

2. If $x_n = c^{1/n}$ then $x_{2n} = \sqrt{x_n}$.
5. If (x_n) does not converge to 0, then there exists $\varepsilon_0 > 0$ and a subsequence (x_{nk}) with $|x_{nk}| > \varepsilon_0$ for all $k \in \mathbf{N}$.
7. (a) 1 (b) $\sqrt{e}$
(c) e (d) $(n + 2)/n = [(n + 2)/(n + 1)] \cdot [(n + 1)/n]$
9. Show that $\lim ((-1)^n x_n) = 0$.
11. Choose a sequence (x_n) in F with $|x_n| \leqslant \inf \{|x| : x \in F\} + 1/n$.

SECTION 3.5

3. (a) $|(-1)^n - (-1)^{n+1}| = 2$ for all $n \in \mathbf{N}$.
6. Let $x^* := \sup \{x_n : n \in \mathbf{N}\}$. If $\varepsilon > 0$, let $H \in \mathbf{N}$ be such that $x^* - \varepsilon < x_H \leqslant x^*$. If $m \geqslant n \geqslant H$, then $x^* - \varepsilon < x_n \leqslant x_m \leqslant x^*$ so that $|x_m - x_n| < \varepsilon$.
7. If $L := x_2 - x_1$, show that $|x_{n+1} - x_n| = L/2^{n-1}$. Note that (x_{2n+1}) is an increasing sequence. Express x_{2n+1} in terms of x_{2n-1}.
9. Show that $|x_{n+1} - x_n| < \frac{1}{4}|x_n - x_{n-1}|$. The limit is $\sqrt{2} - 1$.

SECTION 3.6

3. Note that $|x_n - 0| < \varepsilon$ if and only if $x_n > 1/\varepsilon$.
5. No; as in Example 3.4.5(e), there is a subsequence (n_k) such that $n_k \sin n_k > n_k/2$, and there is a subsequence (m_k) such that $m_k \sin m_k < -m_k/2$.

7. (a) There exists N_1 such that if $n \geq N_1$, then $0 < x_n < y_n$.
9. (b) Let $0 < x_n < M$. If (y_n) does not converge to 0, there exists $\varepsilon_0 > 0$ and a subsequence (y_{nk}) such that $0 < \varepsilon_0 < y_{nk}$. Since $\lim (x_{nk}/y_{nk}) = \infty$, there exists K such that if $k \geq K$ then $M/\varepsilon_0 < x_{nk}/y_{nk}$, which provides a contradiction.

SECTION 4.1

1. (a–c) If $|x - 1| < 1$, then $|x + 1| < 3$ so that $|x^2 - 1| \leq 3|x - 1|$.
   (d) If $|x - 1| < 1$, then $|x^2 + x + 1| < 7$ so that $|x^3 - 1| \leq 7|x - 1|$.
3. If $\lim_{y \to 0} f(y) = L$ then for every $\varepsilon > 0$ there exists $\delta(\varepsilon) > 0$ such that if $0 < |y - 0| < \delta(\varepsilon)$ then $|f(y) - L| < \varepsilon$. Now let $x := y - c$ so that $y = x + c$ to conclude that $\lim_{x \to 0} f(x + c) = L$.
5. Note that the restriction of sgn to $[0,1]$ has a limit at 0.
7. Take $\delta(\varepsilon) := \varepsilon/K$.
9. If $c = 0$, take $\delta(\varepsilon) := \varepsilon^2$. If $c > 0$, show that $|\sqrt{x} - \sqrt{c}| \leq (1/\sqrt{c})\,|x - c|$.
11. (a) If we take $x_n := 1/n$, then $(1/x_n^2) = (n^2)$ which does not converge in $\boldsymbol{R}$.
    (c) Consider the sequences $(x_n) = (1/n)$ and $(y_n) = (-1/n)$.
13. If $f(x) = \operatorname{sgn}(x)$, then $(f(x))^2 = 1$ for all $x \neq 0$. Hence $\lim_{x \to 0} (f(x))^2 = 1$, but $\lim_{x \to 0} f(x)$ does not exist.

SECTION 4.2

1. (a) 10 (b) -3
   (c) 1/12 (d) 1/2
3. Multiply numerator and denominator by $\sqrt{1 + 2x} + \sqrt{1 + 3x}$ and simplify.
5. If $|f(x)| < M$ for $x \in U$, then $|f(x)g(x) - 0| \leq M\,|g(x) - 0|$ for $x \in U$.
9. (a) Note that $g(x) = (f + g)(x) - f(x)$.
   (b) Take $f(x) := x^2$ and $g(x) := 1/\sqrt{x}$ for $x > 0$.
11. (a) No limit (b) 0
    (c) No limit (d) 0

SECTION 4.3

3. Given $\alpha > 0$, if $0 < x < 1/\alpha^2$, then $\sqrt{x} < 1/\alpha$ and so $f(x) > \alpha$. Since $\alpha > 0$ is arbitrary, $\lim_{x \to 0+} f(x) = \infty$.
5. (a) If $\alpha > 1$ and $1 < x < \alpha(\alpha - 1)^{-1}$, then show that $\alpha < x(x - 1)^{-1}$ whence we have $\lim_{x \to 1+} x(x - 1)^{-1} = \infty$.
   (c) If $\alpha > 0$ and $0 < x < 4/\alpha^2$, then $\alpha < 2/\sqrt{x}$ and so $\alpha < (x + 2)/\sqrt{x}$.
   (e) If $x > 0$, then $1/\sqrt{x} < (\sqrt{x + 1})/x$ so $\lim_{x \to 0+} (\sqrt{x + 1})/x = +\infty$.
   (g) 1
7. Use Theorem 4.3.11.
9. There exists $\alpha > 0$ such that if $x > \alpha$, then $|xf(x) - L| < 1$. Hence $|f(x)| < (|L| + 1)/x$ for $x > \alpha$.
13. No. If $h(x) := f(x) - g(x)$, then $\lim_{x \to \infty} h(x) = 0$ and $f(x)/g(x) = 1 + h(x)/g(x) \to 1$.

SECTION 4.4

4. (a) Continuous if $c \neq 0, \pm 1, \pm 2, \ldots$.
   (b) Continuous if $c \neq \pm 1, \pm 2, \ldots$.
   (c) Continuous if $\sin c \neq 0, 1$.
   (d) Continuous if $c \neq 0, \pm 1, \pm 1/2, \ldots$.
5. Yes, define $f(2) := \lim_{x \to 2} f(x) = 3$.
7. Let $V := (0, \infty)$ and let U be a neighborhood such that $f(U) \subseteq V$.
9. (b) Let $f(x) := 1$ for $x \geqslant 0$, $f(x) := -1$ for $x < 0$, and $A := [0,1]$.
10. Note that $||x| - |c|| \leqslant |x - c|$.
12. If x is irrational, then there exists a sequence (r_n) of rational numbers converging to x.
13. The function g is continuous only at 3.
15. Let $I_n := [-1/n, 1/n]$ for $n \in \boldsymbol{N}$. Show that $(\sup f(I_n))$ is a decreasing sequence and $(\inf f(I_n))$ is an increasing sequence. If $\lim (\sup f(I_n)) = \lim (\inf f(I_n))$, then $\lim_{x \to 0} f$ exists.

SECTION 4.5

3. Let f be the Dirichlet discontinuous function [Example 4.4.5(g)] and let $g(x) := 1 - f(x)$.
5. The function g is not continuous at $1 = f(0)$.
7. Let $f(x) := 1$ if x is rational, and $f(x) := -1$ if x is irrational.
9. Show that an arbitrary number is the limit of a sequence of numbers of the form $m/2^n$, where $m \in \boldsymbol{Z}$, $n \in \boldsymbol{N}$.
11. If $h(x) := f(x) - g(x)$, note that $S = \{x \in \boldsymbol{R} : h(x) > 0\}$. Compare this with Exercise 4.4.7.
13. First show that $f(0) = 0$ and (by induction) that $f(x) = cx$ for $x \in \boldsymbol{N}$, and hence also for $x \in \boldsymbol{Z}$. Next show that $f(x) = cx$ for $x \in \boldsymbol{Q}$. Finally, if $x \notin \boldsymbol{Q}$, let $x = \lim (r_n)$ for some sequence in $\boldsymbol{Q}$.

SECTION 4.6

1. Either apply the Boundedness Theorem 4.6.2 to $1/f$, or apply the Maximum-Minimum Theorem to conclude that $\inf f(I) > 0$.
3. Construct a sequence (x_n) in I with $\lim (f(x_n)) = 0$ and apply the Bolzano-Weierstrass Theorem 3.4.6 to (x_n).
5. In the intervals $[1.035, 1.040]$ and $[-7.026, -7.025]$.
7. In $[0.7390, 0.7391]$.
9. Since $f(\pi/4) < 1$ and $f(0) = 1 \leqslant f(\pi/2)$, it follows that $x_0 \in (0, \pi/2)$. If $\cos x_0 > x_0^2$, there exists a neighborhood U of x_0 on which $f(x) = \cos x$, so that x_0 is not an absolute minimum point for f.
10. Consider $f(x) := (1 + x^2)^{-1}$ for $x \in \boldsymbol{R}$.
12. Let $f(x) := \sin(1/x)$ for $x \neq 0$ and $f(0) := 0$.
15. Yes.
17. Let $g(x) := 1/x$ for $x \in (0,1)$.

SECTION 4.7

1. Note that $1/x - 1/y = (y - x)/xy$ so that $|1/x - 1/y| \leq |y - x|$ for x and y in $[1,\infty)$.
3. Note that $f(x) - f(y) = (x + y)(y - x)/(1 + x^2)(1 + y^2)$. Show that $|(x + y)/(1 + x^2)(1 + y^2)| \leq 2$ for all $x,y \in \boldsymbol{R}$.
5. If M is a bound for both f and g on $\boldsymbol{R}$, show that $|f(x)g(x) - f(a)g(a)| \leq M\,|f(x) - f(a)| + M\,|g(x) - g(a)|$.
9. Show that $\lim_{x\to a} f(x)$ and $\lim_{x\to b} f(x)$ exist.
10. There exists $\delta > 0$ such that if $|x - y| \leq \delta$, $x, y \in A$, then $|f(x) - f(y)| < 1$. If A is bounded, it is contained in the union of a finite number of intervals of length δ.
13. Since f is bounded on $[0, p]$, it follows that it is bounded on $\boldsymbol{R}$. Since f is continuous on $J := [-1, p + 1]$, it is uniformly continuous on J. Now show that this implies that f is uniformly continuous on $\boldsymbol{R}$.
17. Show that $|f_2(x) - B_n(x)| \leq 1/4n$ and that equality holds at $x = 1/2$. Hence we must take $n \geq 250$.

SECTION 4.8

1. If $x \in [a,b]$ then $f(a) \leq f(x)$.
5. If f is increasing on $[a,b]$, then $\lim_{x\to a} f(x) = \inf\{f(x) : x \in (a,b]\}$.
7. Show that $\inf\{f(y) - f(x) : x < c < y,\ x,\ y \in I\} = \inf\{f(y) : c < y,\ y \in I\} - \sup\{f(x) : x < c,\ x \in I\}$.
10. Use the Bolzano Intermediate Value Theorem 4.6.6.
11. The function f is not continuous at $x = 1$.

SECTION 4.9

1. Let $I_n := (1 + 1/n, 3)$ for $n \in \boldsymbol{N}$.
3. Let $\mathcal{G}$ be a collection of open sets whose union contains $K_1 \cup K_2$. Then the union of some finite number of these sets contains K_1 [respectively, K_2].
5. Use the Heine-Borel Theorem.
7. Since K is bounded, then $\inf K$ exists. For $n \in \boldsymbol{N}$, let $K_n := \{k \in K : k \leq (\inf K) + 1/n\}$; now apply Exercise 5. [Alternatively, use Exercise 8.]
9. For $n \in \boldsymbol{N}$ let $x_n \in K$ be such that $|c - x_n| \leq \inf\{|c - x| : x \in K\} + 1/n$; now apply Exercise 8.

SECTION 5.1

1. (b) $g'(x) = \lim_{h\to 0} \dfrac{1}{h}\left(\dfrac{1}{x + h} - \dfrac{1}{x}\right) = \lim_{h\to 0} \dfrac{-1}{x(x + h)} = -\dfrac{1}{x^2}$

   (c) $h'(x) = \lim_{h\to 0} \dfrac{1}{h}(\sqrt{x + h} - \sqrt{x}) = \lim_{h\to 0} \dfrac{1}{\sqrt{x + h} + \sqrt{x}} = \dfrac{1}{2\sqrt{x}}$
2. $\lim_{h\to 0} (1/h^{2/3})$ does not exist.
4. Use the inequality $0 \leq f(x) \leq x^2$ for $x \in \boldsymbol{R}$ and the Squeeze Theorem 4.1.7 to obtain $f'(0) = 0$.

5. (a) $f'(x) = (1 - x^2)/(1 + x^2)^2$ (c) $h'(x) = mk\, x^{k-1} (\cos x^k)(\sin x^k)^{m-1}$
6. The function f' is continuous for $n \geqslant 2$, and is differentiable for $n \geqslant 3$.
8. (a) $f'(x) = 2$ for $x > 0$, $f'(x) = 0$ for $-1 < x < 0$, and $f'(x) = -2$ for $x < -1$
   (c) $h'(x) = 2\,|x|$ for all $x \in \boldsymbol{R}$
10. $g'(x) = 2x \sin(1/x^2) - (1/x)\cos(1/x^2)$ for $x \neq 0$, and $g'(0) = 0$. Since $g'(1/\sqrt{2n\pi}) = \sqrt{2n\pi}$ for $n \in \boldsymbol{N}$, then g' is unbounded on any neighborhood of 0.
11. (a) $f'(x) = 2/(2x + 3)$ (b) $g'(x) = 6(L(x^2))^2/x$
    (c) $h'(x) = 1/x$ (d) $k'(x) = 1/(xL(x))$
12. $r > 1$

SECTION 5.2

1. (a) Increasing on $(3/2, \infty)$ (b) Increasing on $(-\infty, -1)$ and $(1, \infty)$
2. (a) Relative minimum at $x = 1$; relative maximum at $x = -1$.
   (c) Relative maximum at $x = 2/3$.
3. (a) Relative minima at $x = \pm 1$; relative maxima at $x = 0, \pm 4$.
   (c) Relative minima at $x = -2, 3$; relative maximum at $x = 2$.
4. $x = (1/n)(a_1 + \ldots + a_n)$
6. If $x < y$, then there exists c in (x,y) such that $|\sin y - \sin x| = |\cos c|\,|y - x|$.
7. There exists c with $1 < c < x$ such that $\log x = (x - 1)/c$. Now use the inequality $1/x < 1/c < 1$.
9. $f'(1/2n\pi) < 0$ for $n \geqslant 2$, and $f'(2/(4n + 1)\pi) > 0$ for $n \geqslant 1$
10. $g'(1/2n\pi) < 0$ and $g'((2/(4n + 1)\pi) > 0$ for $n \in \boldsymbol{N}$
11. For example, $f(x) := \sqrt{x}$
12. Apply Darboux's Theorem 5.2.12. If $g(x) = a$ for $x < 0$, $g(x) = x + b$ for $x \geqslant 0$, where a,b are any constants, then $g'(x) = h(x)$ for $x \neq 0$.
14. Apply Darboux's Theorem 5.2.12.
16. (b) If $b \neq 0$, then for $n \in \boldsymbol{N}$ sufficiently large, if $x \geqslant n$, then there is an $x_n > n$ such that $|(f(x) - f(n))/x| = |(x - n)/x|\,|f'(x_n)| \geqslant |(x - n)/x|\,|b|/2$.
17. Apply the Mean Value Theorem to $g - f$ on $[0,x]$.
20. (c) Apply Darboux's Theorem to the results of (a) and (b).

SECTION 5.3

4. Note that $f'(0) = 0$, but that $f'(x)$ does not exist if $x \neq 0$.
6. (a) 1 (b) 1 (c) 0 (d) 1/3
7. (a) 1 (b) ∞ (c) 0 (d) 0
8. (a) 0 (b) 0 (c) 0 (d) 0
9. (a) 1 (b) 1 (c) e^3 (d) 0
10. (a) 1 (b) 1 (c) 1 (d) 0

SECTION 5.4

1. $f^{(2n-1)}(x) = (-1)^n a^{2n-1} \sin ax$ and $f^{(2n)}(x) = (-1)^n a^{2n} \cos ax$ for $n \in \boldsymbol{N}$
2. $g'(x) = 3x^2$ for $x \geqslant 0$, $g'(x) = -3x^2$ for $x < 0$, and $g''(x) = 6\,|x|$ for $x \in \boldsymbol{R}$

5. $1.095 < \sqrt{1.2} < 1.1$ and $1.375 < \sqrt{2} < 1.5$
6. $R_2(0.2) < 0.0005$ and $R_2(1) < 0.0625$
7. $R_2(x) = (1/6)(10/27)(1 + c)^{-8/3}x^3 < (5/81)x^3$ where $0 < c < x$
9. $|R_n(x)| \leq |x - x_0|^n/n! \rightarrow 0$ as $n \rightarrow \infty$
11. With $n = 4$, $\log 1.5 = 0.40$; with $n = 7$, $\log 1.5 = 0.405$
13. $e = 2.7182818$ with $n = 10$
14. (a) No (b) Relative maximum (c) No (d) Relative minimum
16. Consider $f(x) := x|x|$ at $a = 0$.
19. Since $f(2) < 0$ and $f(2.2) > 0$, there is a zero of f in $[2.0, 2.2]$. The value of x_4 is 2.0945515.
20. $r_1 = 1.45262688$ and $r_2 = -1.16403514$
21. $r = 1.32471796$
22. $r_1 = 0.15859434$ and $r_2 = 3.14619322$
23. $r_1 = 0.5$ and $r_2 = 0.80901699$
24. $r = 0.73908513$

SECTION 6.1

1. Show that if P is any partition of $[a,b]$, then $L(P;f) = U(P;f) = c(b - a)$.
3. (a) Show that if P_n is the partition of $[0,1]$ into n equal subintervals, then $L(P_n;g) = (n - 1)/2n$ and $U(P_n;g) = 1/2$.
   (b) Yes.
5. Show that $L(P_n;f) = (1 - 1/n)^2/4$ and $U(P_n;f) = (1 + 1/n)^2/4$.
7. If P is a partition of $[a,b]$ containing c and $P_c := P \cap [a,b]$, show that $0 \leq U(P_c;f_c) - L(P_c;f_c) \leq U(P;f) - L(P;f)$.
9. If $f(c) > 0$, then $f(x) > f(c)/2$ for x in some interval centered at c. (Why?) Show that this implies that $L(f) > 0$.
11. Let $\varepsilon > 0$ be given and let $n \in N$ be such that $1/n < \varepsilon$. Note that there are only a finite number of points $0 = p_1 < p_2 < \ldots < p_r = 1$ where $h(p_k) \geq 1/n$. Construct disjoint closed intervals $I_1, \ldots, I_r$ such that (i) the sum of the lengths of the I_k is at most ε, and (ii) the point p_k is an interior point of I_k for $1 < k < r$. Let P_ε be the partition obtained by using the end points of these intervals. Show that $U(P_\varepsilon;h) < 2\varepsilon$.
13. Let f_1 be as in Example 6.1.7(d) and let $f_2 := 1 - f_1$.

SECTION 6.2

3. If $P := (x_0, x_1, \ldots, x_n)$ is a partition of I, then $m \leq m_j \leq M_j \leq M$ for $j = 1, \ldots, n$.
4. Show that $L(P;f) \leq L(P;g) \leq L(P;h)$.
6. First show that $\sup\{L(P;f) : P \in \mathcal{P}_c\} \leq L$. Then use Lemma 6.1.2.

SECTION 6.3

1. $H'(x) = -f(x)$
3. (a) $F'(x) = \sin(x^2)$ (c) $F'(x) = (1 + x^2)^{1/2} - 2x(1 + x^4)^{1/2}$
4. $F(x) = x^2/2$ for $0 \leq x < 1$, $F(x) = x - 1/2$ for $1 \leq x < 2$, and $F(x) = (x^2 - 1)/2$ for $2 \leq x \leq 3$. For $x \neq 2$.

5. (b) For fixed y, show that if $g(x) := L(xy) - L(x)$, then $g'(x) = 0$ for all $x > 0$.
   (c) Consider $h(x) := L(x^n) - nL(x)$ for $x > 0$.
7. If $F(x) := \int_0^x f(t)\,dt$, then $F'(x) = f(x) = -f(x)$ so that $F'(x) = 0$ for all $x \in I$.
8. First note that $f'(x) = 1$ for all $x > 0$.
9. Note that if $\varepsilon > 0$, then $M - \varepsilon < f(x)$ for x in some interval $[c,d] \subseteq [a,b]$. Therefore $(M - \varepsilon)^n(d - c) \leq \int_a^b (f(x))^n\,dx \leq M^n(b - a)$.
10. (a) $(2^{3/2} - 1)/3$ (c) $2(1 - \pi/4)$
11. (a) $2\sqrt{2} + 4 \log (2 - \sqrt{2})$ (c) $\log (3 + 2\sqrt{2}) - \log 3$
13. If $|g(x)| \leq M$ for all $x \in I$, show by induction that $|g(x)| \leq MK^n(x - a)^n/n!$.

SECTION 6.4

2. (a) Consider $f(x) := (1 + x)^{-1}$
3. Show that the limit equals $\int_0^1 (1 + x^2)^{-1}\,dx = \text{Arctan } 1$.
5. Given $\varepsilon > 0$, choose $\xi_k \in [x_{k-1}, x_k]$ such that $M_k - \varepsilon/(b - a)n \leq f(\xi_k)$.
8. (a) $\log x \to -\infty$ as $x \to 0+$. The value of the integral is -1.
   (c) $x/(x - 1)^{1/2} \to \infty$ as $x \to 1+$. The value of the integral is $8/3$.
9. (a) The interval is unbounded; the value of the integral is 1.
   (c) The interval is unbounded; the integral is divergent since $x^{-1} \log x > x^{-1}$ for $x > 3$.
10. (a) $(1 - x^2)^{-1/2} \to \infty$ as $x \to \pm 1$; the value of the integral is π.
    (c) The integrand is unbounded on any neighborhood of $x = 0$, and the interval is infinite. The integral is divergent because $\int_a^1 x^{-2}\,dx = 1/a - 1 \to \infty$ as $a \to 0+$.

SECTION 6.5

7. Note that $p^{(4)}(x) = 0$ for all x.
11. Use Exercise 10.
13. Interpret K as an area. Show that $h''(x) = -(1 - x^2)^{3/2}$ and that $h^{(4)}(x) = (-3)(1 + 4x^2)(1 - x^2)^{-7/2}$.
14. Approximately 3.6534845
15. Approximately 4.821159
16. Approximately 0.835649
17. Approximately 1.851937
18. 1
19. Approximately 1.198140
20. Approximately 0.904524

SECTION 7.1

1. If $x \geq 0$ and $\varepsilon < 1$ are given, then $x/(x + n) < \varepsilon$ for all $n > x(1 - \varepsilon)/\varepsilon$.
3. If $x = 0$, the limit is 0; if $x > 0$, the limit is 1.
5. The limit is 0 for all $x \geq 0$.
6. If $x > 0$ and $0 < \varepsilon < \pi/2$, then $\tan(\pi/2 - \varepsilon) > 0$ so that $nx \geq \tan(\pi/2 - \varepsilon)$ for $n \geq n_x$, and thus $\pi/2 - \varepsilon < \text{Arctan } nx \leq \pi/2$.
7. If $x > 0$, then $e^{-x} < 1$, whence it follows that the limit is 0. If $x = 0$, the limit is 1.
9. If $0 \leq c < 1$, then $\lim (n^2c^n) = 0$.
11. If $0 \leq x \leq a$, then $x/(x + n) \leq a/n$. However, if $n \in \mathbf{N}$ and $x \geq n$, then $x/(x + n) \geq 1/2$.

13. Use the sequence $(1/n)$ to show that the convergence is not uniform.
16. Consider the sequence $(1/n)$.
18. If $x \geqslant 0$, then $xe^{-nx} \leqslant 1/n$.

SECTION 7.2

1. The limit function is $f(x) := 0$ for $0 \leqslant x < 1$, $f(1) := 1/2$, $f(x) := 1$ for $1 < x \leqslant 2$.
3. Let $f_n(x) := 1/n$ if x is rational, and $f_n(x) := 0$ if x is irrational.
5. Given $\varepsilon > 0$, let $\delta > 0$ be such that $|x - u| < \delta$ implies that $|f(x) - f(u)| < \varepsilon$. If $n > 1/\delta$, then $|f(x + 1/n) - f(x)| < \varepsilon$ for all $x \in \boldsymbol{R}$.
6. $g(1) = 1$, but $f'(1) = 0$.
8. The Fundamental Theorem of Calculus implies that $\int_a^x f_n'(t)\,dt = f_n(x) - f_n(a)$ for all $n \in \boldsymbol{N}$. Now apply Theorem 7.2.4.
10. If $a = 0$, then the limit exists and has the value 0.
11. The limit is given by $f(x) := 1$ for $0 < x \leqslant 1$, $f(0) := 0$, and the integral of f has the value 1. Also $\int_0^1 f_n(x)\,dx = \int_0^1 (1 - 1/(nx + 1))\,dx = 1 - (1/n)\log(n + 1)$, which has the limit 0 as $n \to \infty$.

SECTION 7.3

1. Take $x = 1$ and $n = 3$ to see that $8/3 < e < 11/4$.
2. 2.71828
4. If $e = m/n$, then $en! - (1 + 1 + 1/2! + \cdots + 1/n!)n!$ is an integer between 0 and 1, which is impossible.
6. 0.0953, 0.3365
7. 0.6931
10. $1 = L'(1) = \lim\,[L(1 + 1/n) - L(1)]/[1/n] = L(\lim\,(1 + 1/n)^n)$.

SECTION 7.4

1. 0.9801, 0.1987, 0.5403, 0.8415

SECTION 8.1

3. Group the terms in the series $\Sigma(-1)^n$ to produce convergence to -1, to 0.
6. (a) Since $1/(\alpha + n)(\alpha + n + 1) = 1/(\alpha + n) - 1/(\alpha + n + 1)$, it follows that the nth partial sum equals $1/\alpha - 1/(\alpha + n + 1)$.
   (b) $1/n(n + 1)(n + 2) = 1/2n - 1/(n + 1) + 1/2(n + 2)$
7. Consider $\Sigma((-1)^n n^{1/2})$. However, if $a_n \geqslant 0$, then $0 \leqslant a_n^2 \leqslant a_n \leqslant 1$ for sufficiently large n.
8. If $a,b \geqslant 0$, then $2(ab)^{1/2} \leqslant a + b$.
9. Show that $b_1 + b_2 + \cdots + b_n \geqslant a_1(1 + 1/2 + \cdots + 1/n)$.
10. Use Exercise 6(a).
11. Show that $a_1 + a_2 + \cdots + a_{2^n}$ is bounded below by $\frac{1}{2}(a_1 + 2a_1 + \cdots + 2^n a_{2^n})$ and above by $a_1 + 2a_2 + \cdots + 2^{n-1}a_{2^{n-1}} + 2^n a_{2^n}$.
12. $\Sigma(2^n(1/2^{np})) = \Sigma(1/(2^{p-1})^n)$, a geometric series

SECTION 8.2

1. (a), (d), (f), and (g) are convergent.
3. (a) and (e) are divergent; (b) is convergent.
4. (b), (c), and (e) are divergent.
5. Compare the series with $\Sigma(1/n^2)$.
7. (a) Convergent (c) Divergent
9. The convergence is conditional.
12. (a) Convergent (b) Divergent
13. (c) If $\Sigma(a_n)$ is absolutely convergent, so is $\Sigma(b_n)$. If $a_n = 0$ except when $\sin n$ is near ± 1, then we can obtain a counterexample.
    (d) Consider $a_n := 1/n(\log n)^2$.

SECTION 8.3

1. (a) and (c) converge uniformly for all x.
   (b) Converges for $x \neq 0$ and uniformly for x in the complement of any neighborhood of $x = 0$.
   (d) Converges for $x > 1$ and uniformly for $x \geqslant a$ for any $a > 1$.
2. Note that $|a_n \sin nx| \leqslant |a_n|$.
3. If the series converges uniformly, then $|c_n \sin nx + \cdots + c_{2n} \sin 2nx| < \varepsilon$ provided n is sufficiently large. Now restrict x to an interval such that $\sin kx > 1/2$ for $n \leqslant k \leqslant 2n$.
6. (a) ∞ (c) $1/e$ (e) 4
7. Both are 1.
9. Note that $\lim (p^{1/n}) = \lim (q^{1/n}) = 1$.
10. Apply the Uniqueness Theorem 8.3.13.
12. Show that if $n \in N$, then there exists a polynomial P_n such that $f^{(n)}(x) = e^{-1/x^2}P_n(1/x)$ for $x \neq 0$.
15. If s_n is the nth partial sum of the series, then $s_n - xs_n = 1 - x^{n+1}$.
17. Use Exercise 15 to show that $1/(1 + x^2) = \Sigma(-1)^n x^{2n}$, then integrate.
19. Integrate $e^{-t^2} = \Sigma(-1)^n t^{2n}/n!$.

INDEX